Phytomicrobiome Interactions and Sustainable Agriculture

Edited by

Amit Verma
S. D. Agricultural University
Palanpur, Gujarat, India

Jitendra Kumar Saini
Central University of Haryana
Mahendergarh, Haryana, India

Abd El-Latif Hesham
Beni-Suef University
Beni-Suef, Egypt

Harikesh Bahadur Singh
GLA University
Mathura, Uttar Pradesh, India

WILEY Blackwell

This edition first published 2021

Registered Offices
John Wiley & Sons, Inc., 111 River Street, Hoboken, NJ 07030, USA
John Wiley & Sons Ltd, The Atrium, Southern Gate, Chichester, West Sussex, PO19 8SQ, UK

Editorial Office
The Atrium, Southern Gate, Chichester, West Sussex, PO19 8SQ, UK

For details of our global editorial offices, customer services, and more information about Wiley products visit us at www.wiley.com.

Wiley also publishes its books in a variety of electronic formats and by print-on-demand. Some content that appears in standard print versions of this book may not be available in other formats.

Library of Congress Cataloging-in-Publication Data
Names: Verma, Amit, 1983- editor.
Title: Phytomicrobiome interactions and sustainable agriculture / edited by Amit Verma, S.D. Agricultural University, Gujarat, India, Jitendra Kumar Saini, Central University of Haryana, Mahendergarh, India, Abd El-Latif Hesham, Beni-Suef University, Egypt, Harikesh Bahadur Singh, GLA University, Mathura, India.
Description: First edition. | Hoboken, NJ : Wiley-Blackwell, 2021. | Includes bibliographical references and index.
Identifiers: LCCN 2020051129 (print) | LCCN 2020051130 (ebook) | ISBN 9781119644620 (hardback) | ISBN 9781119644811 (adobe pdf) | ISBN 9781119644828 (epub)
Subjects: LCSH: Plant-microbe relationships. | Agricultural microbiology. | Plants–Microbiology. | Plants–Effect of stress on. | Sustainable agriculture.
Classification: LCC QR351 .P525 2021 (print) | LCC QR351 (ebook) | DDC 579/.178–dc23
LC record available at https://lccn.loc.gov/2020051129
LC ebook record available at https://lccn.loc.gov/2020051130

Cover Design: Wiley
Cover Image: © Catherine MacBride/Getty Images

Set in 9.5/12.5pt STIXTwoText by SPi Global, Pondicherry, India
Printed and bound by CPI Group (UK) Ltd, Croydon, CR0 4YY

10 9 8 7 6 5 4 3 2 1

To my father and mother

Contents

List of Contributors

Richa Agnihotri
ICAR-Indian Institute of Soybean Research
Indore
Madhya Pradesh
India

Hina Bansal
Amity Institute of Biotechnology
Amity University
Noida
Uttar Pradesh
India

Abhishek Bharti
ICAR-Indian Institute of Soybean Research
Indore
Madhya Pradesh
India

Kalpana Bhatt
Gurukula Kangri Vishwavidhyalaya
Haridwar
Uttarakhand
India

Pankaj Bhatt
Integrative Microbiology Research Centre
South China Agricultural University
Guangzhou
P.R. China

Sunita Chauhan
Amity Institute of Biotechnology
Amity University
Kant
Rajasthan
India

Rakhi Dhankhar
Maharshi Dayanand University
Rohtak
Haryana
India

Ajinath Dukare
ICAR-Central Institute of Post-Harvest Engineering and Technology (CIPHET)
Ludhiana
Punjab
India

Bandita Dutta
Maulana Abul Kalam Azad University of Technology
Kolkata
West Bengal
India

Saurabh Gangola
Graphic Era Hill University
Bhimtal
Uttarakhand
India

Mudasir Gani
Faculty of Agriculture
Sher-e-Kashmir University of Agricultural Sciences and Technology Kashmir (SKUAST-K)
Wadura, Sopore, Kashmir
Jammu-Kashmir (UT)
India

Sayantani Garai
University of Engineering and Management
Kolkata
West Bengal
India

Tripti Grover
Gargi College
University of Delhi
New Delhi
India

Pooja Gulati
Maharshi Dayanand University
Rohtak
Haryana
India

Aparna Gunjal
Dr. D.Y. Patil
Arts, Commerce and Science College
Pune
Maharashtra
India

Shweta Kulshreshtha
Department of Biotechnology, Kumarappa National Handmade Paper Institute
Jaipur
Rajasthan
India

Arun Kumar
Department of Agronomy
Punjab Agricultural University (PAU)
Ludhiana
Punjab
India

Kapila Kumar
Manav Rachna International Institute of Research and Studies
Faridabad
Haryana
India

Sanjeev Kumar
Department of Basic Science, College of Horticulture and Forestry Dr. YSP University of Horticulture and Forestry
Hamirpur
Himachal Pradesh
India

Reena Kumari
Department of Biotechnology, College of Horticulture and Forestry Dr. YSP University of Horticulture and Forestry
Hamirpur
Himachal Pradesh
India

Rekha Kushwaha
University of Missouri
Columbia
MO, USA

Dibyajit Lahiri
University of Engineering and Management
Kolkata
West Bengal
India

Hemant S. Maheshwari
Ecophysiology of Plants, Faculty of Science and Engineering
GELIFES-Groningen Institute for Evolutionary Life Sciences
Groningen
The Netherlands

and

ICAR-Indian Institute of Soybean Research
Indore
Madhya Pradesh
India

Reema Mishra
Gargi College
University of Delhi
New Delhi
India

Aparajita Mohanty
Gargi College
University of Delhi
New Delhi
India

Moupriya Nag
University of Engineering and Management
Kolkata
West Bengal
India

Sharon Nagpal
Department of Microbiology
Punjab Agricultural University (PAU)
Ludhiana
Punjab
India

Neha Patil
PDEA's Annasaheb Magar
Mahavidyalaya
Pune
Maharashtra
India

Suresh Chandra Phulara
Department of Biotechnology
Koneru Lakshmaiah Education Foundation
Guntur
Andhra Pradesh
India

B. Jeberlin Prabina
Tamilnadu Agriculture University
Coimbatore
Tamilnadu
India

Varsha Rani
Department of Biotechnology
Shoolini University
Solan
Himachal Pradesh
India

Rina Rani Ray
Maulana Abul Kalam Azad University of Technology
Kolkata
West Bengal
India

Balram Sahu
Department of Agricultural Microbiology
Indira Gandhi Krishi Vishwavidhyalaya
Raipur
Chhattisgarh
India

Pawan Saini
Department of Plant Breeding and Genetics
Punjab Agricultural University (PAU)
Ludhiana
Punjab
India

and

CSB-Central Sericultural Research & Training Institute (CSR&TI)
Pampore
Jammu-Kashmir (UT)
India

Pooja Saini
Department of Plant Breeding, Genetics and Biotechnology, Dr. Khem Singh Gill College of Agriculture
Eternal University (EU)
Baru Sahib
Himachal Pradesh
India

Deepka Sharma
Department of Biotechnology, College of Horticulture and Forestry Dr. YSP University of Horticulture and Forestry
Hamirpur
Himachal Pradesh
India

Mahaveer P. Sharma
ICAR-Indian Institute of Soybean Research
Indore
Madhya Pradesh
India

Prachie Sharma
Manav Rachna International Institute of Research and Studies
Faridabad
Haryana
India

Sonali Shinde
MES Abasaheb Garware College
Pune
Maharashtra
India

Ravindra Soni
Department of Agricultural Microbiology
Indira Gandhi Krishi Vishwavidhyalaya
Raipur
Chhattisgarh
India

Nimmy Srivastava
Amity Institute of Biotechnology
Amity University
Ranchi
Jharkhand
India

Deep Chandra Suyal
Department of Microbiology, Akal College of Basic Sciences
Eternal University
Baru Sahib
Sirmour
Himachal Pradesh
India

Neha Trivedi
Indian Agricultural Research Institute
New Delhi
India

Amit Verma
S.D. Agricultural University
Palanpur
Gujarat
India

Madan L. Verma
Department of Biotechnology, Indian Institute of Information Technology
Una
Himachal Pradesh
India

Shulbhi Verma
S.D. Agricultural University
Palanpur
Gujarat
India

Meghmala Waghmode
PDEA's Annasaheb Magar Mahavidyalaya
Pune
Maharashtra
India

Preface

The microorganisms are of utmost importance to the survival of plants, animals, and humans. Microorganisms interact with their hosts in both positive and negative manners. Therefore, the microorganisms and the technologies based on the microbial processes find many useful applications in the sustainable development of environment, energy, health, and agriculture. The improvements in crop productivity and management of plant diseases require a good understanding of the interactions between microorganisms with different parts of the plants. The plant–microbiome interaction denotes the genome of all the microorganisms, including bacteria, viruses, and fungi, living in and on various parts of plants. In recent times, significant technological progress has been made to improve our understanding of the plant-microbiome or phytomicrobiome interactions. In order to make agriculture sustainable, it is indeed necessary to learn about various concepts and advances in this field.

This book includes a discussion on some basic concepts, including rhizosphere and phyllospheremicrobiome, analysis of root exudates, role of phytomicrobiome in biotic and abiotic stress management, etc. The book also covers advanced information about metagenomics and proteomics studies in relation to plant microbial community analysis. Modern sustainable practices, including rhizosphere engineering, have also been touched upon.

After numerous deliberations, we came up with the idea to explore the possibility of developing a book on phytomicrobiome interactions and sustainable agriculture. We had in our mind to develop a book that will be beneficial for teachers and students, as well as researchers. Therefore, we invited a variety of researchers and convinced them to contribute the chapters for this book.

We are thankful to our family, friends, students, teachers, and mentors who acted as a source of inspiration to us. At this point we must appreciate the kind gesture of the entire Wiley team, especially Rebecca Ralf (Commissioning Editor), Athira Menon (Project Editor), and Kerry Powell (Managing Editor) who gave us support and showed trust in our capabilities. They generously extended the book timelines in the hard times of the COVID-19 pandemic and kept supporting us continuously, and as a result, we were able to complete this book in its present form.

This has been our maiden effort to produce a book on phytomicrobiome interactions for the sustainable development of agriculture to help students, teachers, and researchers. We hope that we will get support from the readers of the book. We are always open for criticism, suggestions, and recommendations that can help us to explore other aspects of phytomicrobiome interactions for improving crop productivity.

We dedicate this book to all the persons who are directly or indirectly serving the people affected by COVID-19.

Amit Verma
Jitendra Kumar Saini
Abd El-Latif Hesham
Harikesh Bahadur Singh
India, July 2020

About the Editors

Dr. Amit Verma is currently working as Assistant Professor in the Department of Biochemistry, S D Agricultural University, India. He worked under Dr. D.N. Kamra, National Professor, ICAR, India as Research Associate in the Department of Animal Nutrition, IVRI, Bareilly India and studied different aspects of Rumen enzymology. He received his doctorate from G. B. Pant University of Agriculture and Technology, Pantnagar, and investigated the different industrial applications of the Keratinase enzyme. He was awarded with the University Merit Scholarship and CSIR-JRF, CSIR, India during his doctorate. He qualified for examinations, such as GATE, ICMR JRF, ICAR SRF and ASRB NET. He has keen interest in the field of microbial biotechnology and plant–microbe interactions. His current work is focused on the rhizosphere metabolite investigation of arid plants and phytomicrobiome composition of castor crops under biotic stress. He is a member of many national and international professional bodies. He has written more than 25 research and review articles in reputed peer-reviewed journals along with 3 books, 15 book chapters, and more than 15 conference proceedings. He also received a Young Scientist Award from the VIRA foundation, India. He also serves as regular reviewer for international reputed journals with high-impact factors such as (i) *Scientific Reports* (Impact factor 4.122); (ii) *Frontiers in Microbiology* (Impact factor 4.019); (iii) *Frontiers in Plant Science* (Impact factor 3.678); (iv) *PLOS ONE* (Impact factor 2.77); (v) *Protoplasma* (Impact factor 2.8).

Dr. Jitendra Kumar Saini received his BSc (Industrial Microbiology) and MSc (Microbiology) degrees from Gurukula Kangri University, Haridwar. He obtained his PhD in Microbiology from Gobind Ballabh Pant University of Agriculture and Technology, Pantnagar in 2010 after which he worked as a postdoctoral associate at GADVASU, Ludhiana in a World Bank–funded NAIP project on Rumen microbiology. Later he joined the DBT-IOC Centre for Advanced Bioenergy Research, Indian Oil Corporation Ltd., Research and Development Centre, Faridabad as a scientific officer, where he led a team on enzyme development for advanced biofuels. Dr. Saini's work on cellulosic bioethanol production employing thermotolerant yeast won the best poster award at the "International Conference on Emerging Trends in Biotechnology-2014" held at JNU, New Delhi. He joined the Department of Microbiology, Central University of Haryana, Mahendergarh, in 2016 as an Assistant Professor. His current research focuses on enzyme and microbial technologies for sustainable development of energy and environment. Dr. Saini is a recipient of the Early Carrier Research grant from the Science and Engineering Research Board, Department of Science and Technology, Government of India, and a twinning grant from the Department of Biotechnology, Government of India. He is currently supervising two doctoral and 4 postgrad dissertation students, and has supervised 11 postgrad dissertations in the past, besides co-supervising a postdoc. He has filed one US patent, is an author of 23 articles, and is an active reviewer for many reputed journals in biofuel and bioenergy research. Dr. Saini conducted a one-week Global Initiative of Academic Networks course on "Integrated Lignocellulosic Biorefineries for Sustainable Development." Recently, he organized an International Conference AMI-2019 entitled "Microbial Technologies in Sustainable Development of Energy, Environment, Agriculture & Health" as an organizing secretary. He is the review editor for the journal, *Frontiers in Energy Research*. Dr. Saini is a life member of the Association of Microbiologists of India (AMI) and the Asian Federation of Biotechnology (AFOB).

Abd El-Latif Hesham is the Professor of Microbial Genetics and Environmental Meta-Genome Biotechnology, Genetics Department, Faculty of Agriculture, Beni-Suef University, Egypt. He also served as the Professor of Microbial Genetics and Environmental Meta-Genome Biotechnology in the Genetics Department, Faculty of Agriculture, Assiut University, Egypt. He graduated and got his M.Sc. from the Genetics Department, Faculty of Agriculture, Assiut University, Egypt, and his PhD degree from the Chinese Academy of Sciences in Microbial Genetics and Environmental Meta-Genome Biotechnology. He has been awarded postdoctoral studies from CAS-TWAS. He is an expert in microbial genetics and biotechnology, biodegradation, bioremediation and phytoremediation, microbial community structure, soil microbiology and enzyme activities, biological control, antimicrobial activates, biofertilizer, biofuels, and environmental meta-genome biotechnology. He has authored more than 80 peer-reviewed publications in reputed Thomson Reuters high-impact factor journals and one book and seven book chapters in internationally reputed publishers like Elsevier, Springer-Nature, Taylor & Francis, and John Wiley & Sons. He is a key person in many national and international research projects related to field of microbial genetics and applied biotechnology. He has been a scientific and organizing committee member and invited speaker in various international conferences. He is also a recipient of several prestigious national and international awards including most recently, being named a member of the Egyptian National Biotechnology Network of Expertise (NBNE) and Academy of Scientific Research & Technology (ASRT), and has been appointed as the Country Representative for Egypt and the Arab Counties by the International Biodeterioration & Biodegradation Society (IBBS) UK, which belongs to the Federation of European Microbiological Societies (FEMS). He is an associate editor and editorial board member for journals, such as (i) *Scientific Reports*, (ii) *Frontiers in Microbiology*, (iii) *Current Bioinformatics*, (iv) *PeerJ*, (v) *All Life Journal*, (vi) *International Journal of Agriculture & Biology*, (vii) *Journal of Environmental Biology*, and (viii) *Biocatalysis and Agricultural Biotechnology*.

Dr. Harikesh Bahadur Singh is a Professor in the Department of Mycology and Plant Pathology, Institute of Agricultural Sciences, Banaras Hindu University, Varanasi, India. Over the past 35 years, Professor Singh has served Central Universities and CSIR Institutes with his teaching and research skills. Professor Singh has been decorated with several prestigious awards, including the CSIR Technology Award by the Honorable Prime Minister of India, the Vigan Bharati award, the Mundukar memorial award, the BRSI Industrial Medal, the Vigyan Ratna by CSTUP, the Panchanan Meheshwari Medal, and many more, for his key role in translating agriculturally important microorganisms from lab to land. To his credit, he has 20 US patents, 350 research papers, 17 edited books, 70 book chapters, and 55 review articles. Professor Singh is also a Fellow of the National Academy of Agricultural Sciences. Professor Singh is serving as an associate/academic/board editor with journals of international repute.

1

Plant Root Exudate Analysis

Recent Advances and Applications

Shulbhi Verma and Amit Verma

S.D. Agricultural University, Palanpur, Gujarat, India

1.1 Introduction

Root exudates are a group of chemical compounds released from plant roots, which contains the ability to alter the activity of soil and microbes below the ground. These chemical compounds are the basis of interaction among root, soil, and microbes in the rhizosphere and also assist in root colonization. Such interaction assists in development- and defense-related activity in plants through enhancing the nutritional availability and suppressing disease occurrences by the alleviation of various stress factors. Root exudates pave the way for communication within the rhizosphere around plant roots, play a pivotal role in the soil system, and create phytomicrobiomes. Generally, plant root exudate consists of 10–40% carbon, primary and secondary metabolites, mucilage, protein, strigolactones, phytohormone, steroids, terpenoids, fatty acids, and nucleotide derivatives, which act as modifiers of rhizospheric soil (Figure 1.1).

Moreover, different parts of the root release different chemical compounds into the soil, such as meristem, the root elongation zone exudes asparagine and threonine, the root hair zone releases various hydrophobic and hydrophilic amino acids, including phenylalanine, leucine, valine, and glutamic acid, aspartic acid excretes from all parts of the root zone, and the root tip secretes mucilage (Badri and Vivanco 2009). Root exudates characteristics and amounts get changed according to plant genotype, age, biotic, and abiotic stress state, health and disease conditions of the plant, the surface area of the root system, and root architecture (Badri and Vivanco 2009). Apart from the plant characteristics, the environmental factors also show an impact on the root exudation mechanism and exudate constituents (Canarini et al. 2019). Thus, understanding the mechanism of root exudation, and factors affecting it, is of prime importance to support environmentally sustainable agriculture and to alleviate the large use of agricultural chemicals that create biosphere contaminations.

Interaction of root exudates and microbes can change the face of the soil, and plants flourish under such conditions, which combats the ill effects of various environmental

Phytomicrobiome Interactions and Sustainable Agriculture, First Edition. Edited by Amit Verma, Jitendra Kumar Saini, Abd El-Latif Hesham, and Harikesh Bahadur Singh.

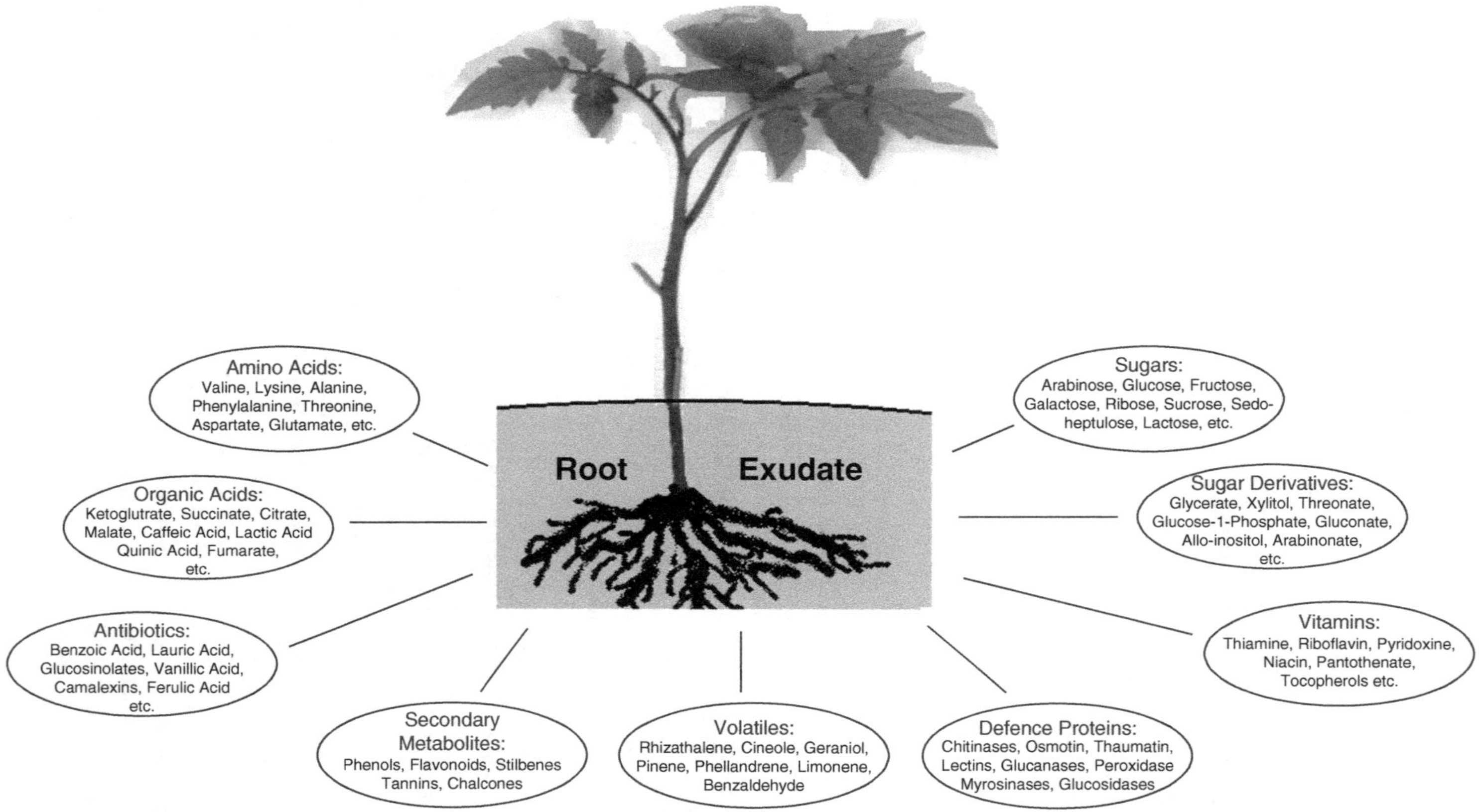

Figure 1.1 Classification of root exudates compounds: various compounds are found in the root exudation, which belongs to wide array of biochemical classes.

factors. Thus, root exudation directly effects plant development, which is beneficial to humans and the environment. It is well established that these interactions can promote the growth and productivity of plants, suppress plant disease, increase immunity and the defense of plants, increase nutrient availability and utilization of plants, and enable better coping mechanisms in stress conditions (Bais et al. 2006). Exudation and its interaction with microbes produces a great impact on soil fertility, which may replace the chemical fertilizer applications and pave a way toward eco-friendly cultivation traditions. It is a well-established fact that a link exists between the root exudate metabolites and the rhizosphere recruitment of different microbial communities (Schulz-Bohm et al. 2018; Sugiyama 2019). Apart from interaction studies, root exudates research is at a culminating point with the development of aptamers specifically for root exudates, which can be utilized for development in a fertilizer system (Emily et al. 2013).

Root exudates exploration revealed the high diversity of chemicals in plant root exudates. Exudation profiling revealed the presence of chemical compounds, and explains the function of concerning chemical in the soil ecosystem and their effects on plant systems. Few technologies, such as NMR and MS, narrate the structural characters of exudate compounds for clearer observations and assist in unravelling the properties of these compounds that are present in root exudates (Verma et al. 2018). Thus, these studies are directed to the chemical synthesis of those compounds, which can be further utilized for conditioning the soil for particular crop cultivation and for paving a way toward "rhizosphere engineering" (Bais et al. 2006). Root exudates profiling may prove to be the strongest technique for elaborating on the root role and overcomes the plant- and soil- related issues that hinders crop productivity and sustainable agriculture.

1.2 Root Exudates Composition: Collection and Analysis

Root exudates chemicals secreted through plant roots are necessary for the soil environment as well as for the plant. Root exudates analysis is quite essential in order to know the function of chemicals present in the exudates. There were several procedures to know the compounds present in the root exudates, such as high performance liquid chromatography (HPLC), UV, visible absorption, transcriptome analysis, metabolomics, proteomics, and gas chromatography–mass spectrometry (GS/MS) utilized to determine various compound present in exudates (Table 1.1).

There are several methods of root exudate collection from mature trees. The characterization of its composition has revealed the hidden capacity of root exudates in stimulating microbial activity and decomposition activity of organic soil matter (Phillips et al. 2008). However, the method depends on a realistic architecture to resemble the natural rhizosphere and doesn't accompany an accumulation of other compounds usually not present in exudate. Root secretion possesses several biological active compounds that reduce the dependency on chemicals in the soil and is safe for environment (Carlsen et al. 2012). To avoid the impact of microbes on root exudates profiles, sterilized water cultivation systems were adopted (Badri et al. 2008; Oburger et al. 2013; Strehmel et al. 2014). Furthermore, hydroponics strategy permits exudate sampling with minimum root damage. Root exudates collected in nutrition solution at three and seven days from model organism

Table 1.1 Root exudation of different plants: biochemical nature of root exudate components and techniques involved in characterization.

Plant System	Exudate Profile	Technique	References
Maize (*Zea mays*)	Wide variety of fatty acids, nitrogenous compounds, organic acids, steroids, and terpenoid derivatives	GC–MS and ^{1}H NMR	Lima et al. (2014)
Lettuce (*Lactuca sativa*)	Exudate comprises amino acids, amides, sugars, sugar alcohols, organic acids	GC–MS	Neumann et al. (2014)
Arabidopsis	Varied composition of sugars, sugar alcohols, amino acids and phenolics at different developmental stages	GC–MS	Chaparro et al. (2013)
Tall fescue (*Lolium arundinaceum*)	Sugars, polyols, growth factors and vitamins, lipids, amines, phenolics, carboxylic acids, nucleosides and others	GC-(TOF) MS (gas chromatography –time of flight–mass spectrometry)	Guo et al. (2015)
Taro plants (*Colocasia esculenta*)	Organic acids like lactic acid, benzoic acid, *m*-hydroxybenzoic acid, *p*-hydroxybenzoic acid, vanillic acid, succinic acid and adipic acid.	GC–MS	Asao et al. (2003)
Potato (*Solanum tuberosum* L.)	Sugar quantification	Colorimetric estimation using glucose and fructose assay kits	Hoysted et al. (2018)
Maize (*Zea mays*)	Benzoxazinoids compounds	HPLC-DAD	Neal et al. (2012)
Beech forest (*Fagus sylvatica* L.)	Low-molecular organic acids	Ion chromatography	Shen et al. (1996)
Sugar Maple (*Acer saccharum* Marsh.)	Organic acids, carbohydrates, amino acids and amides	Thin-layer chromatography	Smith (1970)
Betula alleghaniensis, *Fagus grandifolia*, and *Acer saccharum*	Carbohydrates, amino acids/amides, organic acids, and 9 inorganic ions	–	Smith (1976)
Fagopyrum esculentum	L-tryptophan, fructose-leucine or fructose-isoleucine, fructose-phenylalanine, *N*-acetyl glutamic acid methyl ester	UHPLC-HRMS	Gfeller et al. (2018)
Buckwheat (*Fagopyrumesculentum*)	Caprolactam, an inhibitory allelochemical	1H and 13C NMR spectra	Tin et al. (2009)
Mesocosms having different plants	organic carbon	HPLC-IRMS system	Karlowsky et al. (2018)
Datiscaglomerata	Flavonoids with abundant aglycones	HPLC, UV absorption and LC–MS	Gifford et al. (2018)
Medicago spp.	Flavonoids exclusively isoflavonoids	HPLC, UV absorption and LC–MS	Gifford et al. (2018)

Arabidopsis thaliana for analysis by LC–MS revealed the difference in wild and mutant root exudates composition (Badri et al. 2008). As days increase, root exudates secretion increases, as shown in the seven-day root exudates collection period in a nutrient solution that it is quite sufficient for analysis of chemical composition through ultraperformance liquid chromatographic–electrospray-ionization technique–time of-flight mass spectrometry (UPLC/ESI-QTOFMS). This estimated 103 compounds includes different classes of biological compounds related to nucleic acids, secondary metabolites, amino acids, and defense molecules (Strehmel et al. 2014). In contrast, six hours of exudate of *Zea mays* exposed to different nutrient deficiencies were analyzed through HPLC and statistics were used to reveal the strong influence of a nutritional status of a plant to cope with a limiting nutrient supply (Carvalhais et al. 2010). The limitation of such hydroponic medium collection of exudates is a very different type of plant development as compared to a natural solid substratum growth of a plant. Therefore, a solid medium for exudate collection is designed utilizing vermiculite, sand, perlite, etc., which mimics the rhizospheric conditions more closely. Transcriptome and other omic technologies involved in the analysis of exudate from different microbes reveals the gene and protein involved in the plant microbe's interaction. However, root exudate collection from the natural rhizospheric region is still one of the best methods to collect and analyze root exudate of plants under different conditions. Apart from this, there is a requirement of keeping the physical parameter near to the natural conditions, especially pH and temperature conditions, during the collection of root exudate. Through root exudation analysis we came to know about the impact on microbes and function in rhizosphere, however, its collection is a major constraint during its analysis, due to a lower concentration and presence of many compounds in several times a higher concentration, which is irrelevant to root exudation. So, there is a requirement of using a combination of collection methods to analyze the root components with high accuracy in order to assign the function imparted by the respective component in the growth and development of plants.

1.3 Role of Root Exudates in Shaping Rhizospheric Microbiomes

Root exudate composition resembles the plant constitution and its release of the organic compounds from the root zone allows it to participate in the rhizodeposition process (Jones and Darrah 1995). Root exudation is composed of a tremendous range of chemical compounds, including primary as well as secondary metabolites, ions, mucilage, reactive oxygen molecules, water molecules, amino acids, enzymes, peptides, sugars, vitamins, nucleotides, organic acids, plant inhibitors, growth regulators, sterols, fatty acids, phenolic compounds, flavonoids, and other miscellaneous chemicals (Bais et al. 2006; Huang et al. 2014; Tsuno et al. 2019). Root exudates contain the capacity to modulate the plant microbe interaction in the rhizosphere (De Weert et al. 2002; Zwetsloot et al. 2019). Moreover, these chemical compounds help microbes in the activity in the soil around the root, which is beneficial to plant development as well as to regain soil fertility.

Legume plants are considered an excellent example of shaping rhizospheric microbes through nitrogen assimilation through symbiotic associations of rhizobacteria. The key factor of plant rhizobia interaction in shaping rhizosphere is that flavonoid composition exists in the root exudates of legumes (Abdel Lateif et al. 2012; Weston and Mathesius 2013). Recently, researchers focused on microorganisms other than rhizobiums, which have the capacity of nitrogen fixation, such as cyanobacteria, Frankia, diazotrophsbacteria, etc. Several non-leguminous plants have rhizobia for nitrogen fixation, such as in *Parasponia*sp. (family Cannabaceae), which may assist in the shaping of rhizospheric microbes.

Altering the behavioral changes in microorganisms through the effect of root exudates is proven and has been recently termed as "rhizosphere engineering". In *A. thaliana,* root secretion possesses antimicrobial compounds that are responsible for its resistance to many non-host pathogens that are present in the rhizosphere. Phenolic compounds and antimicrobial protein of root exudates may help in guarding the roots against potential pathogens (Bais et al. 2005). In experimental research, evidence shows that root exudate diversity is a crucial link between plant diversity and soil microorganisms (Steinauer et al. 2016). For better dynamic in-depth interaction knowledge apart from phenolic compounds, flavonoid molecules and several biomolecules possess the capacity to increase favor around the rhizosphere. There are enormous biological molecules in root exudates, such as rosmarinic acid, jasmonic acid, and napthoquinones, which direct microbes toward plants for mutual benefit (Brigham et al. 1999; Bais et al. 2002; Carvalhais et al. 2015). Furthermore, there are few biomolecules, such as flavonoid, strigolactones, and terpenoids, which assist plants in specifically attracting the beneficial counter-partner through root signals (Vranova et al. 2013; Lareen et al. 2016; Massalha et al. 2017). Mutualistic interactions, nodulation and mycorrhizal interactions, signal perception and transduction such as receptor-like kinases (RLKs) are the signal-based interaction of plant and microbes (Lagunas et al. 2015).

In the consecutive discussion, root excaudate's role as a messenger that communicates within rhizosphere is examined. This perspective is well defined in root–root communication involved in an allelopathy phenomenon in the rhizosphere, which contributes in agricultural growth, because rhizospheric allelochemicals are protective in nature for plants. Root microbe communication in the rhizosphere may enhance the plant biomass through an increase in nutrition uptake, secretion of phytohormones, and helps in defense of the plant (Robin et al. 2008).

1.4 Applications of Root Exudation

Root exudate varies across the plant genotype and changes in root exudates of plant can be utilized in the breeding program for the enhancement of nutritional interaction among plants and microbes. For targeted breeding, perspective profiling of root exudates is essential for sustainable agriculture (Kuijken et al. 2015). This phenomenon is well-known through exudate profiles among 19 natural Arabidopsis accessions, which shows a high variation in glycosylated and sulfated metabolites, plant hormones, salicylic acid catabolites, phenylpropanoids, coumarin scopoletin, and polyamine derivatives (Monchgesang et al. 2016). Similar studies were done and demonstrated about the rhizobacterial community composition influenced by varying exudation profiles (Micallef et al. 2009). Root exudate chemical composition changes in the condition of nutrient deprivation, which can further assist said breeding program. Few studies represent the effect of nutrient deficiency in the exudates, and the effect of phosphate limitation investigated in Arabidopsis results in the high abundance of oligolignols, which is responsible for lignifications and coumarins in low quantity (Ziegler et al. 2016). Particularly Fe-deficient strawberry root exudates show a high content of dehydroascorbic acid, trans ferulic acid, galactonic acid, sucrose, and thymidine, whereas P-deficient strawberry root exudates show higher concentration of malic acid, lysine, galactaricacid, butylamine, and simultaneously show a low concentration of ribonic acid, sorbitol 6 phosphate, and proline (Valentinuzzi et al. 2015).

In the environment, plants are exposed to all kinds of friendly and unfriendly microbes with several stresses in the ground. Plant secretion called root exudate has a lot of potential from the perspective of defense. Root exudates possess antimicrobial chemicals, such as phytoanticipins, diterpene rhizathalene A, phytoalexins, phenylpropanoids, t-cinnamic acids, momilacton A, rosmarinic acids, terpenoids, benzoxazinoids, and the defense signaling molecules, salicylic acid, nitric oxide, and methyl jasmonate, which assist in the defense mechanisms of the plant. In addition, these chemical phenolics and terpenoids act as an antimicrobial apart from defense root exudates, which support the stress condition of plants due to their primary and secondary metabolite present in root exudates (Table 1.2).

Plant root exudation indirectly controls resource competition by altering soil chemistry, soil process, and microbial populations, and thus has an important function in plant development. Root exudation possesses the capacity to alter the soil nutrient availability by changing the soil property in the aspect of its chemistry and biology. Root exudate releases mucilaginous substance from root tip of the plant and is a part of exudation quite essential to maintain the water potential (Susan 2018). Root exudate releases plant carbon compounds (border cells and exudates) and primary metabolites into the rhizospheric soil (Canarini et al. 2019). It is important to understand root-mediated communication between plants and other organisms, which assists in the enhancement of agricultural production and can be useful in reduction in the demand for chemical fertilizer, such as in legume plants.

1.5 Conclusion and Future Prospects

Today's main concern of agriculture is the quality of productivity in order to feed the world population. For more production and protection of crops, farmers use chemicals that directly or indirectly effects the fertility of the soil as well as the consumers of the crop. To alleviate the chemical inputs in the agriculture requires detailed information about root exudation and its influence on plant development through soil condition modulation in terms of both biotic as well as abiotic components. Production of the crop depends on plant genotype and soil ecosystem. Root exudate is the key factor in the rhizospheric interaction due to nutrient assets. Researchers should emphasize root exudates, through profiling and their impact on microbes and soil in the rhizosphere for more exploration of root exudate. Still there are several unexplored chemicals of root exudates that can create magnificent changes in the soil and plants in terms of fertility, defense, growth, and yield. A few relevant and essential components of root exudates can be used for developing variety through genetic engineering and breeding, which assists

Table 1.2 Developments in root exudate studies and their action mechanism.

S. No.	Root exudate component(s)	Plant system under study	Mechanism	References
1	Terpenoid class of compound: strigolactone	—	Recruitment of fungal species and establishment of Arbuscular mycorrhiza	Parniske (2008)
2	The isoflavones like daidzein, genistein, and coumestrol	Tribe Phaseoleae plants	Induces the nod gene expression in their rhizobial partners	Dakora (2000)
3	Sugars, sugar acids, amino acids and organic acids	Maize root	Effect of exudate components on the chemosensory systems of *Pseudomonas putida* KT2440	Lopez-Farfan et al. (2019)
4	A class of indole-derived plant chemical, benzoxazinoids	Cereal crops	Antifeedant, insecticidal, antimicrobial, and allelopathic activities are related with this exudate component	Wouters et al. (2016)
5	A phenolic compound luteolin	--	Acts as a potent and specific inducer of *nodABC* gene expression in *Rhizobium meliloti*.	Caetano-Anolles et al. (1988)
6	Catechol and flavonoids catechin, and quercetin	Maize (*Zea mays* L.)	Silicon-induced amelioration of aluminum toxicity	Kidd et al. (2001)
7	Benzoxazinoids, secondary metabolites in grasses	Maize (*Zea mays* L.)	Effects the interaction between maize and *Pseudomonas putida* KT2440.	Neal et al. (2012)
8	Cyclic hydroxamates	Maize (*Zea mays* L.)	Mediate aluminum toxicity resistance in plant	Poschenrieder et al. (2005)
9	Benzoxazinoids, secondary metabolites in grasses	Maize (*Zea mays* L.)	Regulator of innate immunity against aphids and fungi	Ahmad et al. (2005)
10	Carotenoid-derived strigolactones	Pea & *Arabidopsis*	Involved in inhibition of shoot branching in plants	Gomez-Roldan et al. (2008)
11	Soyasaponins	Soybean (*Glycine max*)	Regulates communication with beneficial microorganisms in soil	Tsuno et al. (2018)
12	Phytoestrogens	Clover legume	Male-biased sex ratios and accelerates male metamorphic timing in wood frogs	Lambert (2015)
13	Phenol glycoside: rutin	*Eucalyptus globulus ssp. bicostata* r	Regulate orientation of hyphal elongation toward the root tip, thereby favoring mycorrhizal infection	Lagrange et al. (2001)
14	Artificial root exudate having glucose, organic acids, and serine	—	These components of root exudate accelerate the degradation of pyrene in soil, especially glucose	Lu et al. (2017)

15	Organic acids	Legumes *Lotus corniculatus* L. and *Trifoliumarvense* L. and the grass *Calamagrostis epigeios* (L.)	Influence of these exudation components on plant nutrient acquisition	Boldt-Burisch et al. (2019)
16	Root exudate components: proline and phytohormones	Citrus plants	Alleviation of abiotic stress conditions.	Vives-Peris et al. (2017)
17	Root exudate components: proline and salicylates	Citrus plants	Recruitment of rhizobacteria and alleviation of abiotic stress conditions.	Vives-Peris et al. (2018)
18	1-alanine, 1-proline and oxalic acid of root exudate	*Sedum alfredii*	Phytoremediation of lead contaminated soils	Luo et al. (2017)
19	Organic acids like malic, lactic, acetic, succinic, citric, and maleic acids in root exudates	Corn hybrids (*Zea mays* L.)	These exudate components might contribute to drought tolerance in corn hybrids	Song et al. (2012)
20	Organic acid exudates	White spruce (*Picea glauca*) and subalpine Fir (*Abieslasiocarpa*)	Significant role of exudate organic acids in the transformation of mica and chlorite into smectite in rhizospheric soils	Tuason and Arocena (2009)
21	Root exudate metabolome: primary and secondary metabolites	*Quercus ilex* (holm oak)	Exudate components were changed irreversibly by the lack of water under extreme drought conditions	Gargallo-Garriga et al. (2018)
22	Seed exudate: carbon and nitrogen contents	Chia (*Salvia hispanica*)	Plant exudates greatly affect the physical behavior of soil	Oleghe et al. (2017)
23	Exudate components like total free amino acids, proline, potassium, and some phytohormones	Barley (*Hordeum vulgare* L.)	Root exudate is an important factor in the context of crop performance and carbon balance under conditions of climate change	Olga et al. (2017)
24	Sakuranetin and sorgoleone of root exudate	Sorghum (*Sorghum bicolor* L.)	Biological nitrification inhibition activity	Subbarao et al. (2013)
25	1,9-decanediol	Rice (*Oryza sativa*)	Blocks the ammonia monooxygenase pathway of ammonia oxidation and effects nitrogen use efficiency	Sun et al. (2016)

in a sustainable agriculture system. Several chemical inducers should be explored, which increases the quantity of root exudates for better performance of the rhizosphere. In conclusional, in revealing root exudation secrets, we can support economical agricultural productivity through achievement of environmental sustainability.

References

Abdel Lateif, K., Bogus, D., and Hocher, V. (2012). The role of flavonoids in the establishment of plant roots endosymbioses with arbuscular mycorrhiza fungi, rhizobia and Frankia bacteria. *Plant Signal. Behav.* 7 (6): 636–641.

Ahmad, S., Veyrat, N., Gordon Weeks, R. et al. (2005). Benzoxazinoid metabolites regulate innate immunity against aphids and fungi in maize. *Plant Physiol.* 157 (1): 317–327. https://doi.org/10.1104/pp.111.180224.

Asao, T., Hasegawa, K., Sueda, Y. et al. (2003). Autotoxicity of root exudates from taro. *Sci. Hortic.* 97: 389–396. https://doi.org/10.1016/S0304-4238(02)00197-8.

Badri, D.V. and Vivanco, J.M. (2009). Regulation and function of root excudates. *Plant Cell Environ.* 32: 666–681.

Badri, D.V., Loyola-Vargas, V.M., Broeckling, C.D. et al. (2008). Altered profile of secondary metabolites in the root exudates of *Arabidopsis*. *Plant Physiol.* 146 (2): 762–771.

Bais, H.P., Walker, T.S., Schweizer, H.P., and Vivanco, J.M. (2002). Root specific elicitation and antimicrobial activity of rosmarinic acid in hairy root cultures of sweet basil (*Ocimumbasilicum* L.). *Plant Physiol. Biochem.* 40: 983–995.

Bais, H.P., Prithiviraj, B., Jha, A.K. et al. (2005). Mediation of pathogen resistance by exudation of antimicrobials from roots. *Nature* 434: 217–221.

Bais, H.P., Weir, T.L., Perry, L.G. et al. (2006). The role of root exudates in rhizosphere interactions with plants and other organisms. *Annu. Rev. Plant Biol.* 57: 233–266.

Boldt-Burisch, K., Bernd, U., Annenaeth, M., and Fhüttl, R. (2019). Root exudation of organic acids of herbaceous pioneer plants and their growth in sterile and non-sterile nutrient-poor, sandy soils from post-mining sites. *Pedosphere* 29 (1): 34–44.

Brigham, L.A., Michaels, P.J., and Flores, H.E. (1999). Cell-specific production and antimicrobial activity of naphthoquinones in roots of *Lithospermumerythrorhizon*. *Plant Physiol.* 119: 417–428.

Caetano-Anolles, G., Crist-Estes, D.K., and Bauer, D.W. (1988). Chemotaxis of *Rhizobium meliloti* to the plant flavone luteolin requires functional nodulation genes. *J. Bacteriol.* 170: 3164–3169.

Canarini, A., Kaiser, C., Merchant, A. et al. (2019). Root exudation of primary metabolites: mechanisms and their roles in plant responses to environmental stimuli. *Front. Plant Sci.* 10: 157. https://doi.org/10.3389/fpls.2019.00157.

Carlsen, S.C.K., Pedersen, H.A., Spliid, N.H., and Fomsgaard, I.S. (2012). Fate in soil of flavonoids released from white clover (*Trifoliumrepens* L.). *Appl. Environ. Soil Sci.* 2012: 1–10.

Carvalhais, L.C., Dennis, P.G., Fedoseyenko, D. et al. (2010). Root exudation of sugars, amino acids, and organic acids by maize as affected by nitrogen, phosphorus, potassium, and iron deficiency. *J. Plant Nutr. Soil Sci.* 174: 3–11.

Carvalhais, L.C., Dennis, P.G., Badri, D.V. et al. (2015). Linking jasmonic acid signalling, root excudates, and rhizosphere microbiomes. *Mol. Plant Microb. Interact.* 28 (9): 1049–1058.

Chaparro, J.M., Badri, D.V., Bakker, M.G. et al. (2013). Root exudation of phytochemicals in *Arabidopsis* follows specific patterns that are developmentally programmed and correlate with soil microbial functions. *PLoS One* 8 (2): e55731. https://doi.org/10.1371/journal.pone.0055731.

Dakora, F.D. (2000). Commonality of root nodulation signals and nitrogen assimilation in tropical grain legumes belonging to the tribe *Phaseoleae*. *Aust. J. Plant Physiol.* 27: 885–892.

De Weert, S., Vermeiren, H., Mulders, I.H.M. et al. (2002). Flagella-driven chemotaxis towards exudate components is an important trait for tomato root colonization by *Pseudomonas fluorescens*. *Mol. Plant Microbe Interact.* 15 (11): 1173–1180.

Emily, M., Maureen, M., Carlos, M., and Maria, D.R. (2013). Development and application of crop exudates specific aptamers. *J. Biomol. Struct. Dyn.* 31 (sup1): 89–89.

Gargallo-Garriga, A., Preece, C., Sardans, J. et al. (2018). Root exudate metabolomes change under drought and show limited capacity for recovery. *Sci. Rep.* 8: 12696. https://doi.org/10.1038/s41598-018-30150-0.

Gfeller, A., Glauser, G., Etter, C. et al. (2018). *Fagopyrum esculentum* alters its root exudation after *Amaranthus retroflexus* recognition and suppresses weed growth. *Front. Plant Sci.* 9: 50. https://doi.org/10.3389/fpls.2018.00050.

Gifford, I., Battenberg, K., Vaniya, A. et al. (2018). Distinctive patterns of Flavonoid biosynthesis in roots and nodules of *Datiscaglomerata* and *Medicago* spp. revealed by metabolomic and gene expression profiles. *Front. Plant Sci.* 9: 1463. https://doi.org/10.3389/fpls.2018.01463.

Gomez-Roldan, V., Soraya, F., Philip, B.B. et al. (2008). Strigolactone inhibition of shoot branching. *Nature* 455: 189–194.

Guo, J., McCulley, R.L., and McNear, D.H.J. (2015). Tall fescue cultivar and fungal endophyte combinations influence plant growth and root exudate composition. *Front. Plant Sci.* 6: 183. https://doi.org/10.3389/fpls.2015.00183.

Hoysted, G.A., Bell, C.A., Lilley, C.J., and Urwin, P.E. (2018). Aphid colonization affects potato root exudate composition and the hatching of a soil borne pathogen. *Front. Plant Sci.* 9: 1278. https://doi.org/10.3389/fpls.2018.01278.

Huang, X.F., Chaparro, J.M., Reardon, K.F. et al. (2014). Rhizosphere interactions: root exudates, microbes, and microbial communities. *Botany* 92: 267–275.

Jones, D.L. and Darrah, P.R. (1995). Influx and efflux of organic acids across the soil root interface of *Zea mays* L. and its implications in rhizosphere C flow. *Plant and Soil* 173: 103–109.

Karlowsky, S., Augusti, A., Ingrisch, J. et al. (2018). Drought-induced accumulation of root exudates supports post-drought recovery of microbes in mountain grassland. *Front. Plant Sci.* 9: 1593. https://doi.org/10.3389/fpls.2018.01593.

Kidd, P.S., Llugany, M., Poschenrieder, C. et al. (2001). The role of root exudates in aluminium resistance and silicon-induced amelioration of aluminium toxicity in three varieties of maize (*Zea mays* L.). *J. Exp. Bot.* 52: 1339–1352.

Kuijken, R.C., van Eeuwijk, F.A., Marcelis, L.F., and Bouwmeester, H.J. (2015). Root phenotyping: from component trait in the lab to breeding. *J. Exp. Bot.* 66 (18): 5389–5401.

Lagrange, H., Jay-Allgmand, C., and Lapeyrie, F. (2001). Rutin, the phenolglycoside from eucalyptus root exudates, stimulates *Pisolithus* hyphal growth at picomolar concentrations. *New Phytol.* 149: 349–355.

Lagunas, B., Schäfer, P., and Gifford, M.L. (2015). Housing helpful invaders: the evolutionary and molecular architecture underlying plant root-mutualist microbe interactions. *J. Exp. Bot.* 66 (8): 2177–2186.

Lambert, M.R. (2015). Clover root exudate produces male-biased sex ratios and accelerates male metamorphic timing in wood frogs. *R. Soc. Open Sci.* 2 (12): 1–8. https://doi.org/10.1098/rsos.150433.

Lareen, A., Burton, F., and Schafer, P. (2016). Plant root-microbe communication in shaping root microbiomes. *Plant Mol. Biol.* 90 (6): 575–587.

Lima, L.D.S., Olivares, F.L., de Oliveira, R.R. et al. (2014). Root exudate profiling of maize seedlings inoculated with *Herbaspirillumseropedicae* and humic acids. *Chem. Biol. Technol. Agric.* 23 (1): 1–18.

Lopez-Farfan, D., Reyes-Darias, J.A., Matilla, M.A., and Krell, T. (2019). Concentration dependent effect of plant root exudates on the chemosensory systems of *Pseudomonas putida* KT2440. *Front. Microbiol.* 10: 78. https://doi.org/10.3389/fmicb.2019.00078.

Lu, H., Jianteng, S., and Lizhong, Z. (2017). The role of artificial root exudate components in facilitating the degradation of pyrene in soil. *Sci. Rep.* 7: 7130.

Luo, Q., Wang, S., Sun, L., and Wang, H. (2017). Metabolic profiling of root exudates from two ecotypes of *Sedum alfredii* treated with Pb based on GC-MS. *Sci. Rep.* 7: 39878.

Massalha, H., Korenblum, E., Tholl, D., and Aharoni (2017). Small molecules below-ground: the role of specialized metabolites in the rhizosphere. *Plant J.* 90 (4): 788–807.

Micallef, S.A., Shiaris, M.P., and Colón-Carmona, A. (2009). Influence of *Arabidopsis thaliana* accessions on rhizobacterial communities and natural variation in root exudates. *J. Exp. Bot.* 60 (6): 1729–1742.

Monchgesang, S., Strehmel, N., Schmidt, S. et al. (2016). Natural variation of root exudates in *Arabidopsis thaliana*-linking metabolomic and genomic data. *Sci. Rep.* 6: 29033. https://doi.org/10.1038/srep29033.

Neal, A.L., Ahmad, S., Gordon-Weeks, R., and Ton, J. (2012). Benzoxazinoids in root exudates of maize attracts *Pseudomonas putida* to the rhizosphere. *PLoS One* 7: e35498. https://doi.org/10.1371/journal.pone.0035498.

Neumann, G., Bott, S., Ohler, M.A. et al. (2014). Root exudation and root development of lettuce (*Lactuca sativa* L. cv. Tizian) as affected by different soils. *Front. Microbiol.* 5: 2. https://doi.org/10.3389/fmicb.2014.00002.

Oburger, E., Dell Mour, M., Hann, S. et al. (2013). Evaluation of a novel tool for sampling root exudates from soil-grown plants compared to conventional techniques. *Environ. Exp. Bot.* 87: 235–247.

Oleghe, E., Naveed, M., Baggs, E.M., and Hallett, P.D. (2017). Plant exudates improve the mechanical conditions for root penetration through compacted soils. *Plant and Soil* 421: 19. https://doi.org/10.1007/s11104-017-3424-5.

Olga, C.C., Franzaring, J., Schmid, I. et al. (2017). Atmospheric CO_2 enrichment and drought stress modify root exudation of barley. *Glob. Chang. Biol.* 23: 1292–1304. https://doi.org/10.1111/gcb.13503.

Parniske, M. (2008). Arbuscular mycorrhiza: the mother of plant root endosymbioses. *Nat. Rev. Microbiol.* 6: 763. https://doi.org/10.1038/nrmicro1987.

Phillips, R.P., Erlitz, Y., Bier, R., and Bernhardt, E.S. (2008). New approach for capturing soluble root exudates in forest soils. *Funct. Ecol.* 22: 90–999.

Poschenrieder, C., Tolra, R.P., and Barcelo, J. (2005). A role for cyclic hydroxamates in aluminium resistance in maize? *J. Inorg. Biochem.* 99 (9): 1830–1836.

Robin, A., Vansuyt, G., Hinsinger, P. et al. (2008). Iron dynamics in the rhizosphere: consequences for plant health and nutrition. *Adv. Agron.* 99: 183–225.

Schulz-Bohm, K., Gerards, S., Hundscheid, M. et al. (2018). Calling from distance: attraction of soil bacteria by plant root volatiles. *ISME J.* 12 (5): 1252–1262.

Shen, Y., Strom, L., Jonsson, J.A., and Tyler, G. (1996). Low-molecular organic acids in the rhizosphere soil solution of beech forest (*Fagus sylvatica* L.) cambisols determined by ion chromatography using supported liquid membrane enrichment technique. *Soil Biol. Biochem.* 28: 1163–1169. https://doi.org/10.1016/0038-0717(96)00119-8.

Smith, W.H. (1970). Root exudates of seedling and mature sugar maple. *Phytopathology* 60: 701–703. https://doi.org/10.1094/Phyto-60-701.

Smith, W.H. (1976). Character and significance of forest tree root exudates. *Ecology* 57: 24–331. https://doi.org/10.2307/1934820.

Song, F., Han, X., Zhu, X., and Herbert, S.J. (2012). Response to water stress of soil enzymes and root exudates from drought and non-drought tolerant corn hybrids at different growth stages. *Can. J. Soil Sci.* 92: 501–507.

Steinauer, K., Chatzinotas, A., and Eisenhauer, N. (2016). Root exudate cocktails: the link between plant diversity and soil microorganisms? *Ecol. Evol.* 6: 7387–7396.

Strehmel, N., Bottcher, C., Schmidt, S., and Scheel, D. (2014). Profiling of secondary metabolites in root exudates of *Arabidopsis thaliana*. *Phytochemistry* 108: 35–46.

Subbarao, G.V., Nakahara, T., Ishikawa, H. et al. (2013). Biological nitrification inhibition (BNI) activity in sorghum and its characterization. *Plant and Soil* 366 (1–2): 243–259.

Sugiyama, A. (2019). The soybean rhizosphere: metabolites, microbes, and beyond – a review. *J. Adv. Res.* (In Press) doi: https://doi.org/10.1016/j.jare.2019.03.005.

Sun, L., Lu, Y., Yu, F. et al. (2016). Biological nitrification inhibition by rice root exudates and its relationship with nitrogen-use efficiency. *New Phytol.* 212 (3): 646–656.

Susan, V.F. (2018). Root exudates affect soil stability, water repellency. *Sci. Daily* 04: 809–1833.

Tin, W.W.T., Hayashi, H., Otomatsu, T. et al. (2009). Caprolactam, an inhibitory allelochemical exuded from germinating buckwheat (*Fagopyrum esculentum*) seeds. *Heterocycles* 78: 1217–1222. https://doi.org/10.3987/Com-08-11601.

Tsuno, Y., Teruhisa, F., Keiji, E. et al. (2018). Soyasaponins: a new class of root exudates in soybean (*Glycine max*). *Plant Cell Physiol.* 59 (2): 366–375.

Tuason, M.M.S. and Arocena, J.M. (2009). Root organic acid exudates and properties of rhizosphere soils of white spruce (*Picea glauca*) and sub alpine fir (*Abieslasiocarpa*). *Can. J. Soil Sci.* 89 (3): 287–300. https://doi.org/10.4141/CJSS08021.

Valentinuzzi, F., Cesco, S., Tomasi, N., and Mimmo, T. (2015). Influence of different trap solutions on the determination of root exudates in *Lupinusalbus* L. *Biol. Fertil. Soils* 51 (6): 757–765.

Verma, A., Kumar, S., Hemansi, G.K. et al. (2018). Rhizosphere metabolite profiling: an opportunity to understand plant-microbe interactions for crop improvement. In: *Crop Improvement Through Microbial Biotechnology*, 1e (eds. R. Prasad, S.S. Gill and N. Tuteja), 343–361. Amsterdam: Elsevier.

Vives-Peris, V., Gómez-Cadenas, A., and Perez-Clemente, R.M. (2017). Citrus plants exude proline and phytohormones under abiotic stress conditions. *Plant Cell Rep.* 36 (12): 1971–1984. https://doi.org/10.1007/s00299-017-2214-0.

Vives-Peris, V., Molina, L., Segura, A. et al. (2018). Root exudates from citrus plants subjected to abiotic stress conditions have a positive effect on rhizobacteria. *J. Plant Physiol.* 228: 208–217. https://doi.org/10.1016/j.jplph.2018.06.003.

Vranova, V., Rejsek, K., Skene, K.R. et al. (2013). Methods of collection of plant root exudates in relation to plant metabolism and purpose: a review. *J. Plant Nutr. Soil Sci.* 176: 175–199.

Weston, L.A. and Mathesius, U. (2013). Flavonoids: their structure, biosynthesis and role in the rhizosphere, including allelopathy. *J. Chem. Ecol.* 39: 283–297.

Wouters, F.C., Blanchette, B., Gershenzon, J., and Vassao, D.G. (2016). Plant defence and herbivore counter-defence: benzoxazinoids and insect herbivores. *Phytochem. Rev.* 15: 1127–1151. https://doi.org/10.1007/s11101-016-9481-1.

Ziegler, J., Schmidt, S., Chutia, R. et al. (2016). Non-targeted profiling of semi-polar metabolites in Arabidopsis root exudates uncovers a role for coumarin secretion and lignification during the local response to phosphate limitation. *J. Exp. Bot.* 67 (5): 1421–1432.

Zwetsloot, M.J., Kessler, A., and Bauerle, T.L. (2019). Phenolic root exudate and tissue compounds vary widely among temperate forest tree species and have contrasting effects on soil microbial respiration. *New Phytol.* 218: 530–541.

2

Phytoproteomics

A New Approach to Decipher Phytomicrobiome Relationships

Prachie Sharma and Kapila Kumar

Manav Rachna International Institute of Research and Studies, Faridabad, Haryana, India

2.1 Introduction

The ecosystem thrives with the dynamism of interactions between abiotic and biotic aspects of life. The ecosystem harbors a wide diversity of species ranging from a plethora of microorganisms to macroorganisms. Every living being is predisposed to an assortment of nutritional as well as environmental requirements. These needs, when catered to by the ecosystem, make the living conducive. The most astounding feature of the ecosystem is that every living being is genetically distinct with unique living requirements yet tend to coexist with communal harmony within the same environment. This harmony is met with a balanced equilibrium of benefit-sharing among species for sustained and enhanced living.

Every organism is capable of managing its metabolic niches by first sequestering it through the resources available within the surrounding environment and subsequently subjecting it through metabolic processing for its use. The organisms can only survive in the ecosystem where there are enough resources for its utilization for their healthy living and growth. Nonetheless, at times for a particular metabolic requirement, the mutualistic benefit-sharing occurs between different species when the organism of one species is unable to fulfill its nourishment requirements and depends upon the other. This establishes a harmonious exchange of growth and enhancement benefits without breaching the living integrity of another. However, not all the species contains a harmonious tendency and tries to breach the living system of another to attain a competitive edge in survival. This acts as a detriment to the survival of other species on the whole.

Nonetheless, most of the species reside within an ecosystem with an effort of communal harmony among one another to support each other to sustain and maintain a healthy ecosystem. Hence an ecosystem comprises a consortium of species at the multitudes of genetic diversity existing with each other through a beautiful exchange of mutual benefits while preserving their interests.

Plant–microbe associations best exemplify such mutualistic associations in the ecosystem. The plants act as a meta-organism harboring a consortium of microorganisms living in

Phytomicrobiome Interactions and Sustainable Agriculture, First Edition. Edited by Amit Verma, Jitendra Kumar Saini, Abd El-Latif Hesham, and Harikesh Bahadur Singh.

symbiosis with a variety of plant organ systems. Microbes may colonize the various parts of the plants, be it plant surfaces, areas adjoining the areas around the plant surface, or the interstitial spaces present between the adjoining cells. A phyto-organ system apart from being the sheltered habitat and a future source of nutrition for the surrounding microbial population may attract the tiny creatures for their metabolic benefits by releasing such chemicals that stabilize the colonization of the beneficial microbial colonies. The growth of beneficial microbes, in turn, provides an increment in plant health. Moreover, the microbial signaling system is so developed that upon sensing some stress or threat, the microbial consortium confers changes in metabolism, thus rendering great adaptability and protection by releasing metabolites and antibiotic agents against pathogenic invasion.

Plants are of utmost importance from the view of the ecosystem as well as from the view of commercial viability. Plants form the framework of the ecosystem and provide nourishment and shelter to almost all living beings. Understanding the methods of the best sustenance of the plant within its natural system and providing further enhancement under stressed conditions become the foundation of studying mandatory growth parameters of plants. Since microbes work in close association of plants to enhance and enrich them, thus study of plant microbial associations are of great reputation to humankind as well as to the environment. Therefore, this chapter provides a crisp overview of the plant microbial associations and their manipulation through the proteomics approach, which shall further lead us to touch the proteomic perception of this association deciphered through technical knowhow.

2.2 Phytomicrobiome

Plants are meta-organisms which are the multicellular organisms encompassing a synergistic association with a suite of microorganisms such as fungi, bacteria, archaea, and viruses, as well as several algal symbionts (Bosch and McFall-Ngai 2011). Practically every plant component is colonized by a consortium of microorganisms, which in combination is designated as "plant microbiome" or"phytomicrobiome." Based on their habitation, the microbes are categories as endophytic (within the plant tissue), epiphytic (on the surface of the above-ground part), and rhizospheric (within or on the root surface) (McFall-Ngai 2002, 2008). The microbiome composition and their metabolic activity determines the interaction and sustenance of the host plant within the surrounding environment (Kostic et al. 2013). The rhizospheric zone of the plant hosts a high density of microbial population enthralled with high microbial activity as compared with the other zones of plant habitation even within the surrounding bulk soil (Santoyo et al. 2016). Nonetheless, the microbial diversity is low as compared to the bulk soil and the composition within the bulk soil and rhizosphere are pretty distinct indicating stringency in the selection of microbial population (Pisa et al. 2011).

Years of research have contributed to identifying the associations of microbes with subtly distinguished parts of the plants such as epiphytic and endophytic regions of both stems and leaves of the phyllospheric region; the rhizosphere and the reproductive components such as seeds, fruits, flowers, etc. In the epidermal region of the grape plant along with the xylem of ovaries and ovules, two species of bacillus and the *Pseudomonas* species have specifically been found to colonize endophytically, whereas only the Bacillus species have been found to colonize the seed cell wall of berries (Lugtenberg and and Kamilova 2009; Compant et al. 2010a, 2010b). The plant growth-promoting rhizobacteria such as *Diazotrophicus, Acetobacter*, etc., (Quecine et al. 2012) assist the plant to fix nitrogen from the atmosphere and have an affinity to the plant roots (Pisa et al. 2011) and stem regions of the sugarcane plant and have been found to reside in the apoplastic zone under a low-nitrogen and high-sucrose situation.

The major hub of the interspecies associations that happen is at the rhizosphere which surrounds the roots area of the plants. The population of colonization depends upon the biochemistry of the plant and the microbes plus the available resources within the soil of that locality. One such species that is present in the majority of rhizosphere is the *Streptomyces* sp. They deserve a special mention because they can colonize not only the parts of the plants associated with the soil but also the parts above the ground. These microbes offer a variety of benefits to the host plant where they release antibiotic substances, thereby rendering the plant protected from the malicious conditions present in the surroundings. They are particularly useful as biocontrol agents where they protect various crop plants against pathogenic as well as environmental stresses. They serve as biofertilizers as well owing to their plant growth-promoting abilities. Their sporulating ability aids their survival in the adverse environment rendering them competent over the other microbes. This species releases various lytic enzymes that can metabolize insoluble organic substances to generate nutrients which are imperative for the proper growth and functioning of plants.

The aspect of microbial plant growth promotion can also be utilized for phytoremediation when dealing with a plant sown in a soil laden with high metal contamination. Bioethanol producers such as *Helianthus tuberosus*, a high biomass yielding crop acquires enhancement in the accumulation and sustenance of high concentrations of zinc and cadmium. In another case, the microbial population has been found to be colonizing the internal region of the roots as an endophyte; thereby resulting in increased assimilation of cadmium. The key feature of the presence of metal-solubilizing bacteria is that it makes the host plant tolerant and decreases the effect of metal-induced stress (Montalbán et al. 2017).

Interestingly, such a phenomenon has also been observed for the enhanced salt and drought resistance owing to the effects of plant growth-promoting bacteria as studied in *Lolium perenne,* also known as ryegrass. Unfortunately for this perennially relevant grass, as turf and forage, it is highly sensitive to drought-induced stress and high salinity. However, *Bacillus amyloliquefaciencs* combined with hydrogels to prevent the soil from eroding can help the grass sustain the drought-induced stress by the plant (Su et al. 2017).

In the cotton plant, arbuscular mycorrhizal fungus is a soil-borne mutualistic symbiotic association with the terrestrial planets. One strain, *Rhizophagusirregularis,* acts antagonistically to the growth of *Verticillium dahlia* colonization and induces resistance by releasing the unknown volatiles thereby saving a most important crop plant (Zhang et al. 2018).

A beautiful symphony gets established to carry out mundane functions. The microbiome communities specific to the phyllosphere affect the development of plants along with its function within the ecosystem, whereas the host plant modulates the composition and effector functions of the phytomicrobiome. Environmentally governed factors have been found to amend several metabolite biosynthesis within the plant. Since the roots play a major role in establishing the health of the plant, the rhizomicrobial species participate in modulating plant growth, composition, and development. Interestingly enough, a study was found where just by treating the leaves with a specific consortium of phyllomicrobiome led to the prevention of feeding on the leaves by insect larvae (Badri et al. 2013).

However, it is believed that microbial community composition and their distribution in the phyllospheric region are somewhat arbitrary. Nonetheless, there is high-level specificity in accommodation of the microbial communities within the endospheric and rhizospheric regions (Lebeis 2015).

The study for an in-depth understanding of means of interaction has been attempted through studies involving biochemical, metabolomics, proteomics, and genomics studies. These studies aid in assimilating the knowledge pertaining to the direction of devising methods to enhance the yield of crop produce and biomass production without many additional inputs (Lebeis 2015). Knowledge of this kind underpins the engineering requirements of the phytomicrobiome, thereby catering in the manipulation of the beneficial consortium composition. This, in turn, caters to the plant microbial associations which bear the responsibility of generating signals that enhances the ability of the phytomicrobiome community to overcome various climatic and soil-related stressed or undesirable conditions leading to overall crop productivity (Quecine et al. 2012).

2.3 Phytomicrobiome: The Communication via Signaling

The phytomicrobial associations are very comfortably established between the plant and the microbes through the exchange of signals as a form of communication. The orchestration of signals occurs not only between the host plant and microbial species but also among the microbial species present in a consortium. Appropriate signaling may prevent an unfavorable association by eliciting the defenses against the pathogenic invasions. Signaling governs each other's activity in the consortium as well as the community (Engelmoer et al. 2014). In the event of chemical signals within the community aids positive association for the benefit by enabling the residence of endophytes (Hartmann et al. 2014) and prevents futile or harmful associations.

The microbes in the association of the host have a signature sequence at the molecular level which is highly conserved for that species. The term associated with this phenomenon is known as microbe-associated molecular patterns (MAMP). These molecular receptors present on the microbial species are recognized by pattern recognition receptors (PRR) which are also the ligand-binding ectodomains in plants. The best-known example of this MAMP is where the bacterial species have flagellin and chitin is specific to fungi (Newman et al. 2013). MAMP has one of the highlighting features in establishing plant immunity as well as eliciting antibiotic secretion in microbes. Bacterial species have also been known to interfere with signaling between plants and other strains of microbes. In order to enable pathogenic association, Bacillus strains have been shown to minimize the favorable MAMP-regulated immune response which may also hinder the secretion of antibiotic substances to enable and establish an infection successfully.

These chemical signals may also be termed as the stress signals, which, as the name suggests, are generated in the course of a stressed situation for the growth of the plant. These signals shall in turn elicit the microbial signal which prepares the microbe to provide aid to the stressed plant by its metabolic activity. Signaling compounds produced are either primary metabolic compounds such as carbohydrates, organic acids, and proteins; as well as secondary metabolites such as phenols, phytohormones, and flavonoids, etc. Microbes to host plant signaling compounds include compounds such as phenols, peptides, plant hormones, acyl homoserine lactones, etc., which can also act as microbe to microbe signal. The major contribution in the selection of microbial consortia around the roots happens by the secretions of compounds by the root exudates. *Arabidopsis thaliana* releases malic acid in defense of the attack by the foliage pathogen which stimulates the production of beneficial biofilms in the rhizosphere (Rudrappa et al. 2008).

2.4 Proteomics

Proteins are ubiquitous biomolecules, known to be present fundamentally in some form or another across all living beings. These are the manifestation of all the structural and functional information embedded within the genetic code of the organism. It forms the inevitable component of every bioprocess thus acting as a basic ingredient of a biological pathway. The transformation from one functional form to another to carry out a function is executed by a protein designed for that particular function. Since proteins assemble into functional complexes, every functional complex begs for a varied condition. This makes the protein structure, function, and architecture almost varying at every stage of the process.

The proteome of an organism epitomizes the entire comprehensive range of proteins expressed within an organism. The term also incorporates the analysis of an assortment of proteins expressed within a specified tissue under exceptional conditions. The proteome is derived from the genome of an organism, yet the study of genome contrasts from that of the proteome due to the characteristic stability of the genome under all conditions as a contrast to the dynamism of the proteome expression based on organism's stage of development and time of observation.

Proteomics is the in-depth study of the proteome which implicates the perception of how the proteins fold, interact, and function within a biological system. For instance, several proteins fold to form intricate three-dimensional structures, and several complexes of subunits come together to form a functional complex. Proteins also undergo post-translational modification which decides the fate of their function. This forms an imperative part of the proteomic study as well. The proteins are very sensitive even within the most tightly regulated environment, thus making them vulnerable to distortion once removed from their natural environment. Hence, a proteomic study desires highly sensitive technique.

The major domains in a proteomic study are the in-depth aspects dealing with protein expression levels, effects of post-translational modifications and the behavior of the proteins thereafter; the dynamics of protein interaction with effector molecules and their subsequent effects upon the cellular system on the whole; proteome mining, structural and functional proteomics. Moreover, the proteomic study has been broadly categorized based upon the expression mapping and the interaction mapping pattern. The mapping of protein expression is a global assessment of the changes taking place in a targeted tissue or cell of the amount and type of proteins being expressed. This is done using two-dimensional gel electrophoresis followed by mass spectrometry. For further assessment, in-gel proteolysis is done to form protein spots followed by peptide mass fingerprinting and then mass spectroscopy (Bantscheff et al. 2007). Spots produced on the 2D gel can be used to analyze partial peptide sequences by using the nano-spray method. This flow of technique is utilized to identify and characterize the proteins for further study. Once preliminary data is obtained, then this data is subjected to various bioinformatics assessment using various tools. Such technologies contribute to collating a database consisting of protein expression profiles of specific tissue and cell (Rasmussen et al. 1996).

The proteins within a tissue decide the functional phenotype of the cell. Hence an actual estimate of the progression of the disease and the pathway relevant for the disease manifestation owing to a particular condition and time is determined by proteome analysis. Even though several techniques are available, it is imperative to make a decision as to which technique shall be relevant at what stage of the study. Therefore, it becomes mandatory to take a characteristic overview of every technique and their relevance in brief.

Traditionally, the detection of any plant microbial association like a symbiotic assessment or phytopathogenesis needs a traditional culturing and cultivation method of the plant on the specific media in the laboratory with subsequent biochemical and morphological analysis. This makes the whole process tedious, expensive, and time-consuming (Kav et al. 2007). Until recently, phytomicrobial assessments required distinguishing techniques for the plant as well as microbial characterization. Techniques such as gel electrophoresis, traditional electron microscopy, and serological techniques were needed to ascertain the features of plants under observation while pathogenic microbe detection shall be ascertained using isolation and microscopic analysis, biochemical assays for characterizations, various serological techniques, etc. (Kav et al. 2007).

The next revolution was brought on by the development techniques like enzyme-linked immunosorbent assays (ELISA) and polymerase chain reaction (PCR). With these techniques, the detection and characterizations of phytomicrobial interactions became more reliable and faster. Techniques such as flow cytometry, DNA microarrays, and in situ hybridizations became the standard methods for detection and confirmation of the associations (Kav et al. 2007). Still, these techniques are laden with their own set of disadvantages.

In addition to ELISAs and PCR, currently utilized phytopathogen detection techniques include flow cytometry, fluorescence in situ hybridization and DNA microarrays. While many of the traditional methods for phytopathogen detection have demonstrated their utility as important agricultural tools, there are inherent disadvantages to many of them and the search for novel, rapid and reliable techniques is an ongoing process within the scientific and agricultural communities.

The preparations needed for proteome analysis requires one to understand as to what stage is the analysis required. Proteome studies necessarily begin with whole protein extract which is subsequently subjected to gel-based or non-gel–based separation techniques. Gel-based approaches use two-dimensional gel electrophoresis (2DE) as well as differential gel electrophoresis (DIGE). Then there are few non-gel–based techniques of protein separation such as isobaric technique for relative and absolute quantification (ITRAQ), isotope-coded affinity tag (ICAT), multidimensional protein identification technique (mud PIT) and stable isotope labeling with the aid of amino acids in cell culture

(SILAC). This is succeeded by the method of identification and quantification using tandem mass spectrometry (Quirino et al. 2010). Subsequent to the assimilation of the data on proteins, they are then subjected to a run through the protein databases in order to estimate the proper protein identification which includes the family it belongs to, and the similarity check to get a fair idea about its structural and functional tendencies. The techniques of protein separation can be broadly classified into gel-based and non-gel–based techniques which have been elaborated in the following section.

2.4.1 Gel-Based Protein Separation Techniques

The most frequently used technique with gel-based technology is the 2DE as this method involves utilizing several characteristic features of proteins for separation such as proteins, isoelectric point, molecular mass, as well as solubility (Joshi 2017).

Here in the foremost step, the proteins are subjected to a pH gradation throughout their gel run. Hence proteins separate and precipitate when their isoelectric constant is achieved. At this point at a particular pH, the overall charge of the protein neutralizes thereby precipitating the proteins. This is also known as isoelectric focusing (IEF). IEF is followed by 2DE separation where the proteins are segregated based upon their molecular weight done by the SDS-PAGE technique (Gorg et al. 2004). The protein separation profile is then visualized using certain staining methods that stain the proteins against the gel. Then with the comparison with a standard set of proteins, one can verify and characterize the isolated proteins. The major setback of this technology is limited reproducibility and lower sensitivity, hence only a well abundant and most characterized protein separation is recommended (Bunai and Yamane 2005). 2D fluorescence differential gel electrophoresis (2DDIGE) is an alternative method where the separation of proteins is done on the same gel using differential fluorescent dyes and a comparative is set based upon their spot intensity (Dunn 1993).

2.4.2 Non-Gel Protein Separation Techniques

A non-gel separation technique is usotope coded affinity tag (ICAT) is in principle used to compare the two protein-separated samples. It consists of a tag having one biotin group containing an isotope-coded linker, which forms the heavy and light sections of a tag while the other group is a thiol. This technique is conveniently utilized in the integral membrane protein identification (Gygi et al. 1999). Isobaric tagging of relative and absolute quantification of protein (iTRAQ) uses the N terminus of the peptide chain along with the side chain which is tagged by the isobaric-labeled tags. Within a given sample, this technique permits quantification (relative and absolute) of peptides from varied sources all at once (Ross et al. 2004).

Another non-gel technique is SILAC (stable isotope labeling by amino acid in cell culture). In this, the cell population consisting of protein of interest is cultivated in either N14- or N15-containing cell medium. Cellular proteins are lysed, fractionated, and separated using 2DE. The role of N15 in this technique causes a shift and creates two peaks for each peptide. The ratio of intensities of peaks is substantially indicative of the amount of expression of proteins (Oda et al. 1999; Ong et al. 2002). SILAC is sensitive to the

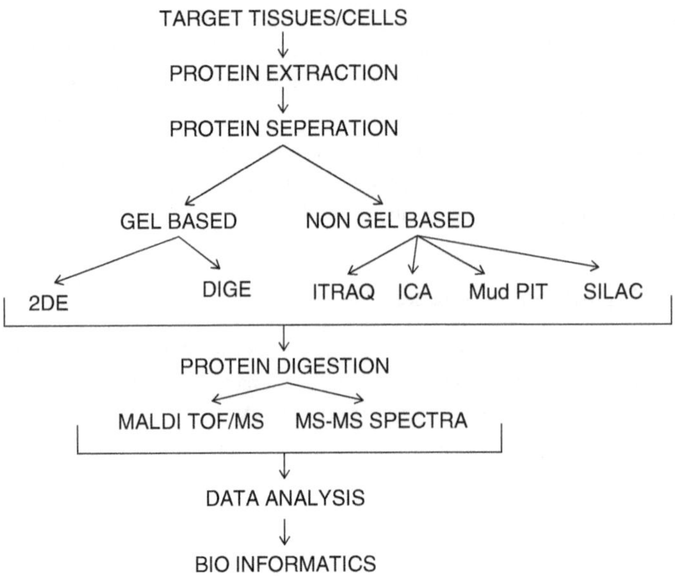

Figure 2.1 A schematic flowchart of proteomic approaches for phytomicrobiome analysis.

detection of the smallest of changes that may occur in a protein thus making this technique precise and a reliable method for quantification within MS. There is a relative assessment of the protein with the slightest change (Blagoev et al. 2004). Mud PIT (multidimensional protein identification technique) is another non-gel–based method which utilizes principles of chromatography to separate the protein are directly associated to tandem mass spectrometry (Washburn et al. 2001; Wolters et al. 2001; McDonald et al. 2002). Quantification of protein is enabled by analysis through mass spectrometry. Upon retrieval of masses of proteins, a comparative of absolute masses is established using bioinformatics tools and computer programs such as OMSSA, MASCOT, and Phenyx (Ross et al. 2004). The path followed using these approaches for the detection of phytomicrobiome relationships is outlined in Figure 2.1.

2.5 Analysis of Phytomicrobial Interactions Using Proteomics Approaches

Thus, microbes can be considered for positive exploitation of the plant growth promotion to enhance the growth of the plant. Microbes also make plants sustainable toward the abiotic as well as biotic stresses looking into progressive climate change. There have been many such types of research, where the positive associations have been well characterized using proteomic analysis and have further been utilized for the enhancement of a natural process.

One such aspect of microbial plant growth promotion is when such an association has been utilized for the purpose of phytoremediation. Phytoremediation is of great help in case of heavy metal remediation. The presence of metal-solubilizing bacteria enhances the ability of the host plant to tolerate metal-induced stress. The best case scenario can be observed in the bioethanol producers, such as *H. tuberosus*, a high biomass yielding crop that acquires enhancement in the accumulation and sustenance of a high concentration of zinc and cadmium. Similarly, the microbial population has been found to be colonizing the internal region of the roots as an endophyte, thereby resulting in increased assimilation of cadmium (Montalbán et al. 2017).

Enhancement of drought and salt resistance observed for the enhanced salt and drought resistance owing to the effects of plant growth–promoting bacteria as studied in *L. perenne*, also known as ryegrass. Unfortunately for this perennially relevant grass as turf and forage, it is highly sensitive to drought-induced stress and high salinity. However, *B. amyloliquefaciens*, combined with hydrogels to prevent the soil from eroding, can help the grass sustain the drought-induced stress by the plant (Su et al. 2017).

In another example, a species of plant growth-promoting bacteria that have been found to be colonizing the roots, stems, and leaf parts of the sugarcane is the species *Gluconacetobacter diazotrophicus*. This is one such example where the bacteria serve as a legume endophyte. However, the proteome interaction studies of this bacterium with the sugar cane are still not very deep. A study was performed to investigate the molecular aspects using MS proteomic analysis using N15 metabolic labeling of bacterial, root samples, and co-cultures. Over 400 proteins were assessed out of which near about 78 combinations of involved proteins appeared relevant among the interaction model. A comparative analysis of the proteomic data thus derived revealed the proteins involved in fundamental roles of protein recognition. Additionally, 30 bacterial active proteins have been identified where 9 were categorically induced by plant signaling pathways. This is the first line of study for the *G. diazotrophicus* and sugar cane (Schirawski and Perlin 2018).

The microbial action on plants can work both ways. Certain interactions work for the enhancement of the plant, however, some have a negative impact owing to such associations. Any plant maybe rendered vulnerable to disease owing to an undesirable microbial association or any abiotic stress situation which throws the plants defenses out of order. In the case of the pathogen associate disease situation, some events are to happen to manifest a plant's invasion by the pathogen. This series of events may be termed as the disease triangle where disease can manifest due to a pathogen if and only if there is appropriate feedback of the responses for the entry of the pathogen into the plant and for that the environmental conditions must be suitable. Plants' obligation as host to a vast array of pathogens such as bacteria, fungi, nematodes, viruses, etc., is a well-understood concept (Dangl and Jones 2001).

The plant's defense system is highly complex and sophisticated. Many microbial species bombard the plant system, some for symbiosis and some for their exploitative tendencies. On the breach by any pathogen, a plant's pathogen-associated molecular patterns (PAMPs) guided basic resistance is activated or cell wall degrading enzymes (CWDE) play their role (Jones and Dangl 2006). Certain pathogens have ascertained methods of evading or suppressing the plant defense which makes the host plant vulnerable to further infection. Sometimes there is a provision of resistance generation of protein expression in certain host plants to safeguard them against the pathogenic attack thereby activating specific resistance mechanisms. Such mechanisms have been deciphered by the proteomic studies. Proteomic analysis of host plant pathogen reveals the protein interaction profile of the proteins expressed during the plant pathogenesis. The plant–pathogen interactions have been categorized into two types depending upon the eventuality of the outcome of the pathogenesis. The first category is the compatible interaction. This is where the pathogen completes its cycle of pathogenesis and the plant is unable to mount an anti-defense response. The second category is the incompatible interaction where the plant mounts a series of complex immune defense responses against the pathogen thereby forestalling the pathogenic progression (Mehta et al. 2008). The progression of the phytopathogenesis can be assessed into three stages: perception or detection, signaling to breach successfully, and response generated by the host of the perceived signals of the pathogens. As the pathogen comes in the vicinity of the host, it generates some signal metabolites which are sensed by the plant system and further communicated to the plant nucleus for feedback. The defense mounted in response to these signals is of two types: the local response which is initiated by the sensation of the elicitor produced by the pathogenic intervention and the other is systemic-acquired resistance (SARs). The intent of the local response is to ameliorate the intervention at the periphery of the first interception. Responses are mediated by innate plant resistance proteins which elicits a rapid yet reliable response. These are high-effect defense mechanisms thereby leading to mobilization of the ion fluxes across the plasma membrane of the host plant, leading to cell wall reinforcement. This subsequently initiates MAP kinase activation and generation of reactive oxygen species thereby causing the programmed cellular death. This type of response is also known as the hypersensitive response (Heath 2000; Jones 2000; Greenberg and Yao 2004).

The second category of defense response is also known as systemic acquired resistance (SARs). The features of this type of response are where the plant gears up the whole system from the localized region to the far distal end against the pathogen entry (Schneider et al. 1996). SARs deals with the phenomenon of raising the immunity guards to defend against the pathogens and also take proactive measures to aid the plant recovery from the damage post an immunogenic response. SARs activation is initiated by the presence of the endogenous salicylic acids (SA) which in turn increases the expression and assimilation of nitric oxides, ethylene, and jasmonic acid thereby eliciting various downstream pathways as cascade (Ryals et al. 1996). Such information regarding the plant-pathogen interactions has been thoroughly substantiated by the genomic, post-genomic, and proteomic analysis. Such studies, therefore, assimilate the validated information pertaining to the pathogenic strategies as well as behavior influenced by the host plant responses. Thus, proteomic studies empower with the most basic functional information regarding the recognition and characterization of the proteins functioning under specific conditions. Proteomics thus assesses the translated protein complement of the gene sequence derived experimentally corresponding to the protein profile of a specified tissue under consideration with a variable growth and treatment conditions (Mehta et al. 2008). Proteomics has helped in elucidating the distinct pathogenic pathways of a plethora of pathogens with their corresponding host plants. Proteomics has helped in elucidating the distinct pathogenic pathways of a plethora of pathogens with their corresponding host plants. Pathogens such as viruses, fungi, and bacteria have special notoriety in causing widespread devastation among the crops and have gained a special place in the study owing to the ease of handling their proteome and trying to establish their methods of pathogenesis. The virus is one of the notorious pathogens which have evolved ways of establishing itself within the host system.

It manipulates the host defenses for its benefits. The prime condition for a viral infection is to succeed in entering the host system; viral particles must first be capable of making it through the innate barriers of the host plant cell either using a carrier or through cell receptors. This is followed by its replication within the specified host cell. Once replication occurs it is then transported to the neighboring cell through plasmodesmata to eventually make its way into the vascular tissue of phloem to infect the host systemically. The viral interactions modulate the host genome and proteome to establish an infection and influence the defense proteins to respond to the infection (Whitham et al. 2006; Mehta et al. 2008).

According to a study where the infection caused by the tobacco mosaic virus (TMV) on the hot pepper plant was analyzed. It was found that this cultivar was resistant against TMV-P0 and susceptible to the TMV-P1.2 strain. Upon proteome analysis using 2DE followed by MALDI-TOF MS, it was found that certain proteins present in hot pepper provided defense against the 26S proteasome RPN7 subunit of TMV=P0. The defenses generated were to initiate Rab 11 GTPase responsible membrane trafficking, i.e. exo and endocytosis, RNA metabolism interference caused to mRNA binding protein, and activation of program cell death pathways (Lee et al. 2009).

The plant–pathogen interaction studies are still quite nascent yet plant fungus interactions have been explored the most through proteomic analysis. Fungal infections are the most prevalent in plant hosts. However, the proteomic study of the fungal proteins encoded by a well-sequenced genome poses one of the greatest challenges. The fungal system regulates several host defenses to enhance its activity as inferred by proteome analysis (Murad et al. 2006, 2007). The rust disease is one of the leading causes of loss to cereals worldwide. One study was conducted on the wheat leaf rust fungus using proteomic tools to assess the disease progression at the molecular level. Susceptible wheat cultivars were exposed to the virulent strain of *Puccinia triticina* where 2DE and MS analysis were used to derive the result. Result analysis revealed 22 distinct proteins involved in pathogenesis with functions of some proteins known and the rest were unknown (Rampitsch et al. 2006; Webb and Feller 2006).

Proteomics has given a glimpse into the involvement of various proteins in plant bioprocesses, like stress response, defense responses, photosynthetic regulation, proteins involved in electron transport, and signal transduction. A study conducted by Houterman et al. (2007) assessed the proteome of tomato xylem sap infected with fungal strain *Fusarium oxysporum*. Thirty-three proteins were relevant to the pathogenesis using 2DE and MS, out of which 13 proteins were found to be specific to infected plants consisting of enzymes chitinases, peroxidases, polygalacturonase, and a subtilisin-like protease. These have been found to be involved in rendering protection to the cell structural integrity and the cell wall. An approach has been specifically assigned to study the proteome of the fungal pathogenic species known to secrete out some protein to facilitate pathogenesis. It is known as the secretome approach. A species of phytopathogenic fungus *Sclerotinia sclerotiorum* secretes the exoproteome. Several CWDE secreted were considered as virulent causative factors, which was derived from liquid culture and further subjected to 2DE and characterized by MALDI-MS/MS (Yajima and Kav 2006). Some of the interactions of phytomicrobiome deciphered using proteomic approaches have been listed in Table 2.1.

Table 2.1 Proteomic approaches available for detection of plant and microbes along with some examples.

Various Types of Proteomics Approaches	List of Interactions Deciphered Using Various Proteomics Approaches	
2 DE (two-dimensional gel electrophoresis)	*Bacillus amyloliquefaciens*	*Lolium perenne*
DIGE (differential gel electrophoresis)	*Pseudomonas* sp.	*Helianthus tuberosus*
ITRAQ (Isobaric technique for relative and absolute quantification)	*Rhizophagusirregularis*	*Gossypium (17 sp.)*
ICAT (isotope-coded affinity tag)	*M. lychnis-dioicae*	*Silene latifolia*
Mud PIT (multidimensional protein identification technique)	*Fusarium proliferatum*	*Solanum lycopersicum*
SILAC (stable isotope labeling with the aid of amino acids in cell culture)	*Gluconacetobacterdiazotrophicus*	*Saccharum officinarum*
2DDIGE (2D fluorescence differential gel electrophoresis)	*Fusarium oxysporum*	*Solanumlycopersicum*
MALDI-TOF (matrix-assisted laser desorption/ionization-time-of-flight) MS Spectra (tandem mass spectrometry)	*Bradyrhizobium japonicum*	*Glycine max*

Bacterial species depict a wide variation of the interaction pathways where pathogenicity is caused by manipulation and translocation of plant host cells. Depending upon the bacterial protein composition, the secretory pathways are classified into various types (Lee and Schneewind 2001). The first secretory pathway, i.e. type-1 secretions are involved in secretion of phytotoxic substances such as rhizobiocin, hemolysins, etc., which are common to all phytopathogens. Though the majority of secretory pathways follow type 3 in which gram-negative bacteria are responsible (Agrios 2004; Puhler et al. 2004). This pathway works on the principle where the effector protein is administered directly to the targeted host cell (Galan and Collmer 1999). The type-2 system of secretion is another imperative pathway for bacterial pathogenicity. It is usually involved in the export of several virulent factors which are usually toxic. Miao and coworkers in their study in 2008 using various transcriptomic and proteomic approaches identified numerous genes and proteins involved in carbon and nitrogen metabolism, plant defense responses, nutrient exchange, and signal transduction that are significantly regulated in *Glycine max* colonized by *Bradyrhizobium japonicum* (Brechenmacher et al. 2010). Examples of a few interactions discussed in the chapter are listed in Table 2.2.

2.6 Conclusion and Future Prospects

The significance of agriculture and sustained crop production to meet the growing need around the world cannot be understated. Specific emphasis in the field of crops is imperative given the rising trends of the population at the global level. Precision, reliability, and

Table 2.2 Examples of phytomicrobiome interactions.

Microorganism	Plant	Habitation	(+/−) Impact of Interaction	Type of Interaction
Bacillus amyloliquefaciens	*Lolium perenne*	rhizospheric	+	Indirect
Pseudomonas sp.	*Helianthus tuberosus*	endophytic	+	Indirect
Rhizophagus irregularis	*Gossypium* (17 sp.)	rhizospheric	+	Direct
M. lychnis-dioicae	*Silene latifolia*	epiphytic	−	Direct
Fusarium proliferatum	*Solanum lycopersicum*	endophytic	−	Direct

accuracy in pathogenic disease detection and diagnosis have become crucial to optimize crop health, yield, and quality. Myriad of biological organisms have been observed to play a positive as well as negative impact on plant health. The positive impact can be enhanced with the aid of technology to increase the overall plant growth. The negative impact can potentially have damaging consequences to crop production worldwide, thereby affecting the overall productivity. Therefore, we need techniques with high-level advancement to assess the phytomicrobial association to derive benefit or precautionary measures well within time. Substantial changes take place within the cell during phytomicrobiome association leading to translation of a novel protein occurring due to plant and microbe genome interactions. Derivation of information on genomics and transcriptomics of plants microbe interactions is available, however, the need of the hour is to validate the data using proteomics. Several classic and recent techniques such as 2DE, MS, ICAT, DIGE, and MudPIT are at our disposal which are now proving to be effective and reliable enough to give us a validated result. The first instinct for the proteome analysis makes use of 2DE and MS for the preliminary extraction and separation of proteins of interest. Subsequently, the techniques are chosen to further delve deep into cellular functions. Owing to their reproducibility of results the data thus generated by such proteomic approaches are proving to be reliable. Identification of protein using peptide mass fingerprinting is filled with constraints owing to the need for the large-scale sequence of the genome.

There is also a necessity to integrate the proteome generated data within the network of bioinformatics software to make the *in-silico* analysis of proteomics more structured. This enhances the ability of the protein-related data to get the validation through in silico analysis. Proteomics has leaped ahead to genomic and transcriptomic studies by making the analysis as close to reality as practically possible. Advancement in proteomics has led the research to leapfrog into the analysis which was simply hypothesized earlier, thereby acting as an aspect that cannot be dispensed of in the field of biological sciences, especially in plant microbial consortium.

References

Agrios, G.N. (2004). *Plant Pathology*, 5e. San Diego: Academic Press.

Badri, D.V., Zolla, G., Bakker, M.G. et al. (2013). Potential impact of soil microbiomes on the leaf metabolome and on herbivore feeding behavior. *New Phytol.* 198: 264–273.

Bantscheff, M., Schirle, M., Sweetman, G. et al. (2007). Quantitative mass spectrometry in proteomics: a critical review. *Anal. Bioanal. Chem.* 389 (4): 1017–1031.

Blagoev, B., Ong, S.E., Kratchmarova, I., and Mann, M. (2004). Temporal analysis of phosphotyrosine-dependent signaling networks by quantitative proteomics. *Nat. Biotechnol.* 22 (9): 1139–1145.

Bosch, T.C.G. and McFall-Ngai, M.J. (2011). Metaorganisms as the new frontier. *Fortschr. Zool.* 114 (4): 185–190.

Brechenmacher, L., Lei, Z., Libault, M. et al. (2010). Soybean metabolites regulated in root hairs in response to the symbiotic bacterium *Bradyrhizobium japonicum*. *Plant Physiol.* 153 (4): 1808–1822.

Bunai, K. and Yamane, K. (2005). Effectiveness and limitation of two dimensional gel electrophoresis in bacterial membrane protein proteomics and perspectives. *J. Chromatogr.* B815 (1-2): 227–236.

Compant, S., Clement, C., and Sessitsch, A. (2010a). Plant growth-promoting bacteria in the rhizo- and endosphere of plants: their role, colonization, mechanisms involved and prospects for utilization. *Soil Biol. Biochem.* 42: 669–678.

Compant, S., Heijden, V.D., and Sessitsch, A. (2010b). Climate change effects on beneficial plant-microorganism interactions. *FEMS Microbiol. Ecol.* 73: 197–214.

Dangl, J.L. and Jones, J.D. (2001). Plant pathogens and integrated defense responses to infection. *Nature* 411 (6839): 826–833.

Dunn, M.J. (1993). Gel electrophoresis: proteins. *Bios*: 41–127.

Engelmoer, D.J.P., Behm, J.E., and Kiers, E.T. (2014). Intense competition between arbuscular mycorrhizal mutualists in an in vitro root microbiome negatively affects total fungal abundance. *Mol. Ecol.* 23: 1584–1593.

Gorg, A., Weiss, W., and Dunn, M.J. (2004). Current two-dimensional electrophoresis technology for proteomics. *Proteomics* 4 (12): 3665–3685.

Greenberg, J.T. and Yao, N. (2004). The role and regulation of programmed cell death in plant-pathogen interactions. *Cell. Microbiol.* 6: 201–211.

Gygi, S.P., Rist, B., Gerber, S.A. et al. (1999). Quantitative analysis of complex protein mixtures using isotope-coded affinity tags. *Nat. Biotechnol.* 17 (10): 994–999.

Hartmann, A., Rothballer, M., Hense, B.A., and Schröder, P. (2014). Bacterial quorum sensing compounds are important modulators of microbe-plant interactions. *Front. Plant Sci.* 5 (131): 10.3389.

Heath, M. (2000). Hypersensitive response-related cell death. *Plant Mol. Biol.* 44: 321–334.

Houterman, P.M., Speijer, D., Dekker, H.L. et al. (2007). The mixed xylem sap proteome of *Fusarium oxysporum*-infected tomato plants. *Mol. Plant Pathol.* 2007 8: 215–221.

Jones, A. (2000). Does the plant mitochondrion integrate cellular stress and regulate programmed cell death? *Trends Plant Sci.* 5: 225–230.

Jones, J.D. and Dangl, J.L. (2006). The plant immune system. *Nature* 444: 323–329.

Joshi, P. (2017). Proteomics. In: *Innovative Approaches in Drug Discovery* (eds. B. Patwardhan and R. Chaguturu), 273–294. Academic Press.

Kav, N., Srivastava, S., Yajima, W., and Sharma, N. (2007). Application of proteomics to investigate plant-microbe interactions. *Curr. Proteomics* 4: 28–4310.

Kostic, A.D., Chun, E., Robertson, L. et al. (2013). *Fusobacterium nucleatum*potentiates intestinal tumorigenesis and modulates the tumor-immune microenvironment. *Cell Host Microbe* 14 (2): 207–215.

Lebeis, S.L. (2015). Greater than the sum of their parts: characterizing plant microbiomes at the community level. *Curr. Opin. Plant Biol.* 24: 82–86.

Lee, V.T. and Schneewind, O. (2001). Protein secretion and the pathogenesis of bacterial infections. *Genes Dev.* 15: 1725–1752.

Lee, K.D., Gray, E.J., Mabood, F. et al. (2009). The class IId bacteriocin thuricin 17 increases plant growth. *Planta* 229: 747–755.

Lugtenberg, B. and and Kamilova, F. (2009). Plant-growth-promoting rhizobacteria. *Annu. Rev. Microbiol.* 63: 541–556.

McDonald, W.H., Ohi, R., Miyamoto, D.T. et al. (2002). Comparison of three directly coupled HPLC MS/MS strategies for identification of proteins from complex mixtures: single-dimension LC-MS/MS, 2- phase MudPIT, and 3-phase MudPIT. *Int. J. Mass Spectrom.* 219 (1): 245–251.

McFall-Ngai, M.J. (2002). Unseen forces: the influence of bacteria on animal development. *Dev. Biol.* 242: 1–14.

McFall-Ngai, M. (2008). Are biologists in 'future shock'? Symbiosis integrates biology across domains. *Nat. Rev. Microbiol.* 6 (10): 789–792.

Mehta, A., Brasileiro, A.C., Souza, D.S. et al. (2008). Plant-pathogen interactions: what is proteomics telling us? *FEBS J.* 275: 3731–3746.

Montalbán, B., Thijs, S., Lobo, M.C. et al. (2017). Cultivar and metal-specific effects of endophytic bacteria in *Helianthus tuberosus* exposed to Cd and Zn. *Int. J. Mol. Sci.* 18: 2026.

Murad, A.M., Laumann, R.A., Lima Tde, A. et al. (2006). Screening of entomopathogenic *Metarhiziumanisopliae* isolates and proteomic analysis of secretion synthesized inresponse to cowpea weevil (*Callosobruchus maculatus*) exoskeleton. *Comp. Biochem.* 142: 365–370.

Murad, A.M., Laumann, R.A., Mehta, A. et al. (2007). Screening and secretomic analysis of entomopathogenic *Beauveria bassiana* isolates in response to cowpea weevil (*Callosobruchus maculatus*) *Journal of Pharmacognosy and Phytochemistryexoskeleton. Comp. Biochem. Physiol. C. Toxicol. Pharmacol.* 145: 333–338.

Newman, M.A., Sundelin, T., Nielsen, J.T., and Erbs, G. (2013). MAMP (microbe-associated molecular pattern) triggered immunity in plants. *Front. Plant Sci.* 4: 139.

Oda, Y., Huang, K., Cross, F.R. et al. (1999). Accurate quantitation of protein expression and site specific phosphorylation. *Proc. Natl. Acad. Sci.* 96 (12): 6591–6596.

Ong, S.E., Blagoev, B., Kratchmarova, I. et al. (2002). Stable isotope labeling by amino acids in cell culture, SILAC, as a simple and accurate approach to expression proteomics. *Mol. Cell. Proteomics* 1 (5): 376–386.

Pisa, G., Magnani, G.S., Weber, H. et al. (2011). Diversity of 16S rRNA genes from bacteria of sugarcane rhizosphere soil. *Braz. J. Med. Biol. Res.* 44: 1215–1221.

Puhler, A., Arlat, M., Becker, A. et al. (2004). What can bacterial genome research teach us about bacteria–plant interactions? *Curr. Opin. Plant. Biol.* 7: 137–147.

Quecine, M.C., Araújo, W.L., Rossetto, P.B. et al. (2012). Sugarcane growth promotion by the endophytic bacterium *Pantoeaagglomerans* 33.1. *Appl. Environ. Microbiol.* 78 (21): 7511–7518.

Quirino, B.F., Candido, E.S., Campos, P.F. et al. (2010). Proteomic approaches to study plant-pathogen interactions. *Phytochemistry* 71: 351–362.

Rampitsch, C., Bykova, N.V., McCallum, B. et al. (2006). Analysis of the wheat and *Puccinia triticina* (leaf rust) proteomes during a susceptible host-pathogen interaction. *Proteomics* 6: 1897–1907.

Rasmussen, H.H., Orntoft, T.F., Wolf, H., and Celis, J.E. (1996). Towards a comprehensive database of proteins from the urine of patients with bladder cancer. *J. Urol.* 155: 2113–2119.

Ross, P.L., Huang, Y.N., Marchese, J.N. et al. (2004). Multiplexed protein quantitation in Saccharomyces cerevisiae using amine-reactive isobaric tagging reagents. *Mol. Cell. Proteomics* 3 (12): 1154–1169.

Rudrappa, T., Czymmek, K.J., Paré, P.W., and Bais, H.P. (2008). Root-secreted malic acid recruits beneficial soil bacteria. *Plant Physiol.* 148: 1547–1556.

Ryals, J., Neuenschwander, U., Willits, M. et al. (1996). Systemic acquired resistance. *Plant Cell* 8 (10): 1809–1819.

Santoyo, G., Moreno-Hagelsieb, G., Del Carmen Orozco-Mosqueda, M., and Glick, B.R. (2016). Plant growth-promoting endophytes. *Microbiol. Res.* 183: 92–99.

Schirawski, J. and Perlin, M.H. (2018). Plant-microbe interaction 2017 – the good, the bad and the diverse. *Int. J. Mol. Sci.* 19 (5): 1374.

Schneider, M., Schweizer, P., Meuwly, P., and Métraux, J.P. (1996). Systemic acquired resistance in plants. *Int. Rev. Cytol.* 168: 303–340.

Su, A.Y., Niu, S.Q., Liu, Y.Z. et al. (2017). Synergistic effects of *Bacillus amyloliquefaciens*(GB03) and water retaining agent on drought tolerance of perennialryegrass. *Int. J. Mol. Sci.* 18: 2651.

Washburn, M.P., Wolters, D., and Yates, J.R. III (2001). Large-scale analysis of the yeast proteome by multidimensional protein identification technology. *Nat. Biotechnol.* 19 (3): 242–247.

Webb, C.A. and Feller, J.P. (2006). Cereal rust fungi genomics and the pursuit of virulence and avirulence factors. *FEMS Microbiol. Lett.* 264: 1–7.

Whitham, S.A., Yang, C., and Goodin, M.M. (2006). Global impact: elucidating plant responses to viral infection. *Mol. Plant Microbe Interact.* 19: 1207–1215.

Wolters, D.A., Washburn, M.P., and Yates, J.R. (2001). An automated multidimensional protein identification technology for shotgun proteomics. *Anal. Chem.* 73 (23): 5683–5690.

Yajima, W. and Kav, N.N. (2006). The proteome of the phytopathogenic fungus Sclerotinia sclerotiorum. *Proteomics* 6: 5995–6007.

Zhang, Q., Gao, X., Ren, Y. et al. (2018). Improvement of *Verticillium* wilt resistance by applying arbuscular mycorrhizal fungi to a cotton variety with high symbiotic efficiency under field conditions. *Int. J. Mol. Sci.* 19: 241.

3

Metagenomics

An Approach to Unravel the Plant Microbiome and Its Function

Ravindra Soni[1], *Deep Chandra Suyal*[2], *Balram Sahu*[1], *and Suresh Chandra Phulara*[3]

[1] *Department of Agricultural Microbiology, Indira Gandhi Krishi Vishwavidhyalaya, Raipur, Chhattisgarh, India*
[2] *Department of Microbiology, Akal College of Basic Sciences, Eternal University, Baru Sahib, Sirmour, Himachal Pradesh, India*
[3] *Department of Biotechnology, Koneru Lakshmaiah Education Foundation, Guntur, Andhra Pradesh, India*

3.1 Introduction

Like the human body, microbes also colonize inside plants (Qin et al. 2010; Zhao 2010; Gevers et al. 2012). The collective colonization of plant-associated microbiota is known as a plant microbiome, which is a key determinant governing plant health and its productivity (Berendsen et al. 2012).

Recent years have seen a remarkable interest in such interactions (Lebeis et al. 2012; Turner et al. 2013). A plant's internal microbiome, also referred to as endophytic microbes, includes members from almost all microbial communities, such as archaea, bacteria, and fungi (Turner et al. 2013; Hardoim et al. 2015). Sometimes these microbial communities colonize inside plants in such a manner that their number surpasses the number of plant's own cells (Mendes et al. 2013). It is interesting to note that soil microbiomes are now touted as a cornerstone of the next green revolution (Parnell et al. 2016). The concept of soil, microbes, and plant interface, i.e. "soil-microbe-plant interface" is not new. However, the "soil-microbe–soil-plant–microbe-plant interface" represents plant microbiome interaction more adequately. The rhizosphere that is affected by several climatic factors influences the plant and microbiome which ultimately utilize the habitat as an information highway (Bais et al. 2004; Roume et al. 2015; Tomer et al. 2017).

A selection effect that is imposed due to the physicochemical changes surrounding roots shapes the microbial composition inside a plant host. One way to this selection is favoring the growth of such opportunistic microbes that are adapted to specific chemical conditions, which is an indirect approach. Alternatively, the microbes that support the growth/development of plants and/or enhance their survival in stress conditions are directly recruited (Bulgarelli et al. 2013; Philippot et al. 2013; Mendes et al. 2014; Suyal et al. 2014). Another habitat where the microbiome interacts with the plant is the rhizoplane, a surface of plant tissues that comes into contact with the soil. Apart from endophytes located inside plant tissues, "epiphytes" is a

Phytomicrobiome Interactions and Sustainable Agriculture, First Edition. Edited by Amit Verma, Jitendra Kumar Saini, Abd El-Latif Hesham, and Harikesh Bahadur Singh.

class of the plant microbiome, which can be found in an adherent form on plant tissues (Bulgarelli et al. 2013). Several studies have explored the structure and function of the microbiome of both model and crop plant species under natural and agricultural environments (Bulgarelli et al. 2012; Rascovan et al. 2016; Hamonts et al. 2018; Kumar et al. 2018a,b). To explore more diversity of plant microbiomes, there is a need to integrate novel molecular technologies, e.g. metagenomics and metatranscriptomics (van der Heijden and Hartmann 2016).

3.2 Metagenomics

Findings have revealed that a gram of soil contains 10^9 microbial cells. This exhibits a great diversity level by attaining about 10^6 taxa. Despite such diversity and number, only 1% of microbes residing in bulk soil and about 10% from plant-influenced zones can culture in standard laboratory conditions. The majority of the soil microbiome remain uncultured. However, they can be detected using advanced molecular-based approaches, which are discussed in later sections (Barret et al. 2013). Metagenomics allows the detection of microbial diversity in environmental samples and enables the construction of community-level gene catalogs (Zengler and Palsson 2012; Franzosa et al. 2015; Guo et al. 2015; Widder et al. 2016). Broadly, metagenomics represents a new advanced approach for genomic analysis. It refers to the simultaneous characterization of genomes in microbial communities from environmental samples, using molecular techniques like PCR, cloning, and DNA sequencing for identifying new sources of plant growth–promoting traits. The approach of metagenomics has also been referred to as environmental genomics, ecogenomics, and community genomics (Song et al. 2013).

Through metagenomics, genomes of organisms in the community can be pooled by entailing the extraction of DNA from the community (Figure 3.1).

Usually, metagenomic analysis conducts 16S rRNA or 18 S rRNA gene surveys to observe microbial community composition while also informing the sequencing complexity required to access high levels of metagenome coverage (Tyson et al. 2004; Lauro et al. 2011). Complete sequencing of clones containing phylogenetic anchors can be performed by using sequence-based metagenomics to indicate the taxonomic group. However, functional metagenomics has identified antibiotics (Hover et al. 2018; Wrighton 2018), antibiotic resistance genes (De Castro et al. 2014; López-Pérez and Mirete 2014), degradative enzymes (Silva et al. 2013; Choi et al. 2018), and plant growth-promoting traits (Tsurumaru et al. 2015). The clones containing metagenomic DNA can also be screened for expression activities by functional analysis. Although it holds great potential in terms of shaping ecological theory; however, due to the slow screening technology it is subsequently lagging behind shotgun sequencing (Zhou et al. 2015).

3.3 Metagenomics of Plant Rhizosphere

The rhizosphere, a narrow zone of soil, is generally altered by factors like plant root secretions. It is rich in the microbiome population and may carry up to 10^8–10^{12} bacterial cells per gram of root or rhizosphere soil (Kennedy and de Luna 2005; Egamberdieva et al. 2008) that comprises more than 30 000 prokaryotic species (Suyal et al. 2016). These

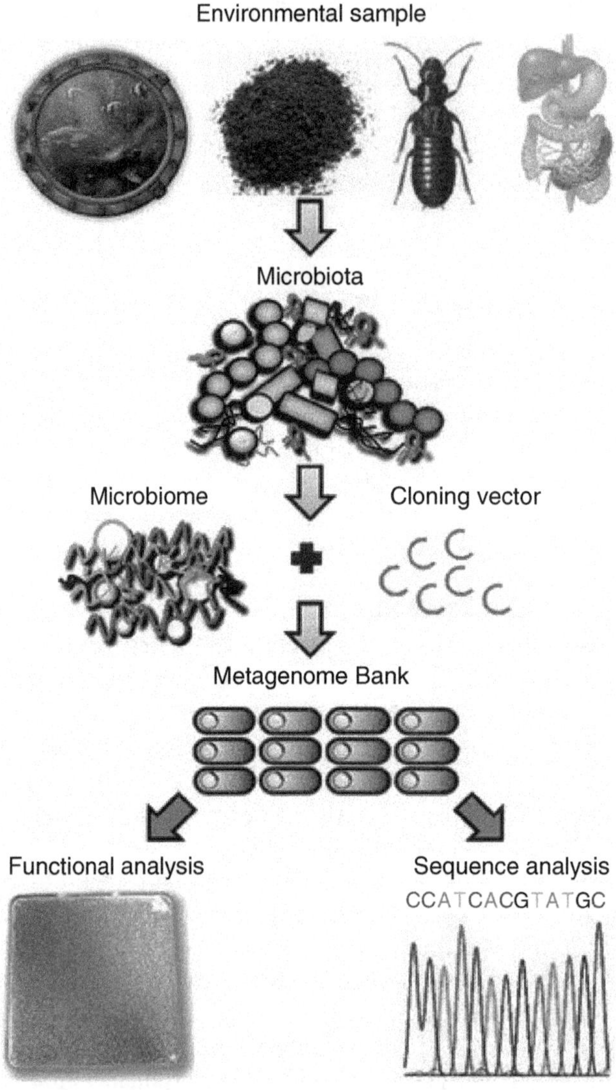

Figure 3.1 Metagenomic approach to reveal the structure and function of a plant microbiome. (*Source:* Sleator et al. (2008) © 2008, Society for Applied Microbiology. Reprinted with permissions of John Wiley & Sons).

rhizospheric microbes have tremendous diversity in terms of their metabolic capabilities, thus playing a vital role in plant health. Therefore, to unveil their roles in the plant system, a proper understanding of their community structure is the need of the hour. Traditionally, genomics has focused on the small proportion, i.e. 1–3% of cultivable microbes; however, due to the advancement in culture-independent molecular tools like metagenomics, high-throughput DNA sequencing, and bioinformatics, our understanding of microbial activities in the rhizosphere has been revolutionized. With the help of advanced DNA—sequencing technologies, the microbial community structures in the rhizosphere can be characterized and compared with the microbial structures in distant

Table 3.1 Advance molecular techniques used for the characterization of various rhizospheric microbial communities.

Techniques used	Rhizosphere	References
Amplicon gene sequencing of conserved marker genes,16S rRNA	Barley and alfalfa	Kumar et al. (2018b)
	Rice roots	Edwards et al. (2018)
Metagenome sequencing	*Taxus* rhizosphere	Hao et al. (2018)
	Wheat and cucumber	Ofaim et al. (2017)
Metatranscriptome sequencing	Wheat	Hayden et al. (2018)
	Seagrass Microbiomes	Crump et al. (2018)
Metaproteomic profiling	*Biscutella laevigata*	Mattarozzi et al. (2017)
	Sugarcane	Lin et al. (2013)
Metabolomic profiling	*Arabidopsis*	Pétriacq et al. (2017)

soil, a part of the soil that does not come under the direct influence of the root (Turner et al. 2013; Bouffaud et al. 2014; Ofek-Lalzar et al. 2014; Roume et al. 2015). Using a high-throughput sequencing-based metagenomic approach, the composition and function of the microbial community in influent water (as a reference matrix) and reed rhizosphere soil of a constructed wetland has been explored (Bai et al. 2014). Recently, microbiome functions in the rhizospheric region were analyzed utilizing a similar approach, in which a metabolic-network–based framework for metagenomics was used (Ofaim et al. 2017). To investigate the microbial diversity in a rhizospheric region of paddy fields, advanced molecular biology tools can be applied. It includes culture-independent molecular techniques, 16S rRNA clone library generation along with RFLP, sequencing, and phylogenetic analysis, etc. (Arjun and Harikrishnan 2011). Furthermore, metagenomic analysis of rhizosphere of plants adapted to acid mine drainage revealed the presence of novel nickel resistance genes (Mirete et al. 2007). These approaches reveal that the soil and rhizosphere microbiomes diversity is highly underrated (Mendes et al. 2013).

However, metagenomic prediction by quantifying the 16S rRNA gene has revealed an abundance of bacteria in the rhizosphere (Lopes et al. 2016). Bacterial diversity associated with the rhizosphere of wheat plants (*Triticum aestivum*) was also explored using metagenomic analysis. It revealed the presence of a highly diverse class of bacteria in the wheat rhizosphere, which includes Actinobacteria, Bacilli, Clostridia, Alphaproteobacteria, Betaproteobacteria, Deltaproteobacteria, Gammaproteobacteria, and few uncultivable bacteria (Velázquez-Sepúlveda et al. 2012). Some of the recent metagenomics techniques used to reveal various rhizospheres are listed in Table 3.1.

3.4 Metagenomics of Plant Phyllosphere

The aerial habitat occupied by microbes is generally termed as the phyllosphere. The estimated colonized microbial area in the terrestrial leaf surface is about 6.4×10^8 km^2 (Morris and Kinkel 2002). Rastogi et al. (2010) found that only 0.1–8.4% of the total bacterial population were cultivable. Phyllosphere microbiome was reported for their involvement in nitrogen fixation, biodegradation of pollutants (Waight et al. 2007), and antibodies production and systemic resistance induction in the host plant (Sundaram et al. 2011; Bodenhausen et al. 2014). Several studies have been conducted to reveal the microbiome of the plant leaf surfaces (Finkel et al. 2016). Methe et al. (2017) identified sets of functional gene annotations that differ between drought and well-watered maize phyllosphere metagenomes from field plots.

The 454-pyrosequencing platform comprising rRNA or ITS amplicon sequencing, whole-genome sequencing, shotgun metagenomics, and transcriptional profiling is widely employed for microbial community analysis (Rastogi et al. 2012; Correa-Galeote et al. 2018). These revolutionary technological innovations facilitate the comparative microbiome analyses and provided a deep insight into the structural and functional characteristics of phyllosphere microbiome. Furthermore, in an interesting study of netaproteogenomic analysis (includes metagenomics and metaproteomics) of rice microbiome revealed that methanol-based methylotrophy linked to the genus *Methylobacterium* dominated within the protein repertoire of the phyllosphere microbiota (Knief et al. 2012). Similarly, a combined metagenomic and metaproteomic approaches were also applied for analyzing the physiology of phyllosphere bacterial communities in situ (Delmotte et al. 2009). Several reports in the public domain reveal phyllosphere microbial diversity through metagenomic approaches (Atamna-Ismaeel et al. 2012; Ritpitakphong et al. 2016; Mukhtar et al. 2017).

3.5 Metagenomics of Plant Endosphere

Microorganisms from the endosphere penetrate and invade into the root internal tissues. A vast array of plants' endosphere is colonized by fungi (Vandenkoornhuyse et al. 2002; Smith and Read 2008), bacteria (Reinhold-Hurek and Hurek 2011) and, some time by Archaea (Sun et al. 2008). The molecular tools to increase our comprehension about the endophytic lifestyle within the host plant are poorly understood (Reinhold-Hurek and Hurek 2011). Metagenomic studies might be helpful to lengthen advancement in this enthralling research area. Furthermore, the metagenomics approach can also be helpful to elucidate the potential of uncultivable endophytic microbial communities, thus revealing the information beyond the genomic information of an individual in the endosphere microbiome (Dinsdale et al. 2008). Moreover, metagenomic studies mainly depend upon the quality of metagenomic DNA extracted from environmental samples. Also, the prevalence of genomic DNA from the plant biomass limits the application of bacterial transcriptome sequencing or metagenome (Utturkar et al. 2016). Here, significant biases in the DNA extraction methodology from endophytic communities may be introduced (Kgomotso et al. 2015). In this context, some workers also observed that the recovery of a diverse class of endophytic communities associated with plant tissues could be enabled by using broad-spectrum SDS-based DNA extraction protocol (Maropola et al. 2015). Separation of the host's genome from the endophytic microbiome's metagenome is a challenging assignment. Due to this, it is rare to achieve full sequencing of an endophytic microbiome by complex procedures like extensive deep genome sequencing (Campisano et al. 2014). However, reports are available which enlighten the endophytic microbiome up to some extents using metagenomics (Tian et al. 2015; Vandenkoornhuyse et al. 2015;

Sengupta et al. 2017). Utilizing metagenomic analysis, Sessitsch et al. (2012) reported the functional characteristics of an endophyte community that colonizes rice roots. Similarly, metagenomics workflow analysis of endophytic bacteria from oil palm fruits has been described by Tanjung and coworkers (Tanjung et al. 2017).

3.6 *In-silico* Tools for Metagenome Analysis

Metagenome sequencing generally ends up with the generation of a huge pool of nucleotide sequences. This vast data set of nucleotide sequences need to be analyzed thoroughly before converting it to meaningful information. The *in-silico* approaches, which include several bioinformatics tools, have helped metagenome researches in data mining. These tools have been developed in such a manner which enable them to examine large preexisting databases for the generation of new information in a short time interval. Majority of the DNA sequences generated in metagenomic studies come from Sanger sequencing and different next-generation sequencing platforms. The pyrosequencing and illumina sequencing are such platforms which produce long and short reads, respectively. Several software packages have been developed for the analysis of data generated by these sequencing techniques.

3.6.1 Mothur

Mothur (https://mothur.org) is one such *in-silico* tool that is now being widely used for the analysis of sequence generated by Sanger and/or 454 ribosomal pyro-tag sequencing (Patrick et al. 2009). Dr. Patrick Schloss and his research group in the Department of Microbiology and Immunology at University of Michigan initiated Mothur in order to develop a single piece of open-source, expandable software that could be able to fill the bioinformatics needs of the microbial ecology community. Since its development, this tool has helped several metagenome researchers and has become among the most cited *in-silico* tools to generate a meaningful information from 16s rRNA sequences.

Similarly, other bioinformatics tools, such as QIIME and MEGEN are also being widely utilized by research community to analyze metagenomic data.

3.6.2 Quantitative Insights into Microbial Ecology (QIIME)

QIIME (http://qiime.org) is also an open-source bioinformatics tool that is designed to generate quality graphics and statistics from raw data generated from Illumina or other platforms (Caporaso et al. 2010). The OTUs (operational taxonomic units) picking is the main output from QIIME pipeline, which illustrates the abundance of sequence variant in quarry samples. Other features such taxonomic assignment, phylogenetic reconstruction, and diversity analyses and visualizations provide additional strength to QIIME. Recently, an advance variant of QIIME, QIIME2 (https://qiime2.org) has been launched (Bolyen et al. 2019). Its q-2-micom plug-in allows users to build and simulate metagenome-scale metabolic community models and they can also predict metabolic fluxes, which are undergoing in microbial consortium.

3.6.3 MEta Genome Analyzer (MEGAN)

The Metagenome Analysis Software (MEGAN) (https://bio.tools/megan) is an *in-silico* tool that allows analysis of a large set of metagenomic data on desktop/laptop. It provides a platform to use several interactive tools under a single application, such as NCBI taxonomy or a more customized taxonomy (e.g. SILVA) for the taxonomic analysis, InterPro2GO, SEED, eggNOG or KEGGforfunctional analysis. In addition to this, one can use bar charts, word count, vronoi tree maps and many other charts to analyze metagenomic dataset.

3.7 Recent Progress in Metagenomic Studies of Plant Microbiome

A lot of metagenomics studies have been conducted in the recent past to characterize the rhizosphere bacterial diversity based on 16S rRNA gene amplification (Soni et al. 2010). Several other advanced molecular techniques such as PCR-RFLP (Singh et al. 2010), DGGE and TTGE (Soni et al. 2010) and real-time PCR (Premalatha et al. 2009; Soni and Goel 2010; Suyal et al. 2015a, 2015b) have also been coupled together with metagenomics to explore the plant microbiome. Researchers have also been able to beautifully paint the dominance of novel genes, such as *csp* (Premalatha et al. 2009) and *nifH* genes (Soni and Goel 2010; Soni et al. 2016) from the Himalayan region in Uttarakhand state of north India. Like *nifH* genes, *nif* might evolve from their nearest genes or adjacent regions and become specific in their functions in due course of time (Soni and Goel 2010). Using metagenomics, the existence of diverse diazotrophic microbial assemblages in the rhizosphere of indigenous red kidney bean (RKB) was also discovered (Suyal et al. 2015a, 2015b). Besides this, several studies related to plant microbiome have been reviewed and compiled to provide comprehensive knowledge about applications of metagenomics in plant microbiome studies (Soni et al. 2012, 2017; Soni 2013; Suyal et al. 2016; Goel et al. 2017a, 2017b, 2017c; Goel et al. 2018; Kumar et al. 2018a).

3.8 Conclusion and Future Prospects

The recent advancement in molecular biology and biotechnology opens several new areas of research especially in plant microbial interactions. Still several unsaid biomes are waiting for their turn. The techniques, such as metagenomics, may provide some new insights to explore these microbiomes. Scientific groups need to focus on these unexplored resources and reservoir of immense potential by using such advanced methodologies.

References

Arjun, J.K. and Harikrishnan, K. (2011). Metagenomic analysis of bacterial diversity in the rice rhizosphere soil microbiome. *Biotechnol. Bioinformatics Bioeng.* 1 (3): 361–367.

Atamna-Ismaeel, N., Finkel, O.M., Glaser, F. et al. (2012). Microbial rhodopsins on leaf surfaces of terrestrial plants. *Environ. Microbiol.* 14: 140–146.

Bai, Y., Liang, J., Liu, R. et al. (2014). Metagenomic analysis reveals microbial diversity and function in the rhizosphere soil of a constructed wetland. *Environ. Technol.* 35 (20): 2521–2527.

Bais, H.P., Park, S.-W., Weir, T.L. et al. (2004). How plants communicate using the underground information superhighway. *Trends Plant Sci.* 9: 26–32.

Barret, M., Tan, H., Egan, F. et al. (2013). Exploiting new systems-based strategies to elucidate plant–bacterial interactions in the rhizosphere. In: *Molecular Microbial Ecology of the Rhizosphere*, vol. 1 (ed. F.J. de Bruijn), 57–68. Hoboken, NJ: Wiley Blackwell.

Berendsen, R.L., Pieterse, C.M.J., and Bakker, P. (2012). The rhizosphere microbiome and plant health. *Trends Plant Sci.* 17: 478–486.

Bodenhausen, N., Bortfeld-Miller, M., Ackermann, M., and Vorholt, J.A. (2014). A synthetic community approach reveals plant genotypes affecting the phyllosphere microbiota. *PLoS Genet.* 10 (4): e1004283.

Bolyen, E., Rideout, J.R., Dillon, M.R. et al. (2019). Reproducible, interactive, scalable and extensible microbiome data science using QIIME2. *Nat. Biotechnol.* 37: 852–857.

Bouffaud, M.L., Poirier, M.A., Muller, D., and Mocenne-Loccoz, Y. (2014). Root microbiome relates to plant host evolution in maize and other Poaceae. *Environ. Microbiol.* 16: 2804–2814.

Bulgarelli, D., Rott, M., Schlaeppi, K. et al. (2012). Revealing structure and assembly cues for Arabidopsis root-inhabiting bacterial microbiota. *Nature* 488 (7409): 91–95.

Bulgarelli, D., Schlaeppi, K., Spaepen, S. et al. (2013). Structure and functions of the bacterial microbiota of plants. *Annu. Rev. Plant Biol.* 64: 9.1–9.32.

Campisano, A., Antonielli, L., Pancher, M. et al. (2014). Bacterial endophytic communities in the grapevine depend on pest management. *PLoS One* 9 (11): e112763.

Caporaso, J.G., Kuczynski, J., Stombaugh, J. et al. (2010). QIIMEallows analysis of high-throughput community sequencing data. *Nat. Methods* 7 (5): 335.

Choi, J.-E., Nguyen, C.M., Lee, B. et al. (2018). Isolation and characterization of a novel metagenomic enzyme capable of degrading bacterial phytotoxintoxoflavin. *PLoS One* 13 (1): e0183893.

Correa-Galeote, D., Bedmar, E.J., and Arone, G.J. (2018). Maize endophytic bacterial diversity as affected by soil cultivation history. *Front. Microbiol.* 9: 484.

Crump, B.C., Wojahn, J.M., Tomas, F., and Mueller, R.S. (2018). Metatranscriptomics and amplicon sequencing reveal mutualisms in Seagrass microbiomes. *Front. Microbiol.* 9: 388.

De Castro, A.P., da Fernandes, G.R., and Franco, O.L. (2014). Insights into novel antimicrobial compounds and antibiotic resistance genes from soil metagenomes. *Front. Microbiol.* 5: 489.

Delmotte, N., Knief, C., Chaffron, S. et al. (2009). Community proteogenomics reveals insights into the physiology of phyllosphere bacteria. *Proc. Natl. Acad. Sci. U.S.A.* 106: 16428–16433.

Edwards, J., Santos-Medellín, C., and Sundaresan, V. (2018). Extraction and 16S rRNA sequence analysis of microbiomes associated with rice roots. *Bio-protocol* 8 (12): e2884.

Egamberdieva, D. et al. (2008). High incidence of plant growth stimulating bacteria associated with the rhizosphere of wheat grown on salinated soil in Uzbekistan. *Environ. Microbiol.* 10: 1–9.

Finkel, O.M., Delmont, T.O., Post, A.F., and Belkin, S. (2016). Metagenomic signatures of bacterial adaptation to life in the phyllosphere of a salt-secreting desert tree. *Appl. Environ. Microbiol.* 82: 2854–2861.

Franzosa, E.A., Hsu, T., Sirota-Madi, A. et al. (2015). Sequencing and beyond: integrating molecular "omics" for microbial community profiling. *Nat. Rev. Microbiol.* 13: 360–372.

Gevers, D., Knight, R., Petrosino, J.F. et al. (2012). The human microbiome project: a community resource for the healthy human microbiome. *PLoS Biol.* 10 (8): e1001377.

Goel, R., Suyal, D.C., Narayan et al. (2017a). Soil Metagenomics: a tool for sustainable agriculture. In: *Mining of Microbial Wealth and MetaGenomics* (eds. V.C. Kalia, Y. Shouche, H.J. Purohit and P. Rahi), 217–225. Springer Nature Singapore.

Goel, R., Suyal, D.C., Kumar, V. et al. (2017b). Stress-tolerant beneficial microbes for sustainable agricultural production. In: *Microorganisms for Green Revolution, Microorganisms for Sustainability* (eds. D.G. Panpatte, Y.K. Jhala, H.N. Shelat and R.V. Vyas), 141–159. Springer Nature Singapore.

Goel, R., Kumar, V., Suyal, D.C. et al. (2017c). Root-associated bacteria: Rhizoplane and endosphere. In: *Plant-Microbe Interactions in Agro-Ecological Perspectives* (eds. D.P. Singh, H.B. Singh and R. Prabha). Springer Nature Singapore Pvt Ltd. https://doi.org/10.1007/978-981-10-5813-4_9.

Goel, R., Kumar, V., Suyal, D.C. et al. (2018). Toward the unculturable microbes for sustainable agricultural production. In: *Role of Rhizospheric Microbes in Soil* (ed. V.S. Meena), 107–123. Springer Nature Singapore Pte Ltd.

Guo, J., Cole, J.R., Zhang, Q. et al. (2015). Microbial community analysis with ribosomal gene fragments from shotgun metagenomes. *Appl. Environ. Microbiol.* 82: 157–166.

Hamonts, K., Trivedi, P., Garg, A. et al. (2018). Field study reveals core plant microbiota and relative importance of their drivers. *Environ. Microbiol.* 20: 124–140.

Hardoim, P.R., van Overbeek, L.S., Berg, G. et al. (2015). The hidden world with in plants: ecological and evolutionary considerations for defining functioning of microbial endophytes. *Microbiol. Mol. Biol. Rev.* 79: 293–320.

Hayden, H.L., Savin, K.W., Wadeson, J. et al. (2018). Comparative meta transcriptomics of wheat Rhizosphere microbiomes in disease suppressive and non-suppressive soils for *Rhizoctoniasolani* AG8. *Front. Microbiol.* 9: 859.

van der Heijden, M.G.A. and Hartmann, M. (2016). Networking in the plant microbiome. *PLoS Biol.* 14 (2): e1002378.

Hover, B. et al. (2018). Culture-independent discovery of the malacidins as calcium-dependent antibiotics with activity against multidrug-resistant Gram-positive pathogens. *Nat. Microbiol.* 3: 415–422.

Kennedy, A.C. and de Luna, L.Z. (2005). Rhiszosphere. In: *Encyclopedia of Soils in the Environment* (pp. 399–406).

Kgomotso, M., Maropola, A., Ramond, J.B., and Trindade, M. (2015). Impact of metagenomic DNAextraction procedures on the identifiable endophytic bacterial diversity in *Sorghum bicolor* (L. Moench). *J. Microbiol. Methods* 112: 104–117.

Knief, C., Delmotte, N., Chaffron, S. et al. (2012). Metaproteogenomic analysis of microbial communities in the phyllosphere and rhizosphere of rice. *ISME J.* 6: 1378–1390.

Kumar, S., Suyal, D.C., Bhoriyal, M., and Goel, R. (2018a). Plant growth promoting potential of psychrotolerant *Dyadobacter* sp. for pulses and finger millet and impact of inoculation on soil chemical properties and diazotrophic abundance. *J. Plant Nutr.* 41 (8): 1035–1046.

Kumar, V., AlMomin, S., Al-Aqeel, H. et al. (2018b). Metagenomic analysis of rhizosphere microflora of oil-contaminated soil planted with barley and alfalfa. *PLoS One* 13 (8): e0202127.

Lauro, F.M., DeMaere, M.Z., Yau, S. et al. (2011). An integrative study of a meromictic lake ecosystem in Antarctica. *ISME J.* 5: 879–895.

Lebeis, S.L., Rott, M., Dangl, J.L., and Schulze-Lefert, P. (2012). Culturing a plant microbiome community at the cross-Rhodes. *New Phytol.* 196: 341–344.

Lin, W., Wu, L., Lin, S. et al. (2013). Metaproteomic analysis of ratoon sugarcane rhizospheric soil. *BMC Microbiol.* 13: 135.

López-Pérez, M. and Mirete, S. (2014). Discovery of novel antibiotic resistance genes through metagenomics. *Rec. Adv. DNA Gene Seq.* 8: 15–19.

Maropola, M.K., Ramond, J.B., and Trindade, M. (2015). Impact of metagenomic DNAextraction procedures on the identifiable endophytic bacterial diversity in *Sorghum bicolor* (L. Moench). *J. Microbiol. Methods* 12: 104–117.

Mattarozzi, M., Manfredi, M., Montanini, B. et al. (2017). A metaproteomic approach dissecting major bacterial functions in the rhizosphere of plants living in serpentine soil. *Anal. Bioanal. Chem.* 409: 2327–2339.

Mendes, R., Paolina, G., and Raaijmaker, J.M. (2013). The rhizosphere microbiome: significance of plant beneficial, plant pathogenic, and human pathogenic microorganisms. *FEMS Microbiol. Rev.* 37: 634–663.

Mendes, R., Garbeva, P., and Raaijmakers, J.M. (2014). The rhizosphere microbiome: significance of plant beneficial, plant pathogenic and human pathogenic microorganisms. *FEMS Microbiol. Rev.* 37: 634–663.

Methe, B.A., Kelvin, Li., Talley, S.P., Gupta, N., Frank, B., Xu, W. et al. (2017). Functional genes that distinguish maize phyllosphere metagenomes in drought and well-watered conditions. BioRxiv. 104331.

Mirete, S., de Figueras, C.G., and González-Pastor, J.E. (2007). Novel nickel resistance genes from the Rhizosphere metagenome of plants adapted to acid mine drainage. *Appl. Environ. Microbiol.* 73 (19): 6001–6011.

Morris, C.E. and Kinkel, L.L. (2002). Fifty years of phylosphere microbiology: significant contributions to research in related fields. In: *Phyllosphere Microbiology* (eds. S.E. Lindow, E.I. Hecht-Poinar and V. Elliott), 365–375. St. Paul, MN: APS Press.

Mukhtar, S., Ishaq, A., Hassan, S. et al. (2017). Comparison of microbial communities associated with halophyte (*Salsolastocksii*) and non-halophyte (*Triticum aestivum*) using culture-independent approaches. *Pol. J. Microbiol.* 66 (3): 353–364.

Ofaim, S., Ofek-Lalzar, M., Sela, N. et al. (2017). Analysis of microbial functions in the rhizosphere using a metabolic-network based framework for metagenomics interpretation. *Front. Microbiol.* 8: 1606.

Ofek-Lalzar, M., Sela, N., Goldman-Voronov, M. et al. (2014). Niche and host-associated functional signatures of the root surface microbiome. *Nat. Commun.* 5: 4950.

Parnell, J.J., Berka, R., Young, H.A. et al. (2016). From the lab to the farm: an industrial perspective of plant beneficial microorganisms. *Front. Plant Sci.* 7: 1110.

Patrick, D.S., Sarah, L.W., Thomas, R. et al. (2009). Introducing mothur: open-source, platform-independent, community-supported software for describing and comparing microbial communities. *Appl. Environ. Microbiol.* 75 (23): 7537–7541.

Pétriacq, P., Williams, A., Cotton, A. et al. (2017). Metabolite profiling of non-sterile rhizosphere soil. *Plant J.* 92 (1): 147–162.

Philippot, L., Raaijmakers, J.M., Lemanceau, P., and vander Putten, W.H. (2013). Going back to the roots: the microbial ecology of the rhizosphere. *Nat. Rev. Microbiol.* 11: 789–799.

Premalatha, K., Soni, R., Khan, M. et al. (2009). Exploration of Csp gene(s) from temperate and glacier soils of Indian Himalaya and in silico analysis of encoding proteins. *Curr. Microbiol.* 58: 343–348.

Qin, J., Li, R., Raes, J. et al. (2010). A human gut microbial gene catalogue established by metagenomic sequencing. *Nature* 464: 59–65.

Rascovan, N., Carbonetto, B., Perrig, D. et al. (2016). Integrated analysis of root microbiomes of soybean and wheat from agricultural fields. *Sci. Rep.* 6: 28084.

Rastogi, G., Tech, J.J., Coaker, G.L., and Leveau, J.H.J. (2010). A PCR-based toolbox for the culture-independent quantification of total bacterial abundances in plant environments. *J. Microbiol. Methods* 83: 127–132.

Rastogi, G., Sbodio, A., Tech, J.J. et al. (2012). Leaf microbiota in an agroecosystem: spatiotemporal variation in bacterial community composition on field-grown lettuce. *ISME J.* 6: 1812–1822.

Reinhold-Hurek, B. and Hurek, T. (2011). Living inside plants: bacterial endophytes. *Curr. Opin. Plant Biol.* 14: 435–443.

Ritpitakphong, U., Falquet, L., Vimoltust, A. et al. (2016). The microbiome of the leaf surface of Arabidopsis protects against a fungal pathogen. *New Phytol.* 210: 1033–1043.

Roume, H., Heintz-Buschart, A., Muller, E.E.L. et al. (2015). Comparative integrated omics: identification of key functionalities in microbial community-wide metabolic networks. *Biofilms Microb.* 1: 15007.

Sengupta, S., Ganguli, S., and Singh, P.K. (2017). Metagenome analysis of the root endophytic microbial community of Indian rice (*O. sativa* L.). *Genomics Data* 12: 41–43.

Sessitsch, A., Hardoim, P., Döring, J. et al. (2012). Functional characteristics of an Endophyte community colonizing Rice roots as revealed by metagenomic analysis. *Mol. Plant Microbe Interact.* 25 (1): 28–36.

Silva, C.C., Hayden, H., Sawbridge, T. et al. (2013). Identification of genes and pathways related to phenol degradation in metagenomic libraries from petroleum refinery wastewater. *PLoS One* 8 (4): e61811.

Singh, C., Soni, R., Jain, S. et al. (2010). Diversification of nitrogen fixing bacterial community using *nif*H gene as a biomarker in different geographical soils of Western Indian Himalayas. *J. Environ. Biol.* 31: 553–556.

Sleator, R.D., Shortall, C., and Hill, C. (2008). Metagenomics. *Lett. Appl. Microbiol.* 47 (5): 361–366.

Smith, S.E. and Read, D.J. (2008). Mycorrhizal symbiosis. *New Phytol.* 137: 563–568.

Song, S., Jarvie, T., and Hattori, M. (2013). Our second genome – human metagenome – how next generation sequencer changes our life through microbiology. *Adv. Microb. Physiol.* 62: 119–124.

Soni, R. (2013). Ecological aspects of metagenomic approach. In: *Reshaping Technology for Agricultural Development* (eds. A. Kumar and A. Prasad), 135–151. New Delhi: Satish Publishing House.

Soni, R. and Goel, R. (2010). Triphasic approach for assessment of bacterial population in different soil systems. *Ekologija* 56 (3 & 4): 99–104.

Soni, R., Shaluja, B., and Goel, R. (2010). Bacterial community analysis using temporal gradient gel electrophoresis of 16 S rDNA PCR products of soil metagenomes. *Ekologija* 56 (3 & 4): 94–98.

Soni, R., Acahrya, C., Premalatha, K., and Goel, R. (2012). Metagenomics technology. In: *Biotechnology in Medicine and Agriculture Principles and Practices* (eds. A. Kumar et al.), 835–843. New Delhi: IKInternational.

Soni, R., Suyal, D.C., Sai, S., and Goel, R. (2016). Exploration of *nif*H gene through soil metagenomes of the western Indian Himalayas. *3 Biotech* 6 (1): 25.

Soni, R., Kumar, V., Suyal, D.C. et al. (2017). Metagenomics of plant Rhizosphere microbiome. In: *Understanding Host-Microbiome Interactions – an Omics Approach* (eds. R.P. Singh, R. Kothari, P.G. Koringa and S.P. Singh), 193–206. Springer Nature Singapore.

Sun, L., Qiu, F., Zhang, X. et al. (2008). Endophytic bacterial diversity in rice (*Oryza sativa* L.) roots estimated by 16S rDNA sequence analysis. *Microb. Ecol.* 55: 415–424.

Suyal, D.C., Shukla, A., and Goel, R. (2014). Growth promotory potential of the psychrophilic diazotroph *Pseudmonasmigulae* S10724 against native *Vigna radiata* (L.) Wilczek. *3 Biotech* 4 (6): 665–668.

Suyal, D.C., Yadav, A., Shouche, Y., and Goel, R. (2015a). Bacterial diversity and community structure of Western Indian Himalayan red kidney bean (*Phaseolus vulgaris*) rhizosphere as revealed by 16S rRNA gene sequences. *Biologia* 70 (3): 305–313.

Suyal, D.C., Yadav, A., Shouche, Y., and Goel, R. (2015b). Diversified diazotrophs associated with the rhizosphere of Western Indian Himalayan native red kidney beans (*Phaseolus vulgaris* L.). *3 Biotech* 5 (4): 433–441.

Suyal, D.C., Soni, R., Sai, S., and Goel, R. (2016). Microbial inoculants as biofertilizer. In: *Microbial Inoculants in Sustainable Agricultural Productivity* (ed. D.P. Singh), 311–318. Springer India.

Tanjung, Z.A., Aditama, R., Sudania, W.M. et al. (2017). Metagenomics workflow analysis of endophytic bacteria from oil palm fruits. *Int. Symp. Bioinform. Chemometrics Metabolism* 835: 012007.

Tian, B.Y., Cao, Y., and Zhang, K.Q. (2015). Metagenomic insights into communities, functions of endophytes, and their associates with infection by root-knot nematode, *Meloidogyne incognita*, in tomato roots. *Sci. Rep.* 5: 17087.

Tomer, S., Suyal, D.C., Rajwar, J. et al. (2017). Isolation and characterization of Phosphate solubilizing bacteria from Western Indian Himalayan soils. *3 Biotech* 7 (2): 1–8.

Tsurumaru, H., Okubo, T., Okazaki, K. et al. (2015). Metagenomic analysis of the bacterial community associated with the taproot of sugar beet. *Microbes Environ.* 30 (1): 63–69.

Turner, T.R., James, E.K., and Poole, P.S. (2013). The plant microbiome. *Genome Biol.* 14: 209.

Tyson, G.W., Chapman, J., Hugenholtz, P. et al. (2004). Community structure and metabolism through reconstruction of microbial genomes from the environment. *Nature* 428: 37–43.

Utturkar, S.M., Cude, W.N., Robeson, M.S. et al. (2016). Enrichment of root Endophytic bacteria from Populusdeltoides and single-cell-genomics analysis. *Appl. Environ. Microbiol.* 82 (18): 5698–5708.

Vandenkoornhuyse, P., Baldauf, S.L., Leyval, C. et al. (2002). Extensive fungal diversity in plant roots. *Science* 295: 2051.

Vandenkoornhuyse, P., Quaiser, A., Duhamel, M. et al. (2015). The importance of the microbiome of the plant holobiont. *New Phytol.* 206: 1196–1206.

Velázquez-Sepúlveda, I., Orozco-Mosqueda, M.C., Prieto-Barajas, C.M., and Santoyo, G. (2012). Bacterial diversity associated with the rhizosphere of wheat plants (*Triticum aestivum*): toward a metagenomic analysis. *FYTON* 81: 81–87.

Waight, K., Pinyakong, O., and Luepromchai, E. (2007). Degradation of phenanthrene on plant leaves by phyllosphere bacteria. *J. Gen. Appl. Microbiol.* 53: 265–272.

Widder, S., Allen, R.J., Pfeiffer, T. et al. (2016). Challenges in microbial ecology: building predictive understanding of community function and dynamics. *ISME J.* 10: 2557–2568.

Wrighton, K.H. (2018). Antibacterial drugs: discovering antibiotics through soil metagenomics. *Nat. Rev. Drug Discov.* 17 (4): 240–241.

Zengler, K. and Palsson, B.O. (2012). A road map for the development of community systems (CoSy) biology. *Nat. Rev. Microbiol.* 10: 366–372.

Zhao, L. (2010). Genomics: the tale of our other genome. *Nature* 465: 879–880.

Zhou, J., He, Z., Yang, Y. et al. (2015). High-throughput metagenomic technologies for complex microbial community analysis: open and closed formats. *MBio* 6 (1): e02288–e02214.

4

Combating the Abiotic Stress Through Phytomicrobiome Studies

Hemant S. Maheshwari[1,2], *Abhishek Bharti*[2], *Richa Agnihotri*[2], *Ajinath Dukare*[3], *B. Jeberlin Prabina*[4], *Saurabh Gangola*[5], *and Mahaveer P. Sharma*[2]

[1] *Ecophysiology of Plants, Faculty of Science and Engineering, GELIFES-Groningen Institute for Evolutionary Life Sciences, Groningen, The Netherlands*
[2] *ICAR-Indian Institute of Soybean Research, Indore, Madhya Pradesh, India*
[3] *ICAR- Central Institute of Post-Harvest Engineering and Technology (CIPHET), Ludhiana, Punjab, India*
[4] *Tamilnadu Agriculture University, Coimbatore, Tamilnadu, India*
[5] *Graphic Era Hill University, Bhimtal, Uttarakhand, India*

4.1 Introduction

4.1.1 Abiotic Stress and Phytomicrobiome

The changing climatic pattern induces irregularities in weather parameters and poses a potential threat for agricultural production, questioning the sustainability of agriculture worldwide. Plants are often exposed to the extreme and adverse effects of global climate change, such as drought, flood, high temperature, salinity, acidity, mineral toxicity, nutrient deficiency, pest and disease incidence (Paul et al. 2017). These stresses affect large agricultural land areas leading to both qualitative and quantitative crop reduction (Cramer et al. 2011). It is projected by FAO that the world population will reach 9.8 billion by 2050 (Harold and Reetz 2016), and the major challenges that the world is going to face are ensuring food security. Productivity could be achieved by the usage of chemical fertilizers and pesticides, but the excessive and irrational use of these chemicals is linked with adverse impacts on soil, environment, and human health. This manmade stress disturbs the eco-balanced nutrient cycling and in turn affects nutrient uptake by crops. Plants have a wide range of protective mechanisms to cope with the stress conditions. Variations in the environmental growth condition due to stress would put the plant metabolism out of homeostasis (Foyer and Noctor 2005), thus creating a condition that would force the plant to create genetic and metabolic changes in the cellular system to some extent only. Breeders are trying to develop stress-tolerant crop varieties using advanced biotechnology tools to make the plants smart to cope with the stress condition. However, an economical way to make the crops sustainable under these stresses is the microbiological intervention–based agriculture. Plants are colonized with diverse microbes constituting the holobionts.

Phytomicrobiome Interactions and Sustainable Agriculture, First Edition. Edited by Amit Verma, Jitendra Kumar Saini, Abd El-Latif Hesham, and Harikesh Bahadur Singh.

It is noted that plants that are harboring the microbes are ecologically more compatible and environmentally sustainable under climate-changing scenarios (Bhattacharyya et al. 2016).

The microorganisms that colonize the plants are collectively termed as a phytomicrobiome, and plants are associated with diverse structural and functional microbial communities (bacteria, fungi, actinobacteria, archea) and constitute holobionts. They can occupy different parts of plants like root (rhizosphere), seed (spermosphere), inside of the plants (endosphere), flower (anthroposphere), and leaf (phyllosphere) (Lakshmanan et al. 2014). Thus, plants represent meta-organisms or holobionts. Carbon-rich root exudates and diverse volatile compounds released from the crop root act as a chemo-attracting molecule for various rhizospheric microorganisms, which in turn perform the beneficial task for the plants. Such rhizospheric microbial communities are implicated in nutrient acquisition, plant growth promotion, alleviation of abiotic and biotic stress via various direct and indirect mechanisms, which protects from plant pathogens. These microorganisms are recruited from bulk soil, and tremendous diversity is associated with bulk soil, which then is passed to the rhizosphere, and then passed to the endosphere. Plants compensate their energy investment through root exudates via filtering beneficial microflora from the bulk soil. Understanding the rhizosphere chemistry and molecular dialogs between plant-microbe interactions are necessary for deciphering the role of rhizosphere microbes and, in turn, interaction in the root zone.

The phytomicrobiome is highly dynamic. It encompasses groups of both beneficial as well as harmful microorganisms, which in equilibrium determines the fate of plant fitness (Raaijmakers et al. 2009). Phytomicrobiome interactions involve interaction among the microbe–microbe, plant–microbe, and either way, and these interactions by way of symbiosis, predation, parasitisms, and antagonism determines the winner in the battlefield. Introduction of new microbes either wild or genetically altered spraying of chemical additives, signal compounds, and the agricultural practices alter the composition of the plant-associated microbiome (Quiza et al. 2015). The composition of the plant microbiome is highly influenced by its genotypes, developmental stages, soil conditions, and abiotic stresses. For example, the studies on the effect of two soil characters and different soybean genotypes on rhizosphere community structure revealed that both affect the microbial community structure, but soil types have a greater impact on shaping the microbial community (Wang et al. 2014). Using advanced techniques such as biolog PCR amplicon pyrosequencing, and real-time PCR, it was shown that metabolic activities of soil microbes were found to be higher in rhizospheric soil than bulk soil. Similarly, pyrosequencing data revealed that soybean rhizospheric–microorganism changes significantly during plant growth, and the plant growth-promoting bacteria like *Bacillus*, *Bradyrhizobium*, and *Rhizobium* were also observed in succession (Sugiyama et al. 2014). In other phytomicrobiome studies, the effect of drought stress on rhizospheric bacterial and fungal community structure in soybean *Glycinesoja* (wild type) and *Glycinemax* (cultivated) revealed more drought-tolerant nature of wild species than cultivated species based on free proline, soluble sugar and gene expression of NCED 1. Furthermore, drought stress was found to be significantly detrimental to the fungal community structure in the plant root zone (Zhang et al. 2017).

4.1.2 Role of Signaling in Phytomicrobiome Interactions

Different kinds of root exudates and volatile compounds act as a chemical signaling molecule in shaping the structural and functional microbial communities of the rhizomicrobiome. Similarly, the plant-associated microbiome also produces volatile, quorum sensing (QS), and phytohormones compounds for their interaction with plants. Interaction of plant and microbe is mediated via various specific signal compounds for better plant growth promotion (Altaf et al. 2017). The role of these signal molecules in phytomicrobiome functioning is discussed in the below section.

4.2 Phytomicrobiome Signaling Compounds

4.2.1 Root Exudates and Plant Volatiles Compounds

Root secret cocktail of diverse chemical compounds and about nearly 20–40% of photosynthetically fixed compounds are released through roots (Badri and Vivanco 2009). These group of compounds mainly includes low-molecular-weight substances like sugars, amino acids, organic acids, and many secondary metabolites (flavonoids, terpenoids, and phenolic), which are exudated passively. Another class is the high-molecular-weight compounds comprising polysaccharides, mucilage, and proteins (Bais et al. 2006). The composition in terms of quality and quantity of rhizo-deposits is determined by the plant variety, genotypes, developmental stage, soil factors, interacting soil microbial community, and prevailing abiotic and biotic stress in the plants (Jones et al. 2004). The nutrient-rich rhizodeposits are utilized by microorganisms for their growth and multiplication in rhizoplane, rhizosphere, and endosphere. Root exudates mediate positive (symbiosis, mycorrhizal association and chemo-attraction of plant growth-promoting rhizobacteria [PGPR]), and negative (allelopathy, competition, herbivory, and parasitism) interactions depending upon the type of microorganisms and host plants (Hayat et al. 2017). The root exudates act as the first line of communication between interacting microbiota for their colonization of various plant parts. Besides root exudates, plant volatiles, which are low-molecular-weight and lipophilic compounds emitted as primary and secondary metabolites emitted from root, stem, leaf, and flowers, also act as a signaling molecule in plant–microbiome interaction (Pichersky et al. 2006). They belong to main terpenoids (cytosolic mevalonic acid (MVA) and plastidial methylerythritol phosphate (MEP), phenylpropanoid/ benzenoid (shikimate pathway), fatty acid and green leaf volatiles (GLVs) derivatives (lipoxygenase pathway) and branched-chain amino acids derivative groups' compounds (tricarboxylic acid cycle) (Dudareva et al. 2013). These volatiles help in priming the plant against various biotic and abiotic stresses, inhibit the growth of phytopathogens, and improve plant growth and productivity (Brilli et al. 2019). Specific plant volatiles such as isoprenoids, emitted from *Arabidopsis* plant, quenches the reactive oxygen species (ROS), provides thermotolerance, and maintains cellular membrane integrity and functionality of thylakoid during high temperature and oxidative stress, which enhances the photosynthetic activity (Velikova et al. 2011). Similarly, isoprenoids formation is linked with carotenoids and cytokinin biosynthetic pathway; therefore, it helps in delaying the senescence of plant parts by scavenging ROS molecules, antioxidative action and prevents programmed cell death or apoptosis during abiotic stresses (Dani et al. 2016).

4.2.2 Microbial Volatile Organic Compounds

Similar to plant volatile, microbial volatile organic compounds (MVOCs) are a low-molecular-weight, high boiling point, high vapor pressure lipophilic compounds, produced via primary and secondary metabolism are involved in the mediation of short- and long-distance innumerable microbe–plant and microbe–microbe interaction in both aerial and soil–plant parts (Kanchiswamy et al. 2015). These MVOCs belong to alcohols, alkenes, ketones, esters, acids, pyrazines, terpenes, and sulfides groups (Piechulla and Degenhardt 2014). The composition of MVOCs depends upon the physiological status and associated environmental factors such as temperature, oxygen, moisture, and pH (Romoli et al. 2014). The MVOCs sometimes have antifungal and antibacterial activities toward the microorganisms occupying the same ecological niche. MVOCs play a role in plant–microbe and microbe–microbe interaction and are implicated in better plant growth promotion (Schmidt et al. 2016). Such compounds mediate long-distance communication and influence population dynamics, behavior, and gene expression of nearby microbial communities (Garbeva et al. 2014).

4.2.3 Quorum Sensing

QS is a population density–dependent regulation of gene expression, which activates after attaining minimum threshold concentration of extracellular chemical signal molecule termed as autoinducers. This mechanism is operating during symbiosis, motility, biofilm formation, colonization, conjugation, antibiotic formation, virulence, and many other plant–microbe interactions in a phytomicrobiome (Miller and Bassler 2001). The cell-to-cell communication in gram-negative bacteria and gram-positive bacteria is mediated by acyl-homoserine lactones (AHL) and autoinducing peptides (AIP), respectively (Ng and Bassler 2009). Other QS signal compounds are cis-11-methyl-2-dodecenoic acid (diffusible signal factor or DSF) produced by *Xanthomonas* (He and Zhang 2008), 2-heptyl-3-hydroxy-4-quinolone (PQS) by *Pseudomonas* (Jimenez et al. 2010), and diketopiperazines (DKZ) in PGPR such as *Pseudomonas* and *Enterobacter* (Holden et al. 1999). Plants also produce some compounds which interfere with QS in bacteria by mimicking AHL molecules, termed as quorum quenching compounds (Teplitski et al. 2000). Plant-associated *Bacillus* sp. produces quorum-quenching compounds, which inactivates AHL (Dong et al. 2002).

The rhizosphere soil has a higher fraction of AHL-producing organisms, thus, signifying their crucial role in root colonization. This also suggests that plants might be taking benefits of a bacterial communication system in modifying and influencing root-colonizing microflora (Lugtenberg and Kamilova 2009). The need for an initial threshold level of microbial inoculum to support plant growth suggests that microbial QS are imperative in plant–microbe interactions in the rhizosphere. The certain bacteria block the expression of many virulence genes of pathogens via quenching autoinducer signals, produced by pathogens. This approach holds a great perspective for crop disease control, as virulence ability of the most plant pathogens is dependent on the QS signaling system. Expression of AHLs producing genes in genetically engineered plants is another approach for enhancing the plant immunity against pathogens.

4.2.4 Underground Phytomicrobiome Signaling

In nature, the well-known examples of the phytomicrobiome signaling system occur in the case of rhizobial-legume symbiosis, mycorrhizal fungi-roots of plants, and with many PGPR (Smith et al. 2015). The key steps involved in the process of a nodule and mycorrhizal colonization are reviewed by Oldroyd (2013) are as follows: (i) plant root releases chemical signal flavonoid to rhizobia, which in turn produces lipochitooligosachharide (LCO)/nod factor, (ii) these LCOs are recognized and binds with Lys domain receptor-like kinases, nod factor receptor 1 and NFR5, (iii) activation of symbiotic signaling pathways followed by calcium spiking in plant cells, (iv) root hair curing, deformation, and infection thread formation for rhizobial movement, and (v) release of rhizobia in the cortical cells, and gets surrounded by plant-derived peribacteroid membrane and form nodule. However, in the case of arbuscular mycorrhizal fungi (AMF), important colonization steps are reviewed by Oldroyd (2013) as follows: (i) plant root releases strigolactones, whereas mycorrhiza produces LCOs or Myc factor, (ii) Myc factor binds with LysM receptor-like kinases and activate the symbiosis signaling event, (iii) calcium spiking in root tissue and AMF invasion into root epidermal cells, and (iv) intercellular hyphal growth inside root cortical tissue and arbuscule formation occur inside the cells (site of nutrient exchange). In addition to both rhizobia and AMF, such signaling also occurs between *Frankia* and plant families' Fagales, Rosales, and Cucurbitales. They also have a common similarity due to the involvement of the receptor-like kinase gene, SymRK (Gherbi et al. 2008).

Many important microorganisms present in bulk soils are chemically attracted and get deposited in the rhizosphere or endosphere by root exudates. PGPR produces numerous phytohormones (auxin, gibberellin, and cytokinin), cell wall lytic enzymes and many secondary metabolites (nitric oxide and 2,4-diacetyl phloroglucinol) that are involved in the modification of root architecture by forming more lateral roots and root hairs (Vacheron et al. 2013). Beneficial as well as pathogenic gram-positive and gram-negative bacteria produce indole-3-acetic acid (IAA) that function as a reciprocal signaling molecule and has a role in the expression of bacterial genes and also in plant–microbe interaction (Duca et al. 2014). PGPR bacteria produce 1-aminocyclopropane-1-carboxylic acid (ACC)-deaminase enzymes, which cleaves ACC, an ethylene precursor into alpha-ketobutyrate and ammonia and therefore reduces the plant ethylene levels. Synergistic action of bacterial IAA- and ACC-deaminase activity is involved in the alleviation of crop abiotic stress (due to drought, salinity, heavy metal tolerance, and flooding) and biotic stress (due to bacterial and fungal) phytopathogens (Glick 2014). A bacterial peptide, thuricin-17, was isolated from PGPR bacterium *Bacillus thuringiensis* NEB 17 (Gray et al. 2006). Spraying of thuricin-17 bacteriocin helps plant growth promotion of soybean and maize (Lee et al. 2009). In banana, among the root exudate, oxalic, malic and fumaric organic acids, the malic acid act as chemoattractants for *Bacillus amyloliquefaciens* NJN-6. However, fumaric acids involved in the formation of biofilm by modulating biofilm-forming gene (yqxM and eps D) (Yuan et al. 2015). *Arabidopsis thaliana* plant roots release malic acid, which selectively chemoattract or recruit PGPR *Bacillus subtilis* FB17 (Rudrappa et al. 2008).

4.3 Mechanisms of Phytomicrobiome Associated with Abiotic Stress Tolerance

Microbial communities colonizing in the rhizosphere, rhizoplane, and endosphere induce physical and chemical changes in the plants and therefore alleviate abiotic stresses termed as induced systemic tolerance (IST) (Yang et al. 2009). Plants exposed to abiotic stresses produce

ROS that damage cellular components and biomolecules, which leads to cellular death. Plant-associated microorganisms produce various metabolites or signals for stress alleviation and plant growth promotion (Ansari 2018). The role of phytomicrobiome in the mitigation of different kinds of plant abiotic stress is discussed in the below section. Furthermore, we explained how different components of phytomicrobiome interact with each other and imparts abiotic stress tolerance in Figure 4.1. And all additional information on the role of different microbes and their mechanisms in the abiotic stress are listed in Table 4.1.

4.3.1 Drought Stress Alleviation

Drought and heat stress are detrimental to crop production and can reduce crop yield by almost 50% (Lamaoui et al. 2018). A plant exposed to drought undergoes various morphological, biochemical, and molecular changes. The key morphological changes are reduction in leaf water potential, stomatal conductance, loss in turgidity, reduction in leaf area and stem length, and decline in photosynthetic yield. Biochemical changes are increasing in levels of antioxidant enzymes (superoxide dismutase, catalase, peroxidases) to scavenge ROS, accumulation of compatible solutes, or metabolites (proline and sugars), and reduced activity of Rubisco enzyme. The molecular changes are increased expression of ABA genes, synthesis of LEA, chaperones, and dehydrin protein and modulation of gene expression of DREB, WRKY, and NAC gene. The activation of regulatory and functional proteins is also observed (Kumar et al. 2018). Root colonizing plant growth–promoting bacteria employ a different direct and indirect mechanism to alleviate drought stress (Mohammadipanah and Zamanzadeh 2019). The key mechanisms are: (i) release of microbial phytohormones (IAA, gibberellic acid, and cytokinin), which modify root system architecture (RSA) for better water and nutrients absorption, (ii) exopolysaccharide (EPS) improves soil aggregation and help in maintaining higher water potential, (iii) alteration of antioxidant enzymes like (SOD, CAT, APX, GR, and POD) for scavenging ROS, (iv) formation of volatile compounds like 2R, 3R-butanediol that helps in the closure of stomata by salicylic acid formation, (v) accumulation of compatible solutes like proline, trehalose, and glycine-betaine for maintaining osmolarity, (vi) alteration in the elasticity of membrane and improve membrane stability, (vii) formation of ACC-deaminase, which cleave ethylene precursors ACC into alpha-ketobutyrate and ammonia and thereby reducing the ethylene levels inside the plants, and (viii) solubilization of macro and micronutrients (P, Zn, K, etc.) and biological nitrogen fixation. The predominant genera involved in abiotic stress alleviation are *Bacillus, Pseudomonas, Azotobacter, Azospirillum, Burkholderia, Klebsiella, Paenibacillus, Serratia, Arthrobacter,* and *Streptomyces* (Yadav and Yadav 2018).

With exposure to drought stress, the composition of root microbiome favorably changes which creates greater plant tolerance and fitness to stress conditions. It was found that under drought enrichment of monoderm lineages (Actinobacteria, Chloroflexi, and Firmicutes) was due to the formation of a positive regulator, glycerol-3-phosphate for monoderms, whereas ROS formation under stress has a negative regulatory role for diderms (Proteobacteria, Bacteroidetes, and Acidobacteria). Following the period of drought stress, the diderms group again dominates in the root microbiome (Xu and Coleman-Derr 2019). Drought stress alters the balance of soil properties, such as soil moisture, nutrients, oxygen content, solute content, and substrate diffusion. Similarly, it also causes decreased photosynthetic activities, alters RSA and depth, reduces carbon fixation and thereby, modifies

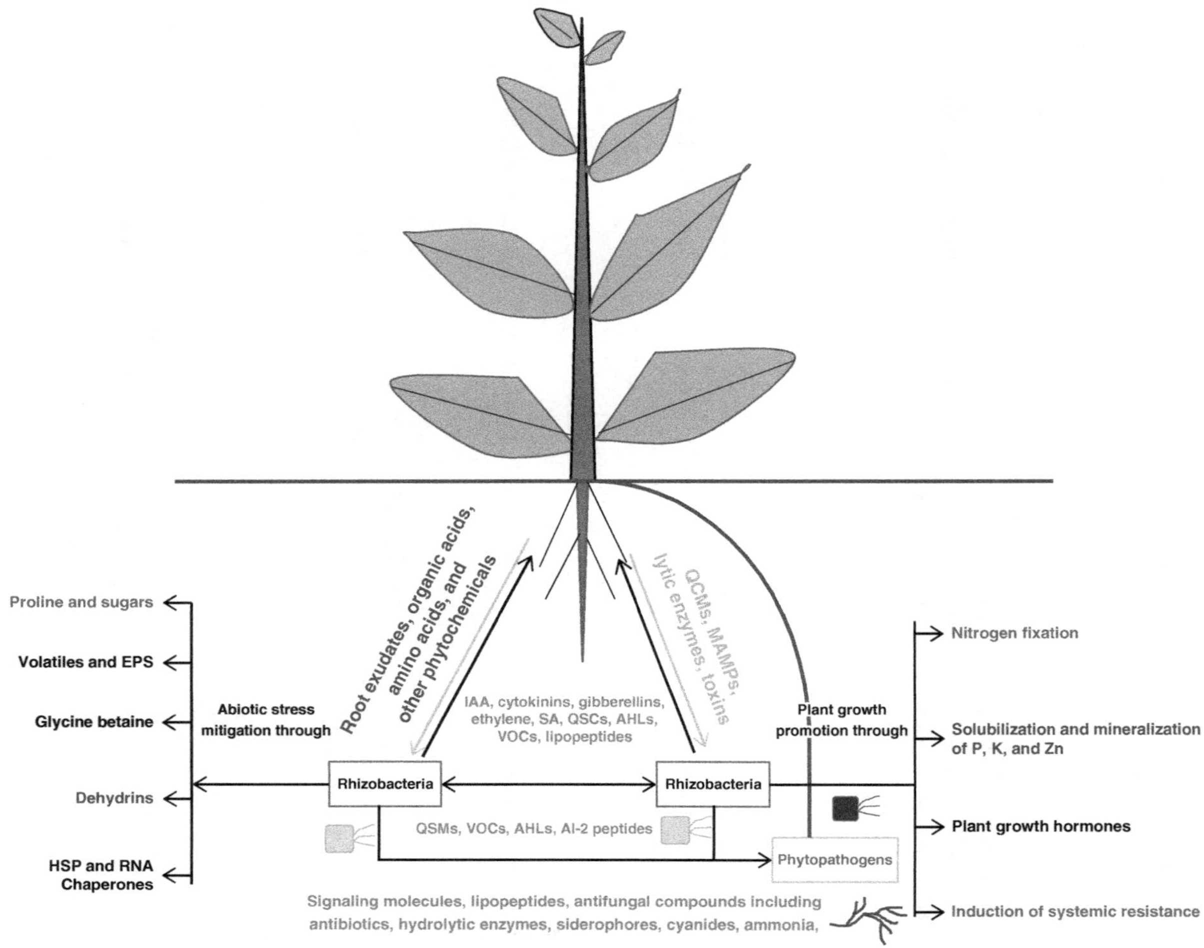

Figure 4.1 Schematic diagram representing the interplay of phytomicrobiome and their interaction with plant and phytopathogens in the vicinity of the root through the network of different chemical signaling.

Table 4.1 Role of phytomicrobiome in abiotic stresses studied recently.

Microorganisms	Remarks	References
Bacillus lentus MKA2, *Bacillus subtilis* MKA and *Bacillus cereus* MKA4	Endophytic bacteria treatment increases plant biomass, relative water content, proline content, stabilizes membrane and scavenges ROS by SOD, CAT and GR enzymes under drought stress	Annapurna et al. (2019)
Bacillus sp.(12D6) and *Enterobacter* sp.(16i)	Modification of root system architecture PGPR secretes IAA and salicylic acid	Jochum et al. (2019)
Seed inoculation of *Azospirillum brasilience* and foliar spray of auxin, gibberellins, and cytokinin	RWC and gas exchange rate were improved and ultimately yield increased	Bulegon et al. (2019)
Halotolerant bacteria like *Arthrobacter woluwensis* AK1, *Microbacterium oxydans* AK2, *Arthrobacter aurescens* AK3, *Bacillus megaterium* AK4, and *Bacillus aryabhattai* AK5	PGPR secretes IAA, siderophore, and solubilize insoluble phosphorus Higher levels of SOD and GSH, uptake of K+, and exclusion of Na+. Salt-tolerant and IAA-producing gene GmST1 and GmLAX3 expressed respectively	Khan et al. (2019)
Tripartite interaction of *Phaseolus vulgaris*-AMF and *Rhizobium*	Dual inoculation increases glomalin production (GRSP) under different levels of salinity Improve physiological parameters and biomass production	Garcia et al. (2019)
Burkholderia phytofirmans strain PsJN	Root growth, enhanced carbon fixation, and O_2 evolution Higher levels of proline, phenomics and starch and imparts cold tolerance	Barka et al. (2006)
Psychrophilic PGPR *B. velezensis* FZB42 and *Bacillus* spp. CJCL2, RJGP41	Regulation of ABA, proline accumulation and lipid peroxidation	Zubair et al. (2019)
Combination of PGPR (*Pseudomonas moraviensis* and *Bacillus cereus*) and carrier-based biofertilizer formulation	Bioremediation of Cd, Co, Cr, Ni, and Mn	Hassan et al. (2017)
B. subtilis OKB105.	Spermidine, a kind of common polyamine, induces growth in *Nicotiana tabacum*	Shan-Shan Xie et al. (2014)

the root exudate profile of crop root. Owing to altered root exudates composition, the shift in soil and root microbiome functions like oligotrophy, sporulation activity, decreased nutrient cycling, decrease in anaerobic respiration, and increase in antibiotic and osmolytes production are observed in general.

The rhizospheric bacteria isolated from the soybean, which produced EPS and IAA, were capable of inducing drought tolerance in a crop plant. Further, it was found that *Bacillus* was key genera involved in crop tolerance to moisture-deficient stress (Susilowati et al. 2018). Similarly, ACC-deaminase producing *Pseudomonas* species, isolated from rainfed plants, alleviated drought stress in *Eleucine coracana* through the prevention of oxidative damage by scavenging ROS (Chandra et al. 2018). The consortium of three ACC-deaminase producing bacteria, *Ochrobactrum pseudogrignonense* RJ12, *B. subtilis* RJ46 and *Pseudomonas* sp. RJ15 were evaluated for their ability to mitigate drought stress in *Vigna mungo* and *Pisum sativum*. The seed treatment with this bacterial consortium noticeably improved seed germination and plant biomass, scavenged toxic ROS, accumulated cellular osmolytes, retained higher chlorophyll and relative water content, reduced the expression of ACC-oxidase (ACO) gene and indirectly provided phosphorus, iron, and nitrogen to plants during stress condition (Saikia et al. 2018). In the similar study, a combination of three *Pseudomonas strains, Bacillus cereus* BSB 38 (14B), and *Arthrobacter nitroguajacolicus* (YB3 and YB5) were inoculated in two rice varieties, and they found that these PGPR scavenges ROS by the formation of SOD, CAT, POD, and APX, and accumulate proline under drought stress (Gusain et al. 2015). A consortium of nine bacteria and one yeast endophytes isolated from poplar plants were inoculated into stem cuttings of *Populus deltoides* x *Populus nigeria* clone OP367 showed increased root and shoot biomass, production of phytohormones (IAA, GA3, JA, and SA), and antioxidant enzymes. The molecular analysis revealed genes for the presence of volatile organic compounds (VOCs), compatible solute, and ACC-deaminase. The improvement of the physiological parameters like the quantum yield of PSII (Fv/Fm), chlorophyll content, and stomatal conductance were also improved (Khan et al. 2016). Priming of wheat seeds by two *B. thuringiensis* AZP2 and *Paenibacillus polymyxa* B, isolated from the harsh environment, improved the drought stress tolerance and survivability. Moreover, the treated plants showed higher plant biomass (78%), photosynthetic activities, and scavenged ROS by upregulation of antioxidant enzyme activities (SOD, GR, MDHAR, and CAT). The inoculation of phosphate solubilizing bacteria *Bacillus polymyxa* on the green gram in drought tolerance was investigated and found that the 1% inoculation improved the physiological and biochemical parameters in plant and alleviated drought stress (Vidya et al. 2016). Besides, plants grown from bacterial primed wheat seeds emitted VOCs like benzaldehyde, b-pinene, and geranyl acetone that were associated with alleviation of drought tolerance (Timmusk et al. 2014).

4.3.2 Salinity Stress Mitigation

Salinity is another most important abiotic stress negatively affecting plant growth and productivity (Isayenkov and Maathuis 2019). The excessive salt content in soil retard plant growth due to excessive uptake of sodium and chloride ions, thus causing ionic imbalance inside plant tissue, interference with mineral homeostasis (mainly calcium and potassium), and causes oxidative stress due to formation of ROS (Isayenkov 2012). Sodium-ion enters the plant cell via nonselective cation channels (NSCCs), which activates Ca^{+2}, ROS, and ABA hormone. These early signals activate the detoxification mechanisms like HKT, NHX, and SOS sodium ion transport for osmotic protections (Deinlein et al. 2014). Salinity causes stomatal closure, retardation of shoot expansion, and the slowdown of metabolic activity due to the accumulation of ion, causing leaf senescence and cell death. Salinity tolerance in the plant is mediated by (i) osmotic tolerance by reducing shoot growth, (ii) ion exclusion by transporters like SOS1 and HKT, and (iii) tissue tolerance due to compartmentalization at the cellular and intercellular level in the vacuole, involvements of proton pumps and ions transporter, and accumulation of compatible solutes, such as proline, glycine-betaine, and sucrose (Roy et al. 2014).

Bacterial and fungal microbial communities associated with plant roots under saline soil are considerably changed. For example, the predominance of Proteobacteria, Acidobacteria, Actinobacteria, and Bacteroidetes were observed in the root microbiome of *Jerusalem artichoke* cultivar Nanyu-1, whereas the population of *Steroidobacter* and *Sphingomonas* were declined considerably (Yang et al. 2016). PGPR has direct and indirect mechanisms for alleviating salinity stress. The key mechanisms are the production of phytohormones (IAA, GA3, and cytokinin), symbiotic and non-symbiotic nitrogen fixation, solubilization of insoluble phosphorus compounds, siderophore production, and production of VOCs (Numan et al. 2018). Other mechanisms are reduction of ethylene via ACC-deaminase enzyme, accumulation of osmolytes (proline, glycine-betaine, glycerol, polysaccharides, and polyamines), maintaining a source-sink relationship, ion-homeostasis, photosynthesis, and increasing plant biomass (Dodd and Pérez-Alfocea 2012). Furthermore, bacterial produced EPS which binds with excess Na^+ ions and prevented its influx to plant roots.

The VOCs produced by PGPR, upregulates expression of HKT 1 transporter in aerial plant parts (shoots) and downregulated in roots, thereby facilitating uptake of K^+ and Ca^{+2} and increasing K^+/Na^+ ratio (Qin et al. 2016). The soil bacterium, *Paenibacillus yonginensis* DCY 84 strain, inoculated in medicinal plant *Panax ginseng*, provided short-term and long-term salinity stress alleviation via the production of IAA and siderophore, phosphorus solubilization, osmolyte production (proline and total soluble sugar), and antifungal activity by hydrolytic enzymes. During long-term salinity exposure, it elicited plant defense-related systems for detoxifying ROS via the production of antioxidant enzymes (SOD, CAT, and peroxidases). Also, ABA synthesis genes and root meristem genes for root hair formation were activated (Sukweenadhi et al. 2018). Similarly, co-inoculation of two PGPR, *Pseudomonas* sp. AK-1 and *Bacillus* sp. SJ-5 in soybean plant showed improved tolerance against salt stress. Treated plants have greater plant biomass and relative water content, higher photosynthetic activity and proline content, and enhanced peroxidase, and lipoxygenase activity for scavenging toxic ROS. These two inoculated bacteria also produced ACC-deaminase, siderophore, EPS, and solubilized phosphorus under salinity stress (Kumari et al. 2015). Seed-born endophytes, *Bacillus amyloliquifaciens* RWL-1, inoculated in rice plant exposed to salinity stress, was found to have a higher accumulation of amino acids (proline, glutamate, aspartate, phenylalanine, and cysteine), and salicylic acids, whereas endogenous ABA levels were decreased in inoculated plants (Shahzad et al. 2017). Seed inoculation of PGPR *B. amyloliquefaciens* NBRI-SN 13 positively improved the physiological parameters, such as proline, total soluble sugar, and lipid peroxidation level, which in turn improved tolerance to drought and salinity in rice plants. Moreover, the well-known drought-tolerant gene, namely DHN, GST, LEA, NAM, GRAM, NRAMP 6, and actin response, were also altered and verified by qRT-PCR analysis (Tiwari et al. 2017).

In addition to bacterial PGPR, AMF employ various mechanisms to the alleviation of salinity stress reviewed by Evelin et al. (2019) as follows: (i) altering the RSA, (ii) maintaining ion homeostasis by Na^+ extrusion in soil solution and K^+ uptake by expression of cations transporters like NHXs, SOS, SKOR, HKT, and AKT, (iii) nutrient acquisition for macro-and macronutrients (nitrogen, phosphorus, potassium, zinc, iron, and copper), (iv) accumulation of osmolytes like proline, trehalose, sucrose, polyamines, sugars, and organic acids for preventing cellular dehydration, (v) ROS scavenging by the production of antioxidant enzymes, (vi) modify the expression of tonoplast and plasma membrane aquaporins (TIPs and PIPs), (vii) increasing efficiency of PS II and other photosynthetic parameters, and (viii) phytohormonal regulation of strigolactones, jasmonate, salicylate, ABA and gibberellic acid. AMF fungi such as *Glomus etunicatum*, *Glomus intraradices,* and *Glomus mosseae*, isolated from the salt-marsh habitat rhizosphere of *Acacia gerrardii* were used to mitigate salt stress in the plants of *Cucumis sativus*. Seed treatment with these AMF increased plant biomass, increased activity of antioxidant enzymes (SOD, CAT, APX, GR, and ascorbic acid) for scavenging ROS and prevented lipid peroxidation and membrane leakage of the cellular membrane in treated plants. Nutrient acquisition of nitrogen, phosphorus, sulfur, and potassium, and reduced uptake of sodium ion was also observed under combined salinity and AMF treatment. Accumulation of osmolytes (proline and total phenol) and phytohormones (salicylic and jasmonic acid) were higher in salinity, and AMF-treated cucumber plants, as compared to the control treatment (Hashem et al. 2018).Polyamines produced by microbes also play an important role in stress management. Bacterial spermidine is a polyamine that contributes to plant protection under stress by biofilm formation and stimulates expansion expression that influences the plant hormone and cell division (Shan-Shan Xie et al. 2014).

4.3.3 Heavy Metal Toxicity

Heavy metal toxicity is another serious abiotic threat to soil ecology and environmental hazard. They are also responsible for altering the existing native microbial communities (their structural and functional diversity) in the soil. It retards plant growth and yield and enters into the food chain and leads to human health hazards too (Etesami 2018). Metal-resistant PGPR has metal detoxifying properties as well as plant growth promotion effects. Mechanisms of heavy metal tolerance by PGPR are through biosorption, biotransformation, bioaccumulation, bioleaching, and precipitation and were diagrammatically explained by Etesami (2018). The key mechanisms involved are as follows: (i) biosorption/metal complexion-heavy metals bind with anionic functional groups like hydroxyl, carboxyl, sulphonate, amide, and amine groups, (ii) production of siderophore, EPS (Bruins et al. 2000), and biosurfactant (Mulligan et al. 2001) that chelate or binds with toxic metal ions (Rajkumar et al. 2012), (iii) metal detoxification by methylation (Rajkumar et al. 2012), (iv) reduction of metal to lower redox state, as in case of chromium (Cr^{VI} into Cr^{III}), (v) accumulation of heavy metal inside the bacterial cell (compartmentalization), and (vi) precipitation in the form of the metal carbonate, phosphate, hydroxide and sulfide, and (vii) biodegradation of metallic compounds. PGPR not only negates the heavy metals but also takes part in plant growth promotion reviewed by Ma et al. 2011. The key mechanism utilized by beneficial microorganisms are as follows: (i) phytohormones production (IAA, gibberellic acid and cytokinin), (ii) ACC-deaminase enzyme production, and (iii) modification of RSA, which help in the nutrient acquisition of macro and micronutrient (N, P, K, Ca, Cu, Fe, Mn, and Zn). The list of PGPR, which imparts heavy metal tolerance and helps in plant growth promotion, and their mechanisms were listed by Etesami (2018).

4.3.4 Low-Temperature Stress

Cold stress induces many adverse physiological changes in plant tissue, such as shrinkage of protoplast volume, a decline in water potential, change in viscosity leading to the rigidity of membrane affecting permeability, ROS formation, a slowdown of metabolic reactions, and ice formation leading to decrease in water potential. The magnitude of the stress depends on crop stages, exposure duration, and level of ice formation (Beck et al. 2004). Under cold stress conditions, ice-nucleating bacteria like *Pseudomonas syringae* and *Erwinia herbicola* can prevent the plant from frost damage (Lindow and Leveau 2002). Cold-tolerant PGPR bacteria impart cold stress tolerance by modification of membrane, RSA, accumulation of osmolytes, enhanced pigment retention, and physiological functions that are involved in photosynthesis and detoxification of ROS by many anti-oxidative enzymes (Barka et al. 2006). A consortium of four psychrotolerant bacteria *Pseudomonas frederiksbergensis* OS261 and OS211, *Pseudomonas vancouverensis* OB155, and *Flavobacterium glaciei* OB146 had shown to improve chilling stress in tomato plants. Upon inoculation, they prevented membrane damage, ROS detoxification by antioxidant enzymes, and accumulated proline (Subramanian et al. 2016). Similarly, seed priming of PGPR *Bacillus velezensis* 5113 in wheat imparted tolerance to low temperature, heat, and drought by reprogramming molecular and metabolic pathways (El-Daim et al. 2019). Biopriming of psychrotolerant PGPR *Pseudomonas* strain isolated from the Himalaya region improved cold stress in wheat plants by increasing chlorophyll content, accumulating proline, amino acid, starch, anthocyanin, and phenolic content. Additionally, increased RWC, reduced membrane damage, and increased Na^+/K^+ ratio were also observed (Mishra et al. 2011).

4.3.5 Nutrient Deficiency

Soil acidity is a major hindrance factor for plant growth due to aluminium and iron toxicity, deficiency of iron, phosphorus, and molybdenum, and toxic concentration of hydrogen ion and phenolic acids (Kidd and Proctor 2001), whereas, alkaline soils are deficient in phosphorus, zinc, copper, iron, and manganese. Soil microorganisms help in the solubilization and mineralization of complex compounds and liberate the available form of the element (N, P, S, and other micronutrients) for the growth and development of plant growth (Richardson et al. 2009).

4.3.6 Flooding or Water Submergence

Flooding stress causes a hypoxic condition in the root zone that induces upregulation of the ACC synthase gene and thereby produces a higher amount of ethylene (Sasidharan et al. 2018). A consequence of increased ethylene is the formation of ROS, which damages macromolecules and photosynthetic machinery (Ahmed et al. 2006). Many ACC-deaminase positive rhizobacteria can cleave ACC, an ethylene precursor into ammonia and alpha-ketobutyrate and lower the ethylene level by almost 60–90% (Saleem et al. 2007). ACC-deaminase positive *Pseudomonas putida* UW4 were treated with *Rumex palustris* plants showed tolerance to complete submergence, whereas mutant does not (Ravanbakhsh et al. 2017). Similarly, inoculation of *Achromobacter xylosoxidans, Serratia ureilytica, Ochrobactrum rhizosphaerae*, and *Herbaspirillum seropedicae* bacteria in *Ocimum sanctum* plants imparted tolerance to flooding stress (Barnawal et al. 2012).

4.4 Importance of Phytomicrobiome Engineering for Crop Stress Alleviation

The phytomicrobiome can be manipulated and modified using either plant-based, meta-organism-based, or microbiome-based strategies. Plant-based strategies include plant-breeding techniques, development of transgenic plants, and selection of cultivar, which allows the growth and multiplication of beneficial phytomicrobiome. Similarly, microbiome-based manipulation is through inoculating microbial strain or genetically modified strains and the addition of soil amendments with microbial signal compounds. The meta-organism–based approaches include coengineering and crop rotation. The purpose of phytomicrobiome manipulation is to enhance the macro- and micronutrient availability to the plants and nutrient cycle, enhanced disease suppressiveness, niche saturation, and to induce resistance to biotic and abiotic stresses (Quiza et al. 2015; Dukare et al. 2018).

The first route of phytomicrobiome engineering is via microbiome alteration. Here, inoculation of PGPR, nitrogen-fixer rhizobia, AMF, ectomycorrhizal fungi (ECM), and bacterial and fungal endophytes or beneficial microorganisms consortia are employed for the better nutrient acquisition, phytohormones production, IST and induced systemic resistance (ISR). Inoculation helps in combined niche exploitation, cross-feeding, and increases the survivability of other associated microorganisms, pathogen exclusion, and niche saturation (Ryan et al. 2009). The main hurdle with this is that an inoculated strain has to compete with native microflora for maintaining optimum population density (Morel and Castro-Sowinski 2013). Inoculations of genetically modified or recombinant strains with beneficial traits have better-surviving potential, adaptation, resilience, and stability (Ryan et al. 2009). The only concern is the loss of desirable genes and risks of releasing genetically modified microorganisms into the environment and undesirable results due to horizontal gene transfer. The phytomicrobiome can be altered by tillage practices or by the application of plant protection chemicals (fungicides, insecticides, weedicides, and fertilizers) (Bakker et al. 2012). The second strategy is plant-based approaches involves the genetic alteration in plant genotype by cultivar selection and breeding approaches for releasing specific root exudate compounds for chemoattracting and recruiting beneficial microbiome and excluding pathogenic- or disease-causing phytopathogens (Hartmann et al. 2009). The disadvantage of this approach is no plant-breeding techniques are designed that have shown a relationship of a cultivar with the root–microbiome composition, and the composition of root exudates is dependent on many edaphic factors also (Bakker et al. 2012). Transgenic plants excreting citrate are suitable for acidic soil having phosphorus deficiency and have higher tolerance against aluminum toxicity (Koyama et al. 2000).

Similarly, transgenic plants expressing ALMT (Al^{3+}-activated malate transporter) and MATE (multidrug and toxin extrusion) grow better in aluminium toxicity and phosphorus-deficient soils (Delhaize et al. 2007). The last route is via the meta-organism approach. Since plants are closely associated with diverse microorganisms termed as meta-organisms or holobiont (Lakshmanan et al. 2014). Here specific crop rotation or mixed cropping system is suggested to enhance the growth of beneficial microorganisms, which help in greater nutrient recycling, improvement of physicochemical soil properties, and development of suppressive soil (Ryan et al. 2009).

4.5 Omics Strategies in Phytomicrobiome Studies

Several omics-based approaches, including genomics, transcriptomics, proteomics, glycomics, interactomics, metabolomics, and phenomics, have a huge potential in advanced studies involving the plant-associated microbiome and their interaction with host plants. These tools can be exploited to unravel the novel genes, proteins, metabolites, and its functions, as well as to discover existing metabolic pathways that are operating under abiotic stress condition. Reverse genetics and gene network mapping techniques reveal a complete picture of given microorganisms. Furthermore, meta-omics techniques like metagenomics and meta-transcriptomics give an idea about a functional nonculturable fraction of microbial communities presently associated with aerial and the below-ground surface parts of host plants that are exposed to the various abiotic stresses (Meena et al. 2017). Genomics, a study of determining the entire DNA sequence of organisms, can be analyzed using next-generation sequencing (NGS), microarray, metagenomics, and HLA genotyping. Furthermore, transcriptomics, being referred to as expression profiling, gives expression level of mRNA in a given microorganism. It includes entire RNA like rRNA, mRNA, tRNA, and other noncoding RNA present in a given microbe or population. It can be studied using RT-PCR, RNA sequencing, small RNA sequencing, and northern blot. Proteomics is a study of the entire complement of proteins, including modifications, identification, and characterization of protein produced by microorganisms. Proteomics can be studied using 2D gel electrophoresis, targeted protein identification, ESI-MS, MS/MS, and MALDI-TOF. Glycomics is a comprehensive study of the entire complement of sugars both in free or in the compound form in a given microorganism. It can be studied using high-resolution mass spectrophotometry (MS), high-performance liquid chromatography (HPLC), lectin microarray, glycan microarray, capillary electrophoresis, LC-MS, and imaging techniques, such as ray crystallography and nuclear magnetic resonance (NMR) with data interpretation using GlycomeDB and Unicarb-DB software. Interact-omics is studying the whole set of molecular interactions in a cell. Furthermore, we can understand protein–protein interaction in a cell. It can be studied using protein complement assay, luminescence-based interactome mapping (LUIER), fluorescence/luminescence resonance energy transfer (F-BRET), and MYTH. Metabolomics, a study of all metabolites like small molecule substrates, intermediates, and products of metabolism, gives an unbiased or targeted analysis of cellular metabolites. It can be studied using GC, MS, HPLC, LC-MS/MS GC-MS/MS, NMR, IR, and X-ray crystallography (Horgan and Kenny 2011). Phenomics is a systemic study of phenotypes, and we can address the phenotype changes under the influence of abiotic stresses that can be studied. It can be studied using imaging techniques, genome-wide association studies, ultrastructural analysis, and high-resolution genome mapping (Großkinsky et al. 2017).

Top-down approach (metagenomics to metabolomics) and bottom-up approach (metabolomics to metagenomics) will help provide the entire blueprint of a given system (Pinu et al. 2019). Using these omics approaches, characterization, and identification of core microbiome, their correlation with crop yield and quality can be done under stressed environmental conditions. Further, identification of signal molecules between crop and microbiome and finally, manipulation of phytomicrobiome for sustainable food production under abiotic stresses can be studied using the omics as mentioned above techniques (Singh and Trivedi 2017). The inclusion of microbiome into the crop breeding programs will give higher plant fitness and tolerance against abiotic stresses. For exploiting the phytomicrobiome concept for sustainable agricultural production, research priorities have been set up on following aspects: (i) developing plant–microbiome models, (ii) identifying core microbiome and metagenome, (iii) elucidation of synthetic and functionally programmable phytomicrobiome assembly and (iv) studying plant genotype phytomicrobiome environment (abiotic and biotic) crop management interactions (Busby et al. 2017).

4.6 Conclusion and Future Prospects

Multi-omics and advanced bioinformatics approaches should be used for studying both culturable and nonculturable fraction of phytomicrobiome. How the member of the phytomicrobiome composition gets altered upon various abiotic stresses may help in the predominance of the microbial community, which helps in tolerance against these stresses. Identification of the key signal compounds exchange between microbe and plant will help in the enrichment of the desired phytomicrobiome. Instead of studying the bidirectional relationship between single plant–microbe, multitrophic interaction should be studied for getting a realistic blueprint of phytomicrobiome for success in combating abiotic stresses. Climate-smart agriculture needs the desired phytomicrobiome manipulation in both above-ground and below-ground for combating stresses.

Acknowledgments

Hemant S. Maheshwari acknowledges the Ministry of Agriculture and Farmer Welfare, Govt of India and Director, ICAR-Indian Institute of Soybean Research, Khandwa Road, Indore-452001, India for providing financial assistance for pursuing doctoral studies at the University of Groningen, The Netherlands (ICAR-NS-IF-2017) under supervision of Dr. J.T.M Elzenga (Professor, Ecophysiology of plants, University of Groningen, The Netherlands) lab.

References

Ahmed, S., Nawata, E., and Sakuratani, T. (2006). Changes of endogenous ABA and ACC, and their correlations to photosynthesis and water relations in mungbean (*Vigna radiata (L.)* Wilczak cv. KPS1) during waterlogging. *Environ. Exp. Bot.* 57 (3): 278–284.

Annapurna, K., Govindasamy, V., Ajit, V. et al. (2019). Mitigation of drought stress in wheat crop by drought tolerant endophytic bacterial isolates. *Vegetos* 32 (4): 486–493.

Ansari, M.I. (2018). Plant microbiome and its functional mechanism in response to environmental stress. *Int. J. Green Pharm.* 12 (1) https://doi.org/10.22377/ijgp.v12i01.1603.

Badri, D.V. and Vivanco, J.M. (2009). Regulation and function of root exudates. *Plant Cell Environ.* 32 (6): 666–681.

Bais, H.P., Weir, T.L., Perry, L.G. et al. (2006). The role of root exudates in rhizosphere interactions with plants and other organisms. *Annu. Rev. Plant Biol.* 57: 233–266.

Bakker, M.G., Manter, D.K., Sheflin, A.M. et al. (2012). Harnessing the rhizosphere microbiome through plant breeding and agricultural management. *Plant and Soil* 360 (1–2): 1–13.

Barka, E.A., Nowak, J., and Clément, C. (2006). Enhancement of chilling resistance of inoculated grapevine plantlets with a plant growth-promoting rhizobacterium, *Burkholderiaphytofirmans* strain PsJN. *Appl. Environ. Microbiol.* 72 (11): 7246–7252.

Barnawal, D., Bharti, N., Maji, D. et al. (2012). 1-Aminocyclopropane-1-carboxylic acid (ACC) deaminase-containing rhizobacteria protect *Ocimum sanctum* plants during waterlogging stress via reduced ethylene generation. *Plant Physiol. Biochem.* 58: 227–235.

Beck, E.H., Heim, R., and Hansen, J. (2004). Plant resistance to cold stress: mechanisms and environmental signals triggering frost hardening and dehardening. *J. Biosci.* 29 (4): 449–459.

Bhattacharyya, P.N., Goswami, M.P., and Bhattacharyya, L.H. (2016). Perspective of beneficial microbes in agriculture under changing climatic scenario: a review. *J. Phytol.* 8: 26–41.

Brilli, F., Loreto, F., and Baccelli, I. (2019). Exploiting plant volatile organic compounds (VOCs) in agriculture to improve sustainable defense strategies and productivity of crops. *Front. Plant Sci.* 10: 264.

Bruins, M.R., Kapil, S., and Oehme, F.W. (2000). Microbial resistance to metals in the environment. *Ecotoxicol. Environ. Saf.* 45 (3): 198–207.

Bulegon, L.G., Guimarães, V.F., Battistus, A.G. et al. (2019). Mitigation of drought stress effects on soybean gas exchanges induced by *Azospirillumbrasilense* and plant regulators. *Pesqui. Agropecu. Trop.* 49: e52807–e52807.

Busby, P.E., Soman, C., Wagner, M.R. et al. (2017). Research priorities for harnessing plant microbiomes in sustainable agriculture. *PLoS Biol.* 15 (3): e2001793.

Chandra, D., Srivastava, R., Glick, B.R. et al. (2018). Drought-tolerant Pseudomonas spp. improve the growth performance of finger millet (*Eleusinecoracana* (L.) Gaertn.) under non-stressed and drought-stressed conditions. *Pedosphere* 28 (2): 227–240.

Cramer, G.R., Urano, K., Delrot, S. et al. (2011). Effects of abiotic stress on plants: a systems biology perspective. *BMC Plant Biol.* 11 (1): 163.

Dani, K.G.S., Fineschi, S., Michelozzi, M. et al. (2016). Do cytokinins, volatile isoprenoids and carotenoids synergically delay leaf senescence? *Plant Cell Environ.* 39 (5): 1103–1111.

Deinlein, U., Stephan, A.B., Horie, T. et al. (2014). Plant salt-tolerance mechanisms. *Trends Plant Sci.* 19 (6): 371–379.

Delhaize, E., Gruber, B.D., and Ryan, P.R. (2007). The roles of organic anion permeases in aluminum resistance and mineral nutrition. *FEBS Lett.* 581 (12): 2255–2262.

Dodd, I.C. and Pérez-Alfocea, F. (2012). Microbial amelioration of crop salinity stress. *J. Exp. Bot.* 63 (9): 3415–3428.

Dong, Y.H., Gusti, A.R., Zhang, Q. et al. (2002). Identification of quorum-quenching N-acyl homoserine lactonases from Bacillus species. *Appl. Environ. Microbiol.* 68 (4): 1754–1759.

Duca, D., Lorv, J., Patten, C.L. et al. (2014). Indole-3-acetic acid in plant–microbe interactions. *Antonie Van Leeuwenhoek* 106 (1): 85–125.

Dudareva, N., Klempien, A., Muhlemann, J.K. et al. (2013). Biosynthesis, function and metabolic engineering of plant volatile organic compounds. *New Phytol.* 198 (1): 16–32.

Dukare, A.S., Paul, S., Nambi, V.E. et al. (2018). Exploitation of microbial antagonists for the control of postharvest diseases of fruits: a review. *Crit. Rev. Food Sci. Nutr.* 58 (2): 1–16.

El-Daim, I.A.A., Bejai, S., and Meijer, J. (2019). *Bacillus velezensis* 5113 induced metabolic and molecular reprogramming during abiotic stress tolerance in wheat. *Sci. Rep.* 9 (1): 1–18.

Etesami, H. (2018). Bacterial mediated alleviation of heavy metal stress and decreased accumulation of metals in plant tissues: mechanisms and future prospects. *Ecotoxicol. Environ. Saf.* 147: 175–191.

Evelin, H., Devi, T.S., Gupta, S. et al. (2019). Mitigation of salinity stress in plants by arbuscular mycorrhizal symbiosis: current understanding and new challenges. *Front. Plant Sci.* 10: 470.

Foyer, C.H. and Noctor, G. (2005). Oxidant and antioxidant signaling in plants: a re-evaluation of the concept of oxidative stress in a physiological context. *Plant Cell Environ.* 28 (8): 1056–1071.

Garbeva, P., Hordijk, C., Gerards, S. et al. (2014). Volatile-mediated interactions between phylogenetically different soil bacteria. *Front. Microbiol.* 5: 289.

Garcia, C.L., Dattamudi, S., Chanda, S. et al. (2019). Effect of salinity stress and microbial inoculations on glomalin production and plant growth parameters of snap bean (*Phaseolus vulgaris*). *Agronomy* 9 (9): 545.

Gherbi, H., Markmann, K., Svistoonoff, S. et al. (2008). SymRK defines a common genetic basis for plant root endosymbioses with arbuscular mycorrhiza fungi, rhizobia, and Frankiabacteria. *Proc. Natl. Acad. Sci. U. S. A* 105 (12): 4928–4932.

Glick, B.R. (2014). Bacteria with ACC deaminase can promote plant growth and help to feed the world. *Microbiol. Res.* 169 (1): 30–39.

Gray, E.J., Lee, K.D., Souleimanov, A.M. et al. (2006). A novel bacteriocin, thuricin 17, produced by plant growth promoting rhizobacteria strain *Bacillus thuringiensis* NEB17: isolation and classification. *J. Appl. Microbiol.* 100 (3): 545–554.

Großkinsky, D.K., Syaifullah, S.J., and Roitsch, T. (2017). Integration of multi-omics techniques and physiological phenotyping within a holistic phenomics approach to study senescence in model and crop plants. *J. Exp. Bot.* 69 (4): 825–844.

Gusain, Y.S., Singh, U.S., and Sharma, A.K. (2015). Bacterial mediated amelioration of drought stress in drought tolerant and susceptible cultivars of rice (*Oryza sativa* L.). *Afr. J. Biotechnol.* 14 (9): 764–773.

Harold, F. and Reetz, J. (2016). *Fertilizers and Their Efficient Use*. International Ferlizers Industry Association www.fertilizers.org.

Hartmann, A., Schmid, M., Van Tuinen, D. et al. (2009). Plant-driven selection of microbes. *Plant and Soil* 321 (1–2): 235–257.

Hashem, A., Alqarawi, A.A., Radhakrishnan, R. et al. (2018). Arbuscular mycorrhizal fungi regulate the oxidative system, hormones and ionic equilibrium to trigger salt stress tolerance in *Cucumis sativus* L. *Saudi J. Biol. Sci.* 25 (6): 1102–1114.

Hassan, T.U., Bano, A., and Naz, I. (2017). Alleviation of heavy metals toxicity by the application of plant growth promoting rhizobacteria and effects on wheat grown in saline sodic field. *Int. J. Phytoremediation* 19 (6): 522–529.

Hayat, S., Faraz, A., and Faizan, M. (2017). Root exudates: composition and impact on plant–microbe interaction. *Biofilms Plant Soil Health.*: 179–193.

He, Y.W. and Zhang, L.H. (2008). Quorum sensing and virulence regulation in *Xanthomonas campestris*. *FEMS Microbiol. Rev.* 32 (5): 842–857.

Holden, M.T., Ram Chhabra, S., De Nys, R. et al. (1999). Quorum-sensing cross talk: isolation and chemical characterization of cyclic dipeptides from *Pseudomonas aeruginosa* and other gram-negative bacteria. *Mol. Microbiol.* 33 (6): 1254–1266.

Horgan, R.P. and Kenny, L.C. (2011). 'Omic'technologies: genomics, transcriptomics, proteomics and metabolomics. *Obstet. Gynaecol.* 13 (3): 189–195.

Isayenkov, S.V. (2012). Physiological and molecular aspects of salt stress in plants. *Cytol. Genet.* 46 (5): 302–318.

Isayenkov, S.V. and Maathuis, F.J. (2019). Plant salinity stress: many unanswered questions remain. *Front. Plant Sci.* 10: 80.

Jimenez, P.N., Koch, G., Papaioannou, E. et al. (2010). Role of PvdQ in *Pseudomonas aeruginosa* virulence under iron-limiting conditions. *Microbiology* 156 (1): 49–59.

Jochum, M., McWilliams, K.M., Borrego, E. et al. (2019). Bioprospecting plant growth-promoting rhizobacteria that mitigate drought stress in grasses. *Front. Microbiol.* 10: 2106.

Jones, D.L., Hodge, A., and Kuzyakov, Y. (2004). Plant and mycorrhizal regulation of rhizodeposition. *New Phytol.* 163 (3): 459–480.

Kanchiswamy, C.N., Malnoy, M., and Maffei, M.E. (2015). Bioprospecting bacterial and fungal volatiles for sustainable agriculture. *Trends Plant Sci.* 20 (4): 206–211.

Khan, Z., Rho, H., Firrincieli, A. et al. (2016). Growth enhancement and drought tolerance of hybrid poplar upon inoculation with endophyte consortia. *Curr. Plant Biol.* 6: 38–47.

Khan, M.A., Asaf, S., Khan, A.L. et al. (2019). Halotolerant rhizobacterial strains mitigate the adverse effects of NaCl stress in soybean seedlings. *Biomed. Res. Int.* 2019: 9530963.

Kidd, P.S. and Proctor, J. (2001). Why plants grow poorly on very acid soils: are ecologists missing the obvious? *J. Exp. Bot.* 52 (357): 791–799.

Koyama, H., Kawamura, A., Kihara, T. et al. (2000). Overexpression of mitochondrial citrate synthase in *Arabidopsis thaliana* improved growth on a phosphorus-limited soil. *Plant Cell Physiol.* 41 (9): 1030–1037.

Kumar, S., Sachdeva, S., Bhat, K.V. et al. (2018). Plant responses to drought stress: physiological, biochemical and molecular basis. In: *Biotic and Abiotic Stress Tolerance in Plants* (ed. S. Vats), 1–25. Singapore: Springer https://doi.org/10.1007/978-981-10-9029-5.

Kumari, S., Vaishnav, A., Jain, S. et al. (2015). Bacterial-mediated induction of systemic tolerance to salinity with expression of stress alleviating enzymes in soybean (*Glycine max* L. Merrill). *J. Plant Growth Regul.* 34 (3): 558–573.

Lakshmanan, V., Selvaraj, G., and Bais, H.P. (2014). Functional soil microbiome: belowground solutions to an aboveground problem. *Plant Physiol.* 166 (2): 689–700.

Lamaoui, M., Jemo, M., Datla, R. et al. (2018). Heat and drought stresses in crops and approaches for their mitigation. *Front. Chem.* 6: 26.

Lee, K.D., Gray, E.J., Mabood, F. et al. (2009). The class IId bacteriocin thuricin-17 increases plant growth. *Planta* 229 (4): 747–755.

Lindow, S.E. and Leveau, J.H. (2002). Phyllosphere microbiology. *Curr. Opin. Biotechnol.* 13 (3): 238–243.

Lugtenberg, B. and Kamilova, F. (2009). Plant-growth-promoting rhizobacteria. *Annu. Rev. Microbiol.* 63: 541–556.

Ma, Y., Prasad, M.N.V., Rajkumar, M. et al. (2011). Plant growth promoting rhizobacteria and endophytes accelerate phytoremediation of metalliferous soils. *Biotechnol. Adv.* 29 (2): 248–258.

Meena, K.K., Sorty, A.M., Bitla, U.M. et al. (2017). Abiotic stress responses and microbe-mediated mitigation in plants: the omics strategies. *Front. Plant Sci.* 8: 172.

Miller, M.B. and Bassler, B.L. (2001). Quorum sensing in bacteria. *Annu. Rev. Microbiol.* 55 (1): 165–199.

Mishra, P.K., Bisht, S.C., Ruwari, P. et al. (2011). Alleviation of cold stress in inoculated wheat (*Triticum aestivum* L.) seedlings with psychrotolerant Pseudomonads from NW Himalayas. *Arch. Microbiol.* 193 (7): 497–513.

Mohammadipanah, F. and Zamanzadeh, M. (2019). Bacterial mechanisms promoting the tolerance to drought stress in plants. In: *Secondary Metabolites of Plant Growth Promoting Rhizomicroorganisms* (eds. H.B. Singh, C. Keswani, M.S. Reddy, et al.), 185–224. Singapore: Springer https://doi.org/10.1007/978-981-13-5862-3.

Morel, M. and Castro-Sowinski, S. (2013). *Plant Microbe Symbiosis: Fundamentals and Advances*. Springer Science & Business Media.

Mulligan, C.N., Yong, R.N., and Gibbs, B.F. (2001). Heavy metal removal from sediments by biosurfactants. *J. Hazard. Mater.* 85 (1–2): 111–125.

Naylor, D. and Coleman-Derr, D. (2018). Drought stress and root-associated bacterial communities. *Front. Plant Sci.* 8: 2223.

Ng, W.L. and Bassler, B.L. (2009). Bacterial quorum-sensing network architectures. *Annu. Rev. Genet.* 43: 197–222.

Numan, M., Bashir, S., Khan, Y. et al. (2018). Plant growth promoting bacteria as an alternative strategy for salt tolerance in plants: a review. *Microbiol. Res.* 209: 21–32.

Oldroyd, G.E. (2013). Speak, friend, and enter: signaling systems that promote beneficial symbiotic associations in plants. *Nat. Rev. Microbiol.* 11 (4): 252.

Paul, S., Dukare, A.S., Manjunatha, B.S., and Annapurna, K. (2017). Plant growth-promoting rhizobacteria for abiotic stress alleviation in crops. In: *Advances in Soil Microbiology: Recent Trends and Future Prospects* (eds. T.K. Adhya, B.B. Mishra, K. Annapurna, et al.), 57–79. Singapore: Springer https://doi.org/10.1007/978-981-10-7380-9.

Pichersky, E., Noel, J.P., and Dudareva, N. (2006). Biosynthesis of plant volatiles: nature's diversity and ingenuity. *Science* 311 (5762): 808–811.

Piechulla, B. and Degenhardt, J. (2014). The emerging importance of microbial volatile organic compounds. *Plant Cell Environ.* 37 (4): 811–812.

Pinu, F.R., Beale, D.J., Paten, A.M. et al. (2019). Systems biology and multi-omics integration: viewpoints from the metabolomics research community. *Metabolites* 9 (4): 76.

Qin, Y., Druzhinina, I.S., Pan, X. et al. (2016). Microbially mediated plant salt tolerance and microbiome-based solutions for saline agriculture. *Biotechnol. Adv.* 34 (7): 1245–1259.

Quiza, L., St-Arnaud, M., and Yergeau, E. (2015). Harnessing phytomicrobiomesignaling for rhizosphere microbiome engineering. *Front. Plant Sci.* 6: 507.

Raaijmakers, J.M., Paulitz, T.C., Steinberg, C. et al. (2009). The rhizosphere: a playground and battlefield for soilborne pathogens and beneficial microorganisms. *Plant and Soil* 321 (1–2): 341–361.

Rajkumar, M., Sandhya, S., Prasad, M.N.V. et al. (2012). Perspectives of plant-associated microbes in heavy metal phytoremediation. *Biotechnol. Adv.* 30 (6): 1562–1574.

Ravanbakhsh, M., Sasidharan, R., Voesenek, L.A. et al. (2017). ACC deaminase-producing rhizosphere bacteria modulate plant responses to flooding. *J. Ecol.* 105 (4): 979–986.

Richardson, A.E., Barea, J.M., McNeill, A.M. et al. (2009). Acquisition of phosphorus and nitrogen in the rhizosphere and plant growth promotion by microorganisms. *Plant and Soil* 321 (1–2): 305–339.

Romoli, R., Papaleo, M.C., De Pascale, D. et al. (2014). GC–MS volatolomic approach to study the antimicrobial activity of the antarctic bacterium *Pseudoalteromonas* sp. TB41. *Metabolomics* 10 (1): 42–51.

Roy, S.J., Negrão, S., and Tester, M. (2014). Salt resistant crop plants. *Curr. Opin. Biotechnol.* 26: 115–124.

Rudrappa, T., Czymmek, K.J., Paré, P.W. et al. (2008). Root-secreted malic acid recruits beneficial soil bacteria. *Plant Physiol.* 148 (3): 1547–1556.

Ryan, P.R., Dessaux, Y., Thomashow, L.S. et al. (2009). Rhizosphere engineering and management for sustainable agriculture. *Plant and Soil* 321 (1–2): 363–383.

Saikia, J., Sarma, R.K., Dhandia, R. et al. (2018). Alleviation of drought stress in pulse crops with ACC deaminase producing rhizobacteria isolated from acidic soil of Northeast India. *Sci. Rep.* 8 (1): 3560.

Saleem, M., Arshad, M., Hussain, S. et al. (2007). Perspective of plant growth promoting rhizobacteria (PGPR) containing ACC deaminase in stress agriculture. *J. Ind. Microbiol. Biotechnol.* 34 (10): 635–648.

Sasidharan, R., Hartman, S., Liu, Z. et al. (2018). Signal dynamics and interactions during flooding stress. *Plant Physiol.* 176 (2): 1106–1117.

Schmidt, R., Etalo, D.W., de Jager, V. et al. (2016). Microbial small talk: volatiles in fungal–bacterial interactions. *Front. Microbiol.* 6: 1495.

Shahzad, R., Khan, A.L., Bilal, S. et al. (2017). Inoculation of abscisic acid-producing endophytic bacteria enhances salinity stress tolerance in *Oryza sativa*. *Environ. Exp. Bot.* 136: 68–77.

Singh, B.K. and Trivedi, P. (2017). Microbiome and the future for food and nutrient security. *Microb. Biotechnol.* 10 (1): 50–53.

Smith, D.L., Praslickova, D., and Ilangumaran, G. (2015). Inter-organismal signaling and management of the phytomicrobiome. *Front. Plant Sci.* 6: 722.

Subramanian, P., Kim, K., Krishnamoorthy, R. et al. (2016). Cold stress tolerance in psychrotolerant soil bacteria and their conferred chilling resistance in tomato (*Solanum lycopersicum* Mill.) under low temperatures. *PLoS One* 11: e0161592–8.

Sugiyama, A., Ueda, Y., Zushi, T. et al. (2014). Changes in the bacterial community of soybean rhizospheres during growth in the field. *PLoS One* 9 (6): e100709.

Sukweenadhi, J., Balusamy, S.R., Kim, Y.J. et al. (2018). A growth promoting bacteria, Paenibacillusyonginensis DCY84T enhanced salt stress tolerance by activating defense-related systems in *Panax ginseng*. *Front. Plant Sci.* 9: 813.

Susilowati, A., Puspita, A.A., and Yunus, A. (2018). Drought resistant of bacteria producing exopolysaccharide and IAA in rhizosphere of soybean plant (*Glycine max*) in Wonogiri Regency Central Java Indonesia. *IOP Conf. Ser.: Earth Environ. Sci.* 142 (1): 012058.

Teplitski, M., Robinson, J.B., and Bauer, W.D. (2000). Plants secrete substances that mimic bacterial N-acyl homoserine lactone signal activities and affect population density-dependent behaviors in associated bacteria. *Mol. Plant-Microbe Interact.* 13 (6): 637–648.

Timmusk, S., El-Daim, I.A.A., Copolovici, L. et al. (2014). Drought-tolerance of wheat improved by rhizosphere bacteria from harsh environments: enhanced biomass production and reduced emissions of stress volatiles. *PLoS One* 9 (5): e96086.

Tiwari, S., Prasad, V., Chauhan, P.S. et al. (2017). *Bacillus amyloliquefaciens* confers tolerance to various abiotic stresses and modulates plant response to phytohormones through osmoprotection and gene expression regulation in rice. *Front. Plant Sci.* 8: 1510.

Vacheron, J., Desbrosses, G., Bouffaud, M.L. et al. (2013). Plant growth-promoting rhizobacteria and root system functioning. *Front. Plant Sci.* 4: 356.

Velikova, V., Várkonyi, Z., Szabó, M. et al. (2011). Increased thermostability of thylakoid membranes in isoprene-emitting leaves probed with three biophysical techniques. *Plant Physiol.* 157 (2): 905–916.

Vidya, P., Shintu, P.V., and Jayaram, K.M. (2016). Impact of phopshate solubilizing bacteria (*Bacillus polymixa*) on drought tolerance of green gram [*Vigna radiata* (L.) Wilczek]. *Ann. Plant Sci.* 5 (4): 1318–1323.

Wang, H., Wang, S.D., Jiang, Y. et al. (2014). Diversity of rhizosphere bacteria associated with different soybean cultivars in two soil conditions. *Soil Sci. Plant Nutr.* 60 (5): 630–639.

Xie, S.S., Wu, H.J., Zang, H.Y. et al. (2014). Plant growth promotion by spermidine-producing *Bacillus subtilis* OKB105. *Mol. Plant Microbe Interact.* 27 (7): 655–663.

Xu, L. and Coleman-Derr, D. (2019). Causes and consequences of a conserved bacterial root microbiome response to drought stress. *Curr. Opin. Microbiol.* 49: 1–6.

Yadav, A.N. and Yadav, N. (2018). Stress-adaptive microbes for plant growth promotion and alleviation of drought stress in plants. *Acta Sci. Agric.* 2 (6).

Yang, J., Kloepper, J.W., and Ryu, C.M. (2009). Rhizosphere bacteria help plants tolerate abiotic stress. *Trends Plant Sci.* 14 (1): 1–4.

Yang, H., Hu, J., Long, X. et al. (2016). Salinity altered root distribution and increased diversity of bacterial communities in the rhizosphere soil of Jerusalem artichoke. *Sci. Rep.* 6: 20687.

Yuan, J., Zhang, N., Huang, Q. et al. (2015). Organic acids from root exudates of banana help root colonization of PGPR strain *Bacillus amyloliquefaciens* NJN-6. *Sci. Rep.* 5: 13438.

Zhang, J., Nasir, F., Kong, Y. et al. (2017). Drought stress shapes the root-associated bacterial and fungal community structure in soybean genotypes. *Pak. J. Bot.* 49 (5): 1933–1942.

Zubair, M., Hanif, A., Farzand, A. et al. (2019). Genetic screening and expression analysis of psychrophilic *Bacillus* spp. reveal their potential to alleviate cold stress and modulate phytohormones in wheat. *Microorganisms* 7 (9): 337.

5

Microbial Diversity of Phyllosphere

Exploring the Unexplored

Rakhi Dhankhar[1]*, Aparajita Mohanty*[2]*, and Pooja Gulati*[1]

[1] *Maharshi Dayanand University, Rohtak, Haryana, India*
[2] *Gargi College, University of Delhi, New Delhi, India*

5.1 Introduction

Phyllosphere represents the aerial plant surfaces that act as habitat for microorganisms. The leaves form a major part of the phyllosphere. As per vegetation modeling, the leaf surface area (upper and lower) of the globe corresponds to 1 017 260 200 km^2, which is almost double the land surface area (Vorholt 2012); thereby suggesting that phyllosphere provides a vast area for inhabitation of the microorganisms. In addition to leaves (commonly referred to as phylloplane), the phyllosphere also includes stem (caulosphere), flowers (anthosphere), and fruits (carposphere) (Bringel and Couée 2015). Various parts of phyllosphere and the multipartite interactions between the phyllosphere, host microorganisms, and environment are depicted in Figure 5.1.

Phyllosphere represents a unique habitat experiencing immense environmental heterogeneity like rapid fluctuations in temperature, changing light intensity due to the diurnal cycle, and variations in water availability. Further, exposure to harsh climatic conditions, UV radiations, nutrient scarcity, and plant metabolites make the phyllosphere a very hostile niche for the microorganisms. Also, the environment is ephemeral (short-lived) as plants shed their leaves regularly. Despite all the adversities, the phyllosphere hosts a plethora of diverse group of microbes, viz. bacteria, fungi, yeast, archaea, cyanobacteria. These microorganisms have evolved several physiological adaptations to cope with the hostile environment of phyllosphere.

The phyllomicrobiome is also beneficial to the host plant in several ways like it protects the plant by pathogen exclusion, release of phytohormones and thereby promotes the growth of plants. The phyllosphere communities have also been reported to influence global biogeochemical cycles and have an ecological role (Fürnkranz et al. 2008). Thus, the study of the phyllosphere is imperative and helps us to understand its microbiota and their interactions to exploit them for plant growth and productivity. Although, the microbiology of phyllosphere is less extensively studied as compared to rhizosphere, the advent of

Phytomicrobiome Interactions and Sustainable Agriculture, First Edition. Edited by Amit Verma, Jitendra Kumar Saini, Abd El-Latif Hesham, and Harikesh Bahadur Singh.

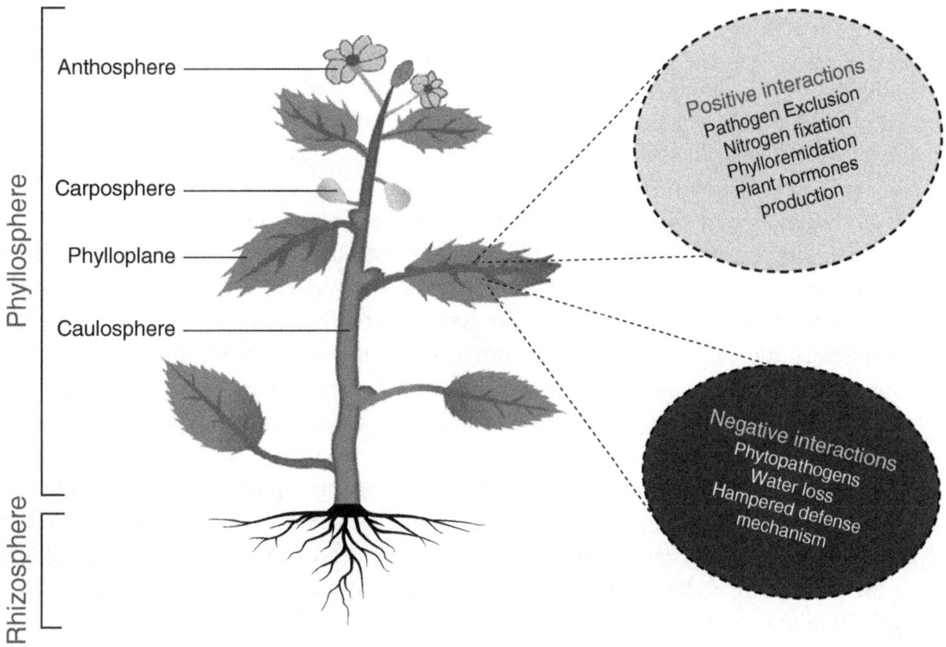

Figure 5.1 Different parts of phyllosphere and the multipartite interactions between phyllosphere, host microorganisms, and environment: The aerial parts of the plant constitute the phyllosphere, and it is divided into various parts; leaf (phylloplane) stem (caulosphere), flowers (anthosphere) and fruits (carposphere). The majority of the phyllosphere is represented by the leaves. Phyllosphere is a dynamic ecosystem, and the microorganisms are in close association with the plants and the environment. They are involved in various beneficial as well as harmful interactions.

high-throughput techniques has opened a new dimension in the field of phyllosphere study. In the last decade, the use of omics and next-generation sequencing (NGS) has provided tremendous insights into phyllomicrobiome. The present chapter focuses on the origin, composition, interactions, and adaptations of the microbial phyllosphere communities. The tools used and challenges in the study of this ecosystem are discussed in detail. The significance and future prospects of phyllosphere microbiota have also been addressed.

5.2 Origin of Phyllosphere Microflora

The origin of microbial communities in phyllosphere is enigmatic as the sources of these microbes can be highly variable. Microorganisms can invade the plant host via vertical transmission (seeds) or horizontal transmission (air, aerosols, water, soil, neighboring plants, animal-borne sources) (Vorholt 2012). It is instinctive to think that air, water, or soil, which are in regular contact with the plants, are the most probable sources of the microorganisms dwelling in the phyllosphere. However, different reports suggest diverse possibilities. The comparison of phyllomicrobiome of nine perennial plants with the bacterial population in the surrounding air using denaturing gradient gel electrophoresis

(DGGE) analysis followed by cloning and sequencing of dominant bands showed that the two environments exhibited a high level of discrepancy in terms of the bacterial population (Vokou et al. 2012). In another study, pyrosequencing of 16S rRNA genes in eudicot trees and their surrounding soil predicted the co-occurrence of very small operational taxonomic units (0.5%) between the two microbiomes (Kim et al. 2012). Contrary to these reports, some evidences support the environmental origin of phyllosphere communities. Metaproteogenomics analysis of phyllosphere and rhizosphere microbiota of paddy field-grown rice was performed and compared with the microbial communities in the flooding water of a paddy field (Knief et al. 2012). Phyllosphere microbiota was found to be more closely related to microbial communities in water as compared to the communities of the rhizosphere, indicating that water is the source for the inoculum of the phyllomicrobiome. Along with soil, water, and air, the organisms might also migrate from the neighboring plants. These migrating organisms have a high chance of survival as they are already adapted to the phyllospheric environment.

In addition to the environment, the plant also plays an important role in driving the composition of its microbial community. A significant correlation exists between plant species and the microbial communities they harbor, which is irrespective of the geographical location. This clearly indicates that the genome of the plant has a significant role in deciding the phyllomicrobiome (Redford et al. 2011; Whipps et al. 2008). Quantitative trait loci (QTL) mapping on recombinant maize has identified the specific chromosomal regions in maize that control the microbial population on the maize phyllosphere (Balint-Kurti et al. 2010). The fact that the pattern of re-establishment of the microbial population remains consistent every year in annual plants further suggests there are local reservoirs of these microorganisms (Knief et al. 2010).

Thus, the phyllosphere has some permanent residents termed as "true or core residents" that are species-specific, whereas a fraction of organisms come from various sources, and these ad hoc colonizers are termed as "transient" residents. However, the relative contribution of various sources for the origin of microbes of phyllosphere is still unclear (Rastogi et al. 2013). Extensive spatial and temporal studies using modern high-throughput molecular approaches and applications like "SourceTracker" are expected to provide a better insight into the area in the coming years.

5.3 Tools to Study Phyllomicrobiome

Initially, culture-dependent basic microbial techniques, such as impression techniques, dilution-plating, and counting the colony-forming units (CFU) were used to analyze the phyllosphere communities (Bainbridge and Dickinson 1972; Wagner et al. 1993). The primary focus was to identify the pathogenic organisms rather than exploring the phyllosphere diversity. This approach was replaced by advanced "first-generation" techniques, viz. clone library, sanger sequencing, DGGE and terminal restriction fragment length polymorphism (RFLP) (Hunter et al. 2010; Izhaki et al. 2013; Vokou et al. 2012). However, these techniques could provide limited insights into the microbial composition of the phyllosphere. Later, the advent of high-throughput next-generation molecular techniques with omics and bioinformatics approaches were a great boon to the understanding of phyllosphere ecology and has opened new horizons of research in this field. These techniques provided a better understanding of various aspects of the phyllomicrobiome, such as the diversity in microbial communities, interaction between plant host and microbe, seasonal, and spatial variation in phyllomicrobiome and adaptations of the microbes. Various tools used to study phyllomicrobiota with the significant findings using these techniques have been discussed in the subsequent sections.

5.3.1 Conventional Methods

The earliest reports on the study of phyllomicrobiome used basic microbiological techniques like impression techniques (leaves pressed on a growth media) or the use of adhesive tapes (microbes removed from leaf surface using tape) and cultured on suitable media (Bainbridge and Dickinson 1972; Langvad 1980). Another culture-dependent approach used was enumerating the bacteria by counting CFU. In this method, the microorganisms were washed from the surface of the plant, and dilution plating was performed on suitable media. But these techniques fail to provide the exact picture of the microbial communities as the fast-growing organisms dominate the plates. Also, most of the phyllosphere microorganisms were viable but nonculturable (VBNC) and hence couldn't be identified using these methods. So, these conventional techniques were biased and could not represent the true diversity of the phyllomicrobiome.

5.3.2 Microscopic Techniques

The ex-situ approaches involving analyses after extraction of the microbes from the leaf surface does not provide the information regarding the relationship of microbial communities with the surface topography. To overcome these issues, in-situ approaches that directly visualize microbes on plant surfaces using different microscopic techniques were developed (Hallett et al. 2010; Lee and Hyde 2002). The leaves were peeled off using bleach and visualized under a bright-field microscope (Dickinson et al. 1974). The major disadvantage of this method was that small and transparent microbes on the leaf surface were difficult to visualize. This problem was addressed by using stains and dyes. Different fluorescent dyes were used to visualize phyllospheric bacteria on environmentally grown silver beech (*Lophozoniamenziesii*) leaves (Remus-Emsermann and Schlechter 2018). In this study, the physiological state of the bacteria was also identified using live-dead staining. The Green fluorescent dye Syto 9, which penetrates living cells, stained the live cells while the red fluorescent dye propidium iodide was used to stain dead cells.

Integrating microscopy with advanced fluorescence techniques such as immuno-labeling and fluorescent in-situ hybridization (FISH) helps in providing clear images of microbial cells in the phyllosphere (Andrews et al. 2002). FISH analyses using fluorescently labeled oligonucleotides showed the wide diversity of microorganisms in the phyllosphere of *Arabidopsis thaliana* (Remus-Emsermann et al. 2013).

In order to understand the distribution pattern of bacteria on the leaf surfaces and the relationship between phyllosphere communities and different plant species, scanning electron microscopy (SEM) was used (Baldotto and Olivares 2008). Leaves of 47 different plant species distributed in 27 botanical families from a Brazilian farm were observed using

SEM. It was found that the bacteria establish themselves preferentially on the trichomes, veins, stomatal opening, and cell wall surfaces and the junction of the epidermis (Baldotto and Olivares 2008).

5.3.3 First-Generation Molecular Techniques

As the actual conditions of the host environment of the phyllosphere are not completely understood and are highly variable; hence, it is difficult to provide suitable media and conditions for the growth of microorganisms present here. Thus, culture-independent approaches were exploited to study phyllomicrobiome. 16SrRNA is the most common house-keeping genetic marker that has been used for accessing the phyllosphere diversity. 16SrRNA sequencing is used to explore the phyllomicrobiota of a variety of plants like rice, apple, lettuce, tomato (Enya et al. 2007; Knief et al. 2012; Williams et al. 2013; Yashiro et al. 2011) (Table 5.1). Another widely used fingerprinting technique to identify bacterial diversity is PCR-DGGE. It generates a profile based on the separation of PCR amplicon of the same size but different sequences (Ercolini 2004). PCR-DGGE technique is also used to analyze the effect of UV radiation on the bacteria assemblages of maize phyllosphere (Kadivar and Stapleton 2003). Other DNA-profiling techniques, like RFLP, amplified rDNA (ribosomal DNA) restriction analysis (ARDRA), were also used. Venkatachalam et al. (2016) performed 16SrRNA, PCR-DGGE and ARDRA analysis to assess the diversity and functional traits of microbes in the rice phyllosphere. They reported the dominance of genus α-proteobacteria (35%) and actinobacteria (38%) in the rice phyllosphere (Venkatachalam et al. 2016). Ribosomal RNA (rRNA) intergenic spacer analysis (RISA) is another approach used for the microbial community analysis in phyllosphere studies. This fingerprinting technique involves PCR amplification of the intergenic spacer region between the small (16S) and large (23S) subunits of rRNA (Gravouil 2012).

5.3.4 Next-Generation Sequencing Methods

NGS has accelerated the pace of phyllosphere research and has made it possible to sequence hundreds of samples in a single run with a drastic reduction in the sequencing cost and time. The 454 and Illumina platform are widely used for microbial community analysis (Degnan and Ochman 2012). The NGS of rRNA amplicons have revealed the community composition of a variety of important plants like soybean, clover, lettuce, spinach, grape, pine, tamarix (Delmotte et al. 2009, Redford et al. 2011) (Table 5.2). Darlison et al. (2019) used both culture-dependent (viable count and identification of bacterial isolates) and culture-independent (illumina sequencing) approaches to access the impact of nitrogen on phyllosphere communities. The alpha diversity (average species diversity in a habitat) decreased after increasing the nitrogen fertilizer dose.

Pantoea agglomerans 299R is a frequent foliar colonizer and is considered as model bacterium for phyllosphere studies (Remus-Emsermann et al. 2013). The genome sequence of *P. agglomerans* was drafted using Illumina sequencing platform. The data unfolded various genes for the adaptive strategies that help the bacteria to strive in the phyllosphere environment. The epiphytic fitness of the organism is due to the presence of genes that are responsible for high-affinity, uptake, and utilization of the photosynthates sucrose, fructose,

Table 5.1 Major findings obtained from the study of the phyllomicrobiome of different plants using various techniques.

S. No.	Approach	Method	Plant	Findings	References
1.	Conventional methods	Impression techniques	Potato	Potato leaves were colonized by filamentous fungi at senescence stage. Green leaves produce phytoalexins, which were inhibitory to spore germination.	Bainbridge and Dickinson (1972)
		Use of adhesive tapes	Fagus and Tilia	Phylloplane fungi *Epicoccum, Cephalosporium* were visualized	Langvad (1980)
2.	Microscopy based methods	Fluorescent microscope	Silver beech	DNA intercalating fluorescent dyes like acridine orange was used to study phyllospheric bacteria in situ. Dye propidium iodide and Syto 9 were used to know the physiological state of bacteria (live/dead staining)	Remus-Emsermann et al. (2018)
		Fluorescence in situ hybridization (FISH)	*Arabidopsis*	FISH analyses using fluorescently labeled oligonucleotides showed the wide diversity of microorganisms in the phyllosphere of this plant	Remus-Emsermann et al. (2018)
		Scanning electron microscope (SEM)	Different plant species (47) belonging to 27 botanical families	Preferred sites for the establishment of bacteria are trichomes, veins, stomatal opening, and cell wall surfaces and junction of the epidermis	Baldotto and Olivares (2008)
3.	First-generation molecular techniques	16S rRNA sequencing	Apple, Arabidopsis, Rice, Tomato, Wheat	Phyllosphere community analysis	Yashiro et al. (2011); Horton et al. (2015); Knief et al. (2012); Enya et al. (2007); Gu et al. (2010)
		PCR-DGEE	Maize, rice, tomato, wheat	Phyllosphere community analysis	Kadivar and Stapleton (2003); Thapa et al. (2018); Enya et al. (2007); Gu et al. (2010)
		RFLP, ARDRA	Rice	Phyllosphere community composition	Venkatachalam et al. (2016)

(Continued)

Table 5.1 (Continued)

S. No.	Approach	Method	Plant	Findings	References
4.	Next-generation sequencing techniques	16S rRNA pyrosequencing	Soybean, clover, Arabidopsis	Unique metabolic traits like high expression in TonB-dependent receptors were reported, which were involved in carbohydrate transport.	Delmotte et al. (2009)
				Methanol was used as a source of energy, which contributes to the successful colonization of *Sphingomonas* and *Methylobacterium* in the phyllosphere	
			Pine	The phyllosphere bacteria community is determined by the host genome	Redford et al. (2010)
			Tamarix	Phyllosphere diversity is determined by the geographical location of the plant	Finkel et al. (2011)
			Lettuce	Geographical location plays a key role in determining the bacterial community on lettuce foliage	Rastogi et al. (2012)
			Grape	The phylloplane and carposphere community composition significantly differ within the same plant	Leveau and Tech (2011)
			Lettuce	The phyllosphere diversity is influenced by seasonal variation, irrigation, leaf age, and artificial *E. coli* inoculation	Williams et al. (2013)
		18S rRNA pyrosequencing	Tamarix	Both geographic location and tree species were determinants of microbial community structures, with the former being more dominant	Finkel et al. (2011)
		ITS pyrosequencing	Oak	The fungal member composition in the phyllosphere is affected by urban and rural management practices	Jumpponen and Jones (2009)
			Beech	The fungal diversity in the phyllosphere is decided by the host plant genome.	Cordier et al. (2012)
			Balsam poplar	Host genotype is a major determinant of phyllosphere fungal community composition	Balint et al. (2013)

5.	Omics approaches	Transcriptomics	African violet	Upregulation of genes related to anaerobiosis, iron uptake, transporters, or permeases, xenobiotic resistance, chemotaxis, and stress responses to reactive oxygen species and heat in the phyllosphere environment	Okinaka et al. 2002
		Proteomics	Arabidopsis	Induction of stress proteins and enzymes involved in methanol utilization in bacteria during epiphytic growth. Two-domain stress regulator, namely PhyR, was discovered in *Alphaproteobacteria*	Gourion et al. (2006)
		Metaproteogenomics	Soybean, clover, Arabidopsis	High consistency w.r.t the proteomes of the communities on the three different plant species	Delmotte et al. (2009)
			Rice	4600 proteins were identified in the rice phyllosphere. Proteins related to methanol-based methylotrophy dominated within the protein repertoire of the phyllosphere microbiota	Knief et al. (2012)
6.	Promoter trapping	In vivo expression technology (IVET)	Arabidopsis	Genes that were induced in *Pseudomonas syringae* pv. tomato during infection of *Arabidopsis thaliana* were identified. Over 500 *in planta*-expressed (ipx) promoter fusions were reported	Boch et al. (2002)
		Recombinase-based in vivo expression technology (RIVET)	Tomato	The genes which were activated in *Xanthomonas campestris* pv. vesicatoria, during its interaction with its host tomato, were studied. The citH homolog gene, encoding a citrate transporter, was found to be important for the interaction of *X. campestris* with tomato	Tamir-Ariel et al. (2007)
		Habitat-inducible rescue of survival (HIRS)	Bean	Plant-inducible genes (pigs) encoding diverse cellular functions, including virulence, transcription regulation, transport, nutrient acquisition, etc., were identified in *Pseudomonas syringae* pv. *Syringae* B728a that makes the bacteria a successful foliar colonizer	Marco et al. (2005)
7.	Community-level physiological profiling (CLPP)		Cucumber and Japanese spindle	Powdery mildew infection changes bacterial community composition in the phyllosphere. Functional diversity, richness, and evenness of bacterial communities were significantly greater in powdery mildew infected diseased plants than uninfected ones	Suda et al. (2009)
8.	Phospholipid derived fatty acid analysis (PLFA)		Tobacco	N-(3-oxohexanoyl) homoserine lactone (occurs in tobacco phyllosphere) acts as inducer for quorum sensing in phyllosphere bacteria	Lv et al. (2012)

Table 5.2 Microbial diversity in the phyllosphere of various crop plants.

S. No.	Crop	Phyllosphere residents	Technique employed	References
1.	Apple (*Malus domestica*)	*Sphingomonadales, Actinomycetales, Rhizobiales, Pseudomonadales, Burkholderiales*	Culture-dependent and 16S rRNA analysis	Yashiro et al. (2011)
2.	Barley (*Hordeum vulgare*)	γ –*Proteobacteria*	PCR, T-RFLP, RISA	Gravouil (2012)
3.	Grapevine (*Vitis vinifera*)	*Sphingomonas, Janthinobacterium, Methylobacterium, Pelomonas Ascomycota, Basidiomycota*	Fungal ITS and 16S rRNA	Perazzolli et al. (2014)
		Aureobasidium, Epicoccum	BOX-fingerprinting	Grube et al. (2011)
4.	Lettuce (*Lactuca sativa*)	*Proteobacteria, Firmicutes, Actinobacteria*	Pyrosequencing and 16S rRNA analysis	Williams et al. (2013)
		Cladosporium, Sporobolomyces	T-RFLP	Hunter et al. (2015)
		Gammaproteobacteria (*Enterobacteriaceae* and *Moraxellaceae* family), Betaproteobacteria (*Comamonadaceae* and *Burkholderiacea* family)	16S rRNA	Williams and Marco (2014)
		Phyla – Proteobacteria, Firmicutes, Bacteroidetes, Actinobacteria Genera – *Pseudomonas, Bacillus, Massilia, Arthrobacter, Pantoea*	16S rRNA	Rastogi et al. (2012)
5.	Rice (*Oryza sativa*)	*Pantoea, Exiguobacterium, Bacillus, Pseudomonas, Enterobacter, Erwinia, Streptomyces* Fungi – *Curvularia lunata, Bipolaris ravenelli*	PCR-DGEE ARDRA	Venkatachalam et al. (2016)
		Actinobacteria, Proteobacteria	16 S rRNA, metaproteomics	Knief et al.(2012)
		Sphingomonas, Acidovorax, Methylobacterium, Pseudomonas, Sinorhizobium	Fluorescent nuclear staining technique	Niwa et al. (2011)

		Proteobacteria, Firmicutes, Planctomycetes, Cyanobacteria, Actinobacteria	PCR-DGGE	Thapa et al. (2018)
6.	Rocket salad (*Diplotaxistenuifolia*)	*Exiguobacterium, Pseudomonas, Phyllobacterium, Pantoea*, Enterobacteriaceae, *Bacillus*	Illumina sequencing and viable count	Darlison et al. (2019)
		Davidiellaceae, Cystofilobasidiaceae, Tremellales Sporidiobolales		
7.	Rocket salad (*Diplotaxistenuifolia*) Lettuce (*Lactuca sativa*)	Proteobacteria, Bacteroidetes, Actinobacteria, Firmicutes	16S rRNA	Dees et al. (2015)
		Families – *Pseudomonadaceae*, *Oxalobacteraceae*, *Enterobacteriaceae*		
		Genera – *Pantoea*, *Sphingomonas*		
8.	Soybean (*Glycine max*), Clover (*Trifolium*)	*Alphaproteobacteria, Sphingomonas* sp., *Methylobacterium*	16S rRNA-DGGE and metaproteomics	Delmotte et al. (2009)
9.	Spinach (*Spinaciaoleracea*)	Proteobacteria and Firmicutes	16S rRNA	Lopez-Velasco et al. (2011)
		Phyllobacterium, Pseudomonas, Pantoea, Bacillus, Exiguobacterium,	Illumina sequencing and viable count	Darlison et al. (2019)
		Dothioraceae, Davidiellaceae, Tremellales family, Incertaesedis		
10.	Tomato (*Solanum lycopersicum*)	*Bacillus, Pseudomonadales, Curtobacterium, Sphingomonas*	16S rRNA analysis and DGGE	Enya et al. (2007)
		Pseudomonas, *Xanthomonas*. *Hypocrea, Aureobasidium, Cryptococcus*	16S rRNA and 18S rRNA	Ottesen et al. (2016)
11.	Wheat (*Triticum aestivum*)	Ascomycota, Basidiomycota.	Fungal ITS2	Karlsson et al. (2014)
		γ-Proteobacteria	16S rDNA, PCR-DGGE	Gu et al. (2010)

and glucose, for the repair of UV-damaged DNA, and production of the osmoprotectants betaine and trehalose. In the same study, PA299R was also found to possess *Pantoea*-typical LPP-1 plasmid, coding for the biosynthesis of thiamine and the pigment zeaxanthin. Besides, the presence of the ipdC gene for the production of the plant hormone indole 3-acetic acid was also confirmed by NGS technology (Remus-Emsermann et al. 2013). The fungal members of the phyllosphere are commonly assessed using nuclear ribosomal internal transcribed spacer (ITS) pyrosequencing. It is recommended as a universal fungal barcode sequence to identify fungal species. The composition of the fungal community of oak, beech, and balsam poplar was studied using ITS pyrosequencing. The major findings of these studies are listed in Table 5.1.

5.3.5 Omics and Bioinformatics Approaches

In the last decade, omics exploration, and bioinformatics tools have revolutionized the field of phyllosphere studies. Omics approaches collectively represent high-dimensional biology, which aims to detect the structure and function of genes (genomics), the study of mRNA (transcriptomics), proteins (proteomics), and global metabolite profiles (metabolomics) in a biological sample. These techniques have allowed us to understand the complex molecular systems present in the phyllomicrobiome. A huge amount of data is generated in the omics approaches; hence bioinformatics tools are crucial to manage the overflowing data sets and extract valuable information from it. The development of open-access and indexed databases has facilitated researchers to explore the phyllomicrobiome with great ease; for example, to study the origin of phyllomicrobiota in tomato plants, 16SrRNAillumina sequencing was performed at regular intervals in the growing season. The vast data generated were analyzed using QIIME (open-source bioinformatics pipeline for microbiome analysis from raw sequencing data) with the Green Gene database (http://greengenes.lbl.gov) (Ottesen et al. 2016). Other bioinformatics tools like UCHIME (to detect chimera sequence), PYNAST (for multiple sequence alignment), Fast Tree (for the phylogenetic tree) and UCLUST (for sequence clustering) have aided in the phyllomicrobiome analysis.

Okinaka et al. (2002) used a transcriptomics approach to find the genes that are specifically up- and downregulated during the infection of plants by *Erwinia chrysanthemi*. Several plants up- and downregulated clones were end sequenced and the resulting data were analyzed by BLASTX, BLASTP, and BLASTN searches. The microarray profiling, along with bioinformatics tools, identified that the virulence genes, the genes responsible for iron uptake and transport, chemotaxis, and stress response in the bacteria are upregulated in the phyllosphere of African violet leaves (Okinaka et al. 2002). Proteomics study of phyllosphere bacteria *Methylobacterium extorquens* discovered PhyR as a response regulator, essential for epiphytic growth (Gourion et al. 2006). The proteome study revealed that proteins and enzymes necessary for methylotropy are upregulated in the phyllosphere inhabiting *M. extorquens*.

The association of metagenomic and metaproteomic approaches known as metaproteogenomics is also used in the ecological analysis of the phyllosphere. Microbial community analysis of soybean, clover, and *A. thaliana* was carried out using this approach. Temporal and spatial stability in the phyllosphere community and the proteomes of the three species were observed (Delmotte et al. 2009). The metaproteogenomics analysis of the rice phyllosphere revealed an abundance of the protein linked to methanol-based methylotrophy (C_1 metabolism) in the rice phyllomicrobiome. This observation indicates that genus *Methylobacterium* dominates the rice phyllosphere and it derives energy from methanol, which is a byproduct of cell wall metabolism in rice plants (Knief et al. 2012). The metagenomic data obtained through various studies are submitted at open source and open submission web application server MG-RAST (Metagenomic Rapid Annotations using Subsystems Technology). This pipeline carries out automated phylogenetic and functional analysis of metagenomes and the data can be retrieved for further studies.

5.3.6 Other Molecular Methods

In addition to the above-discussed techniques, several other approaches are employed to study the phyllosphere. Promoter trapping methods like in vivo expression technology (IVET) and habitat-inducible rescue of survival (HIRS) are used in various phyllospheric studies to identify the unique genes that are induced or expressed during leaf colonization by various bacteria. In the promoter trapping method, promoter regions for the genes are used as bait to trap and identify the protein of interest. Boch et al. (2002) identified the genes that were induced during the infection of *A. thaliana* by *Pseudomonas syringae* pv. tomato using IVET (Boch et al. 2002). IVET identifies transcriptionally active portions of a microbial genome during interaction of the microorganism with a particular environment or host organism. In a similar study, recombinase-based in vivo expression technology (RIVET) was used to study the interaction between other phytopathogenic bacteria *Xanthomonas campestris* pv. *vesicatoria* and its host tomato plant (Tamir-Ariel et al. 2007). A green fluorescent protein-based IVET leaf array was used to identify genes in *Erwinia chrysanthemi* 3937 that were specifically upregulated during plant infection compared with growth in a laboratory culture medium (Yang et al. 2004). The use of nonselective reporter (GFP) is laborious; hence a novel variant of IVET was developed, namely HIRS. This technique uses endogenous factors (like methionine prototrophy) for complementation to identify genes that are specifically induced in *P. syringae* pv. *syringae* B728a during growth on bean leaves (Marco et al. 2005).

The functional diversity and richness of bacterial assemblages in the phyllosphere of powdery mildew infected leaves of cucumber and Japanese spindle were compared with noninfected leaves using community-level physiological profiles (CLPP) (Suda et al. 2009). CLPP studies the catabolic diversity of bacteria based on the substrate utilization pattern. The ability to utilize different carbon and nitrogen source present in microtitre plate format is analyzed and the data is utilized to study diversity. Phospholipid fatty acid analysis (PLFA) is another useful technique that is used to study the composition of bacterial communities in the tobacco phyllosphere (Lv et al. 2012). PLFAs are found in cellular membranes and act as markers for a particular taxonomic group (Thapa and Prasanna 2018).

5.4 Biodiversity of Phyllosphere

Phyllosphere act as a niche for a variety of microorganisms. The inhabitants of the phyllosphere are referred to as epiphytes. The species richness of a particular plant's phyllosphere is less than its rhizosphere, yet if the phyllomicrobiome of the earth is considered, it is huge and highly diverse (Vorholt 2012). The population size of these epiphytes considerably varies over a smalltime scale and even over the growing seasons due to large fluctuations in the nutritional and environmental conditions of the phyllosphere. The variations also occur within or among the plants of the same species. Furthermore, the species of the plant also affects the microbial load of the leaf. The host plant genome is identified as a key factor in influencing the phyllosphere community composition.

Phyllosphere is inhabited by almost all major groups of microorganisms like bacteria, cyanobacteria, fungus, algae, and some protozoa. The microbial communities of economically important crops are described in Table 5.2. The most common resident of the phyllomicrobiome are bacteria (Thapa and Prasanna 2018). Studies on the phyllosphere of various crops (wheat, apple lettuce spinach, rice) and trees showed that phylum Actinobacteria, Bacteroidetes, Firmicutes, and Proteobacteria are most abundant residents of phyllosphere (Bulgarelli et al. 2013; Thapa and Prasanna 2018). The community structure at the genus level is dominated by *Pseudomonas, Sphingomonas, Methylobacterium, Bacillus, Massilia, Arthrobacter*, and *Pantoea,* etc. Some phytopathogens such as *Agrobacterium, Burkholderia, Clavibacter, Leifsonia*, and *Xanthomonas* are also found in abundance in the phyllosphere (Vorholt 2012). Most of the phyllosphere bacteria do not occur as single cells but as aggregates embedded in the extracellular polymeric substances (EPS). EPS provides a hydrated environment around the cells and protects them from harsh climatic conditions (Rigano et al. 2007).

Filamentous fungi are regarded as the transient residents of the leaves, while yeasts, and rapidly sporulating fungal species are the core residents. These exist as phytopathogens, saprophytes, or sometimes in association with algae in the form of lichens (Frey-Klett et al. 2011). The population of filamentous fungi ranges between 10^2 and 10^8 CFU/g leaf, while 10–10^{10} CFU/g leaf yeast are present in the phyllosphere (Thapa and Prasanna 2018). Some commonly found fungi in the phyllomicrobiome are *Cladosporium, Alternaria, Penicillium, Acremonium, Mucor,* and *Aspergillus*, whereas yeasts like *Cryptococcus, Sporobolomyces,* and *Rhodotorula* predominate (Table 5.2). The ITS pyrosequencing studies of some microbiomes reported that members from division Ascomycota and Basidiomycota dominate the phyllosphere (Balint et al. 2013). The study also reported that the genotype of the plant host shapes the foliar fungal community composition. The previous study on the phyllosphere of European beech (*Fagus sylvatica*) also supports the finding that the genetic distance predominates over geographic distance. Some bio-control agents like *Aureobasidium pullulans* (yeast-like fungus) are also reported in the phyllosphere. Other than fungus and bacteria, the presence of cyanobacteria and algae are also observed in the phyllosphere, but their habitat is restricted to the phyllosphere of forest trees. They rarely contribute to crop phyllomicrobiome.

5.5 Microbial Adaptation to Phyllosphere

Phyllosphere offers a challenging environment for the epiphytes to grow. Thus, successful survival and reproduction in this habitat require certain adaptations by the microorganisms (Levy et al. 2018). Microbes exhibit various adaptation strategies to enhance their epiphytic fitness to combat various biotic and abiotic stresses, and nutrient scarcity (Figure 5.2). These adaptive strategies are discussed below.

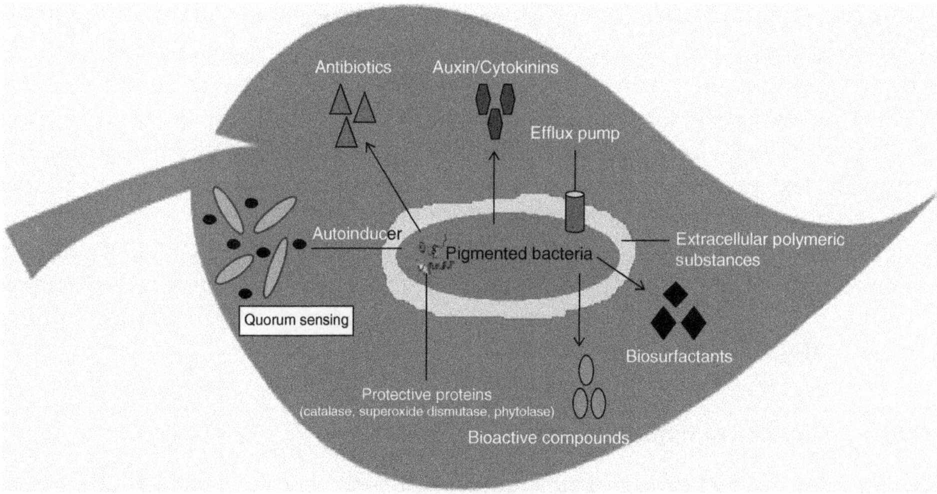

Figure 5.2 Classification of root exudates compounds: various compounds are found in the root exudation, which belongs to a wide array of biochemical classes.

5.5.1 Adaptation to Abiotic Stresses

The phyllosphere bacteria are exposed to abiotic stresses like harmful UV light, extreme temperatures, low humidity, and osmotic stress. These conditions also vary throughout the day. Phyllosphere microbes cope up with these challenges. The phyllomicrobiome is dominated by UV tolerant microbes. A study of the peanut phyllosphere exposed to high UV fluxes revealed that the phyllosphere microbes tolerate high UV fluctuations (Sundin and Jacobs 1999). A large number of bacteria in the phyllosphere, such as *Methylobacterium, Sphingomonas,* and *Pseudomonas,* are pigmented; pigmentation provides protection against UV radiation (Lindow and Brandl 2003). Phyllosphere bacteria also possess special DNA repair mechanisms like phytolases. In *P. syringe,* which is a frequent colonizer of the phyllosphere, the DNA repair mechanism is associated with *rulAB* genes of plasmid, which are induced on UV exposure.

Desiccation and osmotic stress are other abiotic factors that affect the survival of epiphytes. Bacteria exhibit two major strategies to counteract these stresses: aggregation and production of bioactive compounds. The formation of cell aggregates is mediated by the production of extracellular polymeric (EPS). EPS helps the bacteria to aggregate in biofilms and increase the water availability to bacterial cells. Xanthum gum is one of the predominant EPS secreted by *Xanthomonas* species on the surface of leaves of citrus plants, which significantly enhances the epiphytic fitness of the bacteria (Rigano et al. 2007). Certain specific proteins like haemagglutin in family proteins in *Dickeya dadantii* (formerly *Erwini achrysanthemi*) and *Xanthomonas oryzae* also assist the bacteria in adhering to the surface of the leaf (Rojas et al. 2002; Das et al. 2009). Adhesion favors colonization as it protects bacteria from being washed off from the plant surface with raindrops (Vorholt 2012). In addition to the aggregation, phyllosphere bacteria produce bio surfactants that keep the surface wet and avoid desiccation. In order to manage osmotic stress, phyllosphere microbes also produce compounds like choline or trehalose (Chen and Beattie 2008;

Freeman et al. 2010). Some microbes derive osmoprotectants from host plants. *P. syringae* is known to produce bioactive molecules like coronatine (a toxin) or syringolin A, which enhances their survival on plants (Melotto et al. 2006; Schellenberg et al. 2010).

The phyllosphere microbes are also reported to exhibit the phenomenon of quorum sensing as an adaptive measure to the challenging environments. Quorum sensing is a density-dependent behavior in which microbes communicate with each other with the help of small signaling molecules called autoinducer (Thapa and Prasanna 2018). Quorum sensing in *P. syringae* is well studied, and it is found that the signaling molecule acyl-homoserine lactones (AHLs) are important for the epiphytic fitness of the bacteria. It induces EPS production, resistance to desiccation, and suppresses motility (Vorholt 2012).

5.5.2 Adaptation to Biotic Stresses

The microbes also face various biotic hurdles in the process of successful colonization on the phyllosphere. Plants are known to produce a variety of natural products, particularly antimicrobials, which play a role in defense, communication, and development (Fan et al. 2011). However, microbes have evolved with various mechanisms like the upregulation of certain genes and the presence of antibiotic-resistant transport systems to overcome these defenses. *Arabidopsis* plant produces aliphatic glucosinolates derived natural product, sulforaphane (4-methylsulfinylbutyl isothiocyanate) that inhibits the growth of *Pseudomonas* on the plant. Still, the organism became resistance by the expression of multiple sax genes (saxCAB/F/D/G) (Fan et al. 2011). Deletion of the sax operon impaired the growth of *P. syringae* on *Arabidopsis*.

The presence of membrane-associated multidrug efflux (MDE) transporters is another defense strategy employed by bacteria against plant antimicrobial compounds. A wide array of these efflux systems is reported in different phyllosphere dwelling microbes and their role in bacterial self-protection is recognized. The RND-type transporter IfeAB exports isoflavonoid and provides protection to *Agrobacterium tumefaciens* against this antimicrobial plant metabolite (Palumbo et al. 1998). Another drug transporter pump MexAB-OprM has been identified in *P. syringae*. The mutation in MexAB-OprM reduced the tolerance of bacteria against various antimicrobial compounds and the survival rate in the phyllosphere was also reduced (Stoitsova et al. 2008).

Sometimes apart from plant-originated metabolites, the antimicrobial compounds from bacteria that are already present in the phyllosphere also inhibit the growth of other microbes. The phyllosphere microbes have developed protection mechanisms against these microbial-originated compounds as well (Vorholt 2012). For example, multi-drug transporter NorM from MATE family enhances the epiphytic fitness of *Erwinia amylovora* on apple flowers by conferring resistance against toxin produced by *P. agglomerans*.

Another challenge for phyllosphere microbes is to protect their nucleic acids from damage by reactive oxygen species (ROS) as they are exposed to a high amount of oxygen produced by the leaves during photosynthesis. The ROS detoxification is carried out by enzymes, catalases, and superoxide dismutase. These enzymes, along with some chaperons and DNA-binding proteins, are overexpressed in phyllosphere bacteria, as evident from the metaproteogenomic study of soybean, clover, and Arabidopsis phyllomicrobiome (Delmotte et al. 2009).

5.5.3 Adaptation to Nutrient Scarcity

Phyllosphere is not favorable for the growth of microbes in terms of nutrient availability. The successful phyllosphere inhabitants can derive their energy by utilizing metabolites produced by the host plants. The plants like soybean, clover, and Arabidopsis produced methanol as a by-product during the metabolism of the plant cell walls (Sy et al. 2005). These plants are dominantly colonized by *Methylobacterium* as the members of this group are facultative methylotrophs and can utilize methanol as an energy source (Delmotte et al. 2009). Studies revealed that the phyllosphere-inhabiting microbes express methylotrophic enzymes (methanol dehydrogenase, formaldehyde-activating enzyme) during colonization (Knief et al. 2012). Similarly, *Flavobacteria* adapt them to the phyllosphere by producing enzymes of glycoside hydrolase families GH78 and GH106, which help them to utilize the complex polysaccharide and rhamnogalacturonan present in plants (Kolton et al. 2013).

Epiphytic bacteria also produce certain compounds that enhance substrate availability. The bio-surfactants produced by these bacteria not only increase water availability and protect plants from desiccation but also favor substrate leaching. Some phyllosphere bacteria are known to produce hormones like IAA, which loosen the plant cell wall and release saccharides from the plant. Some also secrete cytokinin, which triggers the cell wall of plants to release methanol (Kutschera 2007). The presence of specialized transport systems like ABC transporter in *P. syringae* and TonB-dependent transport system in *Sphingomonas* favors the uptake of various nutrients from leaf surface despite their low concentration (Bulgarelli et al. 2013).

Since the phyllosphere experience wide exposure to the sunlight, the microbial communities present there can also utilize light as an additional source of energy by the expression of bacteriorhodopsins. Reports confirm the presence of aerobic anoxygenic photoautotrophs in the phyllomicrobiome (Atamna-Ismaeel et al. 2012a). The metagenomic study of five different phyllomicrobiomes (rice, soybean, tobacco, tamarix, clover) revealed the presence of homologs of gene *bch*Y and *puf*M, which code for the Y subunit of chlorophyllide reductase and the M subunit of the photosynthetic reaction center, respectively (Atamna-Ismaeel et al. 2012b).

5.6 Interaction of Phyllomicrobiota with Plants

The phyllomicrobiota exhibits diverse biotic interactions like antagonism, parasitism, commensalism, and mutualism with plants (Vorholt 2012). As these relationships are complex, thus, a clear demarcation between these interactions, particularly commensal, and mutualistic symbiont, is difficult. So broadly, we can categorize the interactions as positive or negative depending on the effect of microbiome communities on the host plant.

The common benefit that the host plant provides to the associated microbial communities is the continuous supply of nutrients and habitat for their growth. On the other hand, the effect of microbial communities on the host plant is highly variable.

5.6.1 Positive Interactions

The plant mainly benefits from its microflora as they produce mimics of plant hormones like cytokinins and indole acetic acid (Kutschera 2007). These hormones result in increased longevity of leaves (Jones and Dangl 2006). This trait is also exploited by the pathogenic microbes that sustain on leaves for a longer time. Similar to rhizosphere bacteria, some phyllosphere nitrogen-fixing organisms like cyanobacteria help the plants in providing nitrogen to the leaves (Lambais et al. 2006; Moyes et al. 2016; Rigano et al. 2007). The fungal communities on the leaves help them to survive in a water-scarce environment by facilitating foliar water uptake. The fungal hyphae grow into the stomatal pores and form a direct pathway for water uptake. The growth of fungi also decreases the surface tension on leaves (Rosado et al. 2018).

Another important contribution of the phyllomicrobiome is that they protect the plants against pathogens by their antagonism toward pathogenic organisms. Various reports have supported the idea of plant protection by the phyllosphere bacteria. Strains of *Sphingomonas* have been reported to suppress the growth of pathogenic *P. syringae* pv. syringae DC3000 (causing bacterial speck disease) on leaves of *A. thaliana* plant (Innerebner et al. 2011). This protective feature was found to be restricted only to *Sphingomonas* isolated from the plant's environment. The same organism from the non-host plant lacked this capability signifying that the residents of the phyllosphere modify themselves to favor the host. Reports confirm that colonization by pathogenic *E. amylovora* (fire blight disease in pear and apple) is inhibited by prior colonization of stigma in flowers by nonpathogenic bacteria such as *Pseudomonas fluorescens* strain A506 and *P. agglomerans* (Pusey 2002). Another evidence of pathogen exclusion was shown for leaf spot disease of lettuce, where the population of causative agent *X. campestris* was found to be negatively correlated with the phyllosphere bacterial population of *Bacillus, Erwinia,* and *Pantoea* (Rastogi et al. 2012). In another study of a lettuce plant, it was reported that bacterium *Enterobacter asburiae* negatively impacted the survival of *Escherichia coli* O157:H7 (an enteric pathogen) (Cooley et al. 2006). Recently, Qin et al. (2019) studied the ecological function of phyllosphere communities of tobacco and found that genus *Pantoea* and *Sphingomonas* suppress the wildfire disease in tobacco (Qin et al. 2019). The molecular basis of such cases of pathogenic exclusion is not yet clear, but the state of plant protection arises from an interplay of several traits. Plant phyllomicrobiome creates a competition of space and nutrients among the microbial communities. Besides, the phyllosphere microbes also produce an antimicrobial substance that interferes with the cell signaling systems to antagonize the pathogens. They also induce systemic resistance in the plants and protect the plants.

One of the most significant phyllosphere services to the plant and the environment is its role in bioremediation (phylloremediation). The phyllosphere bacteria are known to metabolize and degrade toxic compounds like nicotine, phenol, polycyclic aromatic hydrocarbons, which are produced by car exhausts. These compounds are harmful to plants and humans (Bringel and Couée 2015).

5.6.2 Negative Interactions

Along with beneficial residents, the phyllosphere also harbors some phytopathogens *viz. P. syringae, X. oryzae, Ustilago maydis, Sclerotinia sclerotiorum*. These phyllosphere bacteria are not directly pathogenic but secrete certain phytotoxins that make the plant susceptible to invasion of other pathogenic bacteria or viruses. For example, *P. syringae* produces a toxin coronatine that mimics the plant hormone jasmonic acid isoleucine that favors stomatal opening, thus facilitating the entry of pathogenic bacteria (Zheng et al. 2012). Some phyllosphere-associated bacteria like *Stenotrophomonas, Pseudomonas,* and *Enterobacter* are reported to influence the defense of plants against insect defoliators and support the growth of colorado potato beetle larvae (Chung et al. 2013).

Another adverse effect of the phyllomicrobiome on plants is increased water loss. The phyllosphere microbes are reported to enhance the rate of cuticular transpiration (Schreiber et al. 2005). The leaf associated fungal hyphae also lead to water loss by providing a pathway for water bypassing stomatal opening (Rosado et al. 2018).

5.7 Significance of Phyllomicrobiome Studies

Phyllosphere represents an ecologically and economically important ecosystem that is amenable for manipulation. Thus, it is regarded as a well-suited model to study the principles of ecology (Vorholt 2012). The study of the phyllosphere is also significant as these microbes play an important role in deciding plant health and ensure food safety to humans. It is often said that "plants wear their guts to the outside" and as our gut flora provides us the protection against various diseases; similarly, the microbes in the phyllosphere act as biological control agent and provide protection to plants by using the principle of pathogen exclusion (Andrews and Harris 2000). The knowledge about the indigenous microbial population of the phyllosphere will help us to design bio-strategies for plant protection. Various reports are available to support the biocontrol potential of phyllosphere microorganisms. In maize phyllosphere, microbes like *Bacillus,* and *Pantoea* inhibit the growth of leaf blight causing fungi, *Exserohilum turcicum* (Sartoria et al. 2015). A significant negative correlation was found between the population of these two bacteria and the growth of the pathogenic fungi. In the apple plant, *Pseudomonas graminis* protected the plant against *E. amylovora,* which causes fire blight disease (Mikiciński et al. 2016). A study on rice phyllosphere reported that the phyllosphere *Bacillus* produces antifungal secondary metabolites to inhibit the growth of fungi *Rhizoctonia solani* and *Burkholderia glumae* that causes sheath blight and panicle blight disease in rice respectively (Shrestha et al. 2016). The phyllomicrobiome enzymes like nonribosomal peptide synthetase and type 1 polyketide synthase gene are reported to be involved in the production of bioactive compounds that inhibit the rice blast disease caused by *Pyricularia oryzae* (Harsonowati et al. 2017). Some bacteria use siderophore for their biocontrol potential. *Pseudomonas protegens* CS1 prevent citrus canker in lemon by generating ROS and using siderophore pyochelin (Michavila et al. 2017).

Understanding phyllomicrobiome is also crucial in the context of food safety for humans. The knowledge about phyllosphere organisms and their interactions with pathogenic microbes in fruits and vegetables will help to control disease outbreaks related to the consumption of the infected food products (Rastogi et al. 2012). Studies revealed that major foodborne pathogen *E. coli* O157:H7 (Shiga toxin-producing) on lettuce leaves can be controlled by another epiphytic bacteria *E. asburiae* (Cooley et al. 2006). Members of Firmicutes and Enterobacteriaceae family also hamper the growth of *E. coli* O157:H7 on spinach leaves (Lopez-Velasco et al. 2011). Newton et al. (2010) suggested that manipulating the phyllomicrobiome using agrochemicals and plant-breeding techniques can ensure food safety of fresh produce.

The epiphytic bacteria are also important agents for phytoremediation. Many harmful organic compounds like phenol, isoprene, nicotine, aromatic hydrocarbons present in vehicle exhausts (acenaphthylene, acenaphthene, fluorene, phenanthrene) are metabolized by microbial communities of phyllosphere (Bringel and Couée 2015). *Arthrobacter* present in the phyllosphere of a variety of plants can degrade aromatic hydrocarbons and pesticides (Scheublin and Leveau 2013). The comparative transcriptome profiling of this organism from leaves of the *Phaseolus vulgaris* (common bean) with nonphyllospheric strains revealed that gene *cph* responsible for degrading chlorophenol is induced in the phyllosphere environment (Scheublin et al. 2014). Thus, the phyllosphere help plants in pollution mitigation. A detailed study in this field will help us to tailor the phyllomicrobiota of trees with microbes that can minimize the extent of the harmful products from our environment.

5.8 Conclusion and Future Prospects

Phyllosphere represents a vital microbial habitat with vast diversity across the globe. An insight into phyllomicrobiome with respect to the structure and function of the microbiota present, their adaptations, and interaction of microbes within this hostile environment is of utmost importance considering their extend and ecological impact. Although phyllosphere was earlier a neglected milieu, the use of high-throughput NGS techniques and omics approaches has opened new avenues in phyllosphere microbiology. A rapid increase in phyllomicrobiome studies has enhanced our knowledge about various aspects, particularly their survival, adaptation strategies, and interactions. Advanced techniques like community-based profiling and metaproteogenomics have further widened our understanding of phyllomicrobiome communities and epiphytic-fitness traits.

However, the knowledge is still in its infancy as compared to rhizosphere research. There are many questions that still remain unanswered; what is the source of phyllomicrobiota? How does the plant genome decide the structure of the phyllosphere microbial community? What is the nature of foliar exudates and how they interact with microbes present on leaves? The phyllosphere composition of major crop plants also needs to be revealed. The dynamics of multipartite interaction between plant–microbes, microbes–microbe, and microbe–environment need a better understanding. The spatial and temporal effect on plant phyllosphere communities also needs to be deciphered. Further research addressing these questions will broaden our paradigm on plant–microbial communities and their practical applications. A better understanding of the phyllomicrobiome and their interactions at the microscale and in-situ levels using complementary approaches will help us to design synthetic communities for their immense biotechnological applications like developing pathogen-resistant plants with improved biomass. The role of phyllosphere microorganism in sustainable agriculture, pollution mitigation, and climate change are other intriguing areas of research.

References

Andrews, J.H. and Harris, R.F. (2000). The ecology and biogeography of microorganisms on plant surfaces. *Annu. Rev. Phytopathol.* 38: 145–180.

Andrews, J.H., Spear, R.N., and Nordheim, E.V. (2002). Population biology of *Aureobasidium pullulans* on apple leaf surfaces. *Can. J. Microbiol.* 48 (6): 500–513.

Atamna-Ismaeel, N., Finkel, O., Glaser, F. et al. (2012a). Bacterial anoxygenic photosynthesis on plant leaf surfaces. *Environ. Microbiol. Rep.* 4 (2): 209–216.

Atamna-Ismaeel, N., Finkel, O., Glaser, F. et al. (2012b). Microbial rhodopsins on leaf surfaces of terrestrial plants. *Environ. Microbiol.* 14 (1): 140–146.

Bainbridge, A. and Dickinson, C.H. (1972). Effect of fungicides on the microflora of potato leaves. *Trans. Br. Mycol. Soc.* 59 (1): 31–41.

Baldotto, L.E.B. and Olivares, F.L. (2008). Phylloepiphytic interaction between bacteria and different plant species in a tropical agricultural system. *Can. J. Microbiol.* 54 (11): 918–931.

Balint, M., Tiffin, P., Olson, M.S. et al. (2013). Host genotype shapes the foliar fungal microbiome of Balsam Poplar (*Populus balsamifera*). *PLoS One* 8 (1): e53987.

Balint-Kurti, P., Simmons, S.J., Blum, J.E. et al. (2010). Maize leaf epiphytic bacteria diversity patterns are genetically correlated with resistance to fungal pathogen infection. *Mol. Plant Microbe Interact.* 23 (4): 473–484.

Boch, J., Joardar, V., Gao, L. et al. (2002). Identification of *Pseudomonas syringae* pv. tomato genes induced during infection of *Arabidopsis thaliana*. *Mol. Microbiol.* 44 (1): 73–88.

Bringel, F. and Couée, I. (2015). Pivotal roles of phyllosphere microorganisms at the interface between plant functioning and atmospheric trace gas dynamics. *Front. Microbiol.* 6: 1–14.

Bulgarelli, D., Schlaeppi, K., Spaepen, S. et al. (2013). Structure and functions of the bacterial microbiota of plants. *Annu. Rev. Plant Biol.* 64 (1): 807–838.

Chen, C. and Beattie, G.A. (2008). *Pseudomonas syringae* BetT is a low-affinity choline transporter that is responsible for superior osmoprotection by choline over glycine betaine. *J. Bacteriol.* 190 (8): 2717–2725.

Chung, S.H., Rosa, C., Scully, E.D. et al. (2013). Herbivore exploits orally secreted bacteria to suppress plant defenses. *Proc. Natl. Acad. Sci.* 110 (39): 15728–15733.

Cooley, M.B., Chao, D., and Mandrell, R.E. (2006). *Escherichia coli* O157:H7 survival and growth on lettuce is altered by the presence of epiphytic bacteria. *J. Food Prot.* 69 (10): 2329–2335.

Cordier, T., Robin, C., Capdevielle, X. et al. (2012). The composition of phyllosphere fungal assemblages of European beech (*Fagus sylvatica*) varies significantly along an elevation gradient. *New Phytol.* 196 (2): 510–519.

Darlison, J., Mogren, L., Rosberg, A.K. et al. (2019). Leaf mineral content govern microbial community structure in the phyllosphere of spinach (*Spinacia oleracea*) and rocket (*Diplotaxis tenuifolia*). *Sci. Total Environ.* 675: 501–512.

Das, A., Rangaraj, N., and Sonti, R.V. (2009). Multiple adhesin-like functions of *Xanthomonas oryzae* pv. Oryzae are involved in promoting leaf attachment, entry, and virulence on rice. *Mol. Plant Microbe Interact.* 22 (1): 73–85.

Dees, M.W., Lysøe, E., Nordskog, B. et al. (2015). Bacterial communities associated with surfaces of leafy greens: shift in composition and decrease in richness over time. *Appl. Environ. Microbiol.* 81 (4): 1530–1539.

Degnan, P.H. and Ochman, H. (2012). Illumina-based analysis of microbial community diversity. *ISME J.* 6 (1): 183–194.

Delmotte, N., Knief, C., Chaffron, S. et al. (2009). Community proteogenomics reveals insights into the physiology of phyllosphere bacteria. *Proc. Natl. Acad. Sci.* 106 (38): 16428–16433.

Dickinson, C.H., Watson, J., and Wallace, B. (1974). An impression method for examining epiphytic microorganisms and its application to phylloplane studies. *Trans. Br. Mycol. Soc.* 63 (3): 616–619.

Enya, J., Shinohara, H., Yoshida, S. et al. (2007). Culturable leaf-associated bacteria on tomato plants and their potential as biological control agents. *Microb. Ecol.* 53 (4): 524–536.

Ercolini, D. (2004). PCR-DGGE fingerprinting: novel strategies for detection of microbes in food. *J. Microbiol. Methods* 56 (3): 297–314.

Fan, J., Doerner, P., and Lamb, C. (2011). *Pseudomonas* sax genes overcome non-host resistance in Arabidopsis. *Science* 1185: 1185–1188.

Finkel, O.M., Burch, A.Y., Lindow, S.E. et al. (2011). Geographical location determines the population structure in phyllosphere microbial communities of a salt-excreting desert tree. *Appl. Environ. Microbiol.* 77 (21): 7647–7655.

Freeman, B.C., Chen, C., and Beattie, G.A. (2010). Identification of the trehalose biosynthetic loci of *Pseudomonas syringae* and their contribution to fitness in the phyllosphere. *Environ. Microbiol.* 12 (6): 1486–1497.

Frey-Klett, P., Burlinson, P., Deveau, A. et al. (2011). Bacterial-fungal interactions: hyphens between agricultural, clinical, environmental, and food microbiologists. *Microbiol. Mol. Biol. Rev.* 75 (4): 583–609.

Fürnkranz, M., Wanek, W., Richter, A. et al. (2008). Nitrogen fixation by phyllosphere bacteria associated with higher plants and their colonizing epiphytes of a tropical lowland rainforest of Costa Rica. *ISME J.* 2 (5): 561–570.

Gourion, B., Rossignol, M., and Vorholt, J.A. (2006). A proteomic study of *Methylobacterium extorquens* reveals a response regulator essential for epiphytic growth. *Proc. Natl. Acad. Sci.* 103 (35): 13186–13191.

Gravouil, C. (2012). Identification of the barley phyllosphere and characterization of manipulation means of the bacteriome against leaf scald and powdery mildew. Ph.D. Thesis. The University of Nottingham.

Grube, M., Schmid, F., and Berg, G. (2011). Black fungi and associated bacterial communities in the phyllosphere of grapevine. *Fungal Biol.* 115 (10): 978–986.

Gu, L., Bai, Z., Jin, B. et al. (2010). Assessing the impact of fungicide enostroburin application on bacterial community in wheat phyllosphere. *J. Environ. Sci.* 22 (1): 134–141.

Hallett, I.C., Boyd-Wilson, K.S.H., and Everett, K.R. (2010). Microscope methods for observation of the phylloplane flora. *N. Z. Plant Prot.* 63: 15–23.

Harsonowati, W., Astuti, R.I., and Wahyudi, A.T. (2017). Leaf blast disease reduction by rice-phyllosphere actinomycetes producing bioactive compounds. *J. Gen. Plant Pathol.* 83 (2): 98–108.

Horton, M.W., Natacha, B., Kathleen, B. et al. (2015). Genome-wide association study of *Arabidopsis thaliana*'s leaf microbial community. *Encycl. Astrobiol.* 1561–1561.

Hunter, P.J., Hand, P., Pink, D. et al. (2010). Both leaf properties and microbe-microbe interactions influence within-species variation in bacterial population diversity and structure in the lettuce (*Lactuca* species) phyllosphere. *Appl. Environ. Microbiol.* 76 (24): 8117–8125.

Hunter, P.J., Pink, D.A., and Bending, G.D. (2015). Cultivar-level genotype differences influence diversity and composition of lettuce (Lactuca sp.) phyllosphere fungal communities. *Fungal Ecol.* 17: 183–186.

Innerebner, G., Knief, C., and Vorholt, J.A. (2011). Protection of *Arabidopsis thaliana* against leaf-pathogenic *Pseudomonas syringae* by *Sphingomonas* strains in a controlled model system. *Appl. Environ. Microbiol.* 77 (10): 3202–3210.

Izhaki, I., Fridman, S., Gerchman, Y., and Halpern, M. (2013). Variability of bacterial community composition on leaves between and within plant species. *Curr. Microbiol.* 66 (3): 227–235.

Jones, J.D. and Dangl, J.L. (2006). The plant immune system. *Nature* 444: 323–329.

Jumpponen, A. and Jones, K.L. (2009). Massively parallel 454 sequencing indicates hyperdiverse fungal communities in temperate Quercus macrocarpa phyllosphere. *New Phytol.* 184 (2): 438–448.

Kadivar, H. and Stapleton, A.E. (2003). Ultraviolet radiation alters maize phyllosphere bacterial diversity. *Microb. Ecol.* 45 (4): 353–361.

Karlsson, I., Friberg, H., Steinberg, C. et al. (2014). Fungicide effects on fungal community composition in the wheat phyllosphere. *PLoS One* 9 (11).

Kim, M., Singh, D., Lai-Hoe, A. et al. (2012). Distinctive phyllosphere bacterial communities in tropical trees. *Microb. Ecol.* 63 (3): 674–681.

Knief, C., Delmotte, N., Chaffron, S. et al. (2012). Metaproteogenomic analysis of microbial communities in the phyllosphere and rhizosphere of rice. *ISME J.* 6 (7): 1378–1390.

Knief, C., Ramette, A., Frances, L. et al. (2010). Site and plant species are important determinants of the *Methylobacterium* community composition in the plant phyllosphere. *ISME J.* 4 (6): 719–728.

Kolton, M., Sela, N., Elad, Y. et al. (2013). Comparative genomic analysis indicates that niche adaptation of terrestrial flavobacteria is strongly linked to plant glycan metabolism. *PLoS One* 8 (9): 1–11.

Kutschera, U. (2007). Plant-associated methylobacteria as co-evolved phytosymbionts: a hypothesis. *Plant Signal. Behav.* 2 (2): 74–78.

Lambais, M.R., Crowley, D.E., Cury, J.C. et al. (2006). Bacterial diversity in tree canopies of the Atlantic Forest. *Science* 312 (5782): 1917.

Langvad, F. (1980). A simple and rapid method for qualitative and quantitative study of the fungal flora of leaves. *Can. J. Microbiol.* 26 (6): 666–670.

Lee, O.H.K. and Hyde, K.D. (2002). Phylloplane fungi in Hong Kong mangroves: evaluation of study methods. *Mycologia* 94 (4): 596–606.

Leveau, J.H.J. and Tech, J.J. (2011). Grapevine microbiomics: bacterial diversity on grape leaves and berries revealed by high-throughput sequence analysis of 16S rRNA amplicons. *Acta Hortic.* 905: 31–42.

Levy, A., Salas Gonzalez, I., Mittelviefhaus, M. et al. (2018). Genomic features of bacterial adaptation to plants. *Nat. Genet.* 50 (1): 138–150.

Lindow, S.E. and Brandl, M.T. (2003). Microbiology of the Phyllosphere. *Appl. Environ. Microbiol.* 69 (4): 1875–1883.

Lopez-Velasco, G., Welbaum, G.E., Boyer, R.R. et al. (2011). Changes in spinach phylloepiphytic bacteria communities following minimal processing and refrigerated storage described using pyrosequencing of 16S rRNA amplicons. *J. Appl. Microbiol.* 110 (5): 1203–1214.

Lv, D., Ma, A., Bai, Z. et al. (2012). Response of leaf-associated bacterial communities to primary acyl-homoserine lactone in the tobacco phyllosphere. *Res. Microbiol.* 163 (2): 119–124.

Marco, M.L., Legac, J., and Lindow, S.E. (2005). *Pseudomonas syringae* genes induced during colonization of leaf surfaces. *Environ. Microbiol.* 7 (9): 1379–1391.

Melotto, M., Underwood, W., Koczan, J. et al. (2006). Plant stomata function in innate immunity against bacterial invasion. *Cell* 126 (5): 969–980.

Michavila, G., Adler, C., De Gregorio, P.R. et al. (2017). *Pseudomonas protegens* CS1 from the lemon phyllosphere as a candidate for citrus canker biocontrol agent. *Plant Biol.* 19 (4): 608–617.

Mikiciński, A., Sobiczewski, P., Puławska, J. et al. (2016). Control of fire blight (*Erwinia amylovora*) by a novel strain 49M of *Pseudomonas graminis* from the phyllosphere of apple (*Malus* sp.). *Eur. J. Plant Pathol.* 145 (2): 265–276.

Moyes, A.B., Kueppers, L.M., Pett-Ridge, J. et al. (2016). Evidence for foliar endophytic nitrogen fixation in a widely distributed subalpine conifer. *New Phytol.* 210 (2): 657–668.

Newton, A.C., Gravouil, C., and Fountaine, J.M. (2010). Managing the ecology of foliar pathogens: ecological tolerance in crops. *Ann. Appl. Biol.* 157 (3): 343–359.

Niwa, R., Yoshida, S., Furuya, N. et al. (2011). Method for simple and rapid enumeration of total epiphytic bacteria in the washing solution of rice plants. *Can. J. Microbiol.* 57 (1): 62–67.

Okinaka, Y., Yang, C.H., Perna, N.T. et al. (2002). Microarray profiling of *Erwinia chrysanthemi* 3937 genes that are regulated during plant infection. *Mol. Plant Microbe Interact.* 15 (7): 619–629.

Ottesen, A.R., Gorham, S., Reed, E. et al. (2016). Using a control to better understand phyllosphere microbiota. *PLoS One* 11 (9): 1–16.

Palumbo, J.D., Kado, C.I., and Phillips, D.A. (1998). An isoflavonoid-inducible efflux pump in Agrobacterium tumefaciens is involved in competitive colonization of roots. *J. Bacteriol.* 180 (12): 3107–3113.

Perazzolli, M., Antonielli, L., Storari, M. et al. (2014). Resilience of the natural phyllosphere microbiota of the grapevine to chemical and biological pesticides. *Appl. Environ. Microbiol.* 80 (12): 3585–3596.

Pusey, P.L. (2002). Biological control agents for fire blight of apple compared under conditions limiting natural dispersal. *Plant Dis.* 86 (6): 639–644.

Qin, C., Tao, J., Liu, T. et al. (2019). Responses of phyllosphere microbiota and plant health to application of two different biocontrol agents. *AMB Express* 9 (1): 42.

Rastogi, G., Coaker, G.L., and Leveau, J.H.J. (2013). New insights into the structure and function of phyllosphere microbiota through high-throughput molecular approaches. *FEMS Microbiol. Lett.* 348 (1): 1–10.

Rastogi, G., Sbodio, A., Tech, J.J. et al. (2012). Leaf microbiota in an agroecosystem: spatiotemporal variation in bacterial community composition on field-grown lettuce. *ISME J.* 6 (10): 1812–1822.

Redford, A., Bowers, R., Knight, R. et al. (2011). Variability in the distribution of bacteria on tree leaves. *Environ. Microbiol.* 12 (11): 2885–2893.

Remus-Emsermann, M.N.P. and Schlechter, R.O. (2018). Phyllosphere microbiology: at the interface between microbial individuals and the plant host. *New Phytol.* 218 (4): 1327–1333.

Remus-Emsermann, M.N.P., Kim, E.B., Marco, M.L. et al. (2013). Draft genome sequence of the phyllosphere model bacterium *Pantoea agglomerans* 299R. *Genome Announc.* 1 (1): 2012–2013.

Rigano, L.A., Siciliano, F., Enrique, R. et al. (2007). Biofilm formation, epiphytic fitness, and canker development in *Xanthomonas axonopodis* pv. *citri*. *Mol. Plant Microbe Interact.* 20 (10): 1222–1230.

Rojas, C.M., Ham, J.H., Deng, W.L. et al. (2002). HecA, a member of a class of adhesins produced by diverse pathogenic bacteria, contributes to the attachment, aggregation, epidermal cell killing, and virulence phenotypes of *Erwinia chrysanthemi* EC16 on *Nicotiana clevelandii* seedlings. *Proc. Natl. Acad. Sci.* 99 (20): 13142–13147.

Rosado, B.H.P., Almeida, L.C., Alves, L.F. et al. (2018). The importance of phyllosphere on plant functional ecology: a phyllo trait manifesto. *New Phytol.* 219 (4): 1145–1149.

Sartoria, M., Nescia, A., Formento, Á. et al. (2015). Selection of potential biological control of *Exserohilum turcicum* with epiphytic microorganisms from maize. *Rev. Argent. Microbiol.* 47 (1): 62–71.

Schellenberg, B., Ramel, C., and Dudler, R. (2010). *Pseudomonas syringae* virulence factor syringolin a counteracts stomatal immunity by proteasome inhibition. *Mol. Plant Microbe Interact.* 23 (10): 1287–1293.

Scheublin, T.R. and Leveau, J.H.J. (2013). Isolation of *Arthrobacter* species from the phyllosphere and demonstration of their epiphytic fitness. *Microbiol. Open.* 2 (1): 205–213.

Scheublin, T.R., Deusch, S., Moreno-Forero, S.K. et al. (2014). Transcriptional profiling of Gram-positive *Arthrobacter* in the phyllosphere: induction of pollutant degradation genes by natural plant phenolic compounds. *Environ. Microbiol.* 16 (7): 2212–2225.

Schreiber, L., Krimm, U., Knoll, D. et al. (2005). Plant-microbe interactions: identification of epiphytic bacteria and their ability to alter leaf surface permeability. *New Phytol.* 166 (2): 589–594.

Shrestha, B.K., Karki, H.S., Groth, D.E. et al. (2016). Biological control activities of rice-associated *Bacillus* sp. strains against sheath blight and bacterial panicle blight of rice. *PLoS One* 11 (1): 1–18.

Stoitsova, S.O., Braun, Y., Ullrich, M.S. et al. (2008). Characterization of the RND-type multidrug efflux pump MexAB-OprM of the plant pathogen *Pseudomonas syringae*. *Appl. Environ. Microbiol.* 74 (11): 3387–3393.

Suda, W., Nagasaki, A., and Shishido, M. (2009). Powdery mildew-infection changes bacterial community composition in the phyllosphere. *Microbes Environ.* 24 (3): 217–223.

Sundin, G.W. and Jacobs, J.L. (1999). Ultraviolet radiation (UVR) sensitivity analysis and UVR survival strategies of a bacterial community from the phyllosphere of field-grown peanut (*Arachis hypogeae* L.). *Microb. Ecol.* 38 (1): 27–38.

Sy, A., Timmers, A.C.J., Knief, C. et al. (2005). Methylotrophic metabolism is advantageous for *Methylobacterium extorquens* during colonization of *Medicago truncatula* under competitive conditions. *Appl. Environ. Microbiol.* 71 (11): 7245–7252.

Tamir-Ariel, D., Navon, N., and Burdman, S. (2007). Identification of genes in *Xanthomonas campestris* pv. vesicatoria induced during its interaction with tomato. *J. Bacteriol.* 189 (17): 6359–6371.

Thapa, S. and Prasanna, R. (2018). Prospecting the characteristics and significance of the phyllosphere microbiome. *Ann. Microbiol.* 68 (5): 229–245.

Thapa, S., Ranjan, K., Ramakrishnan, B. et al. (2018). Influence of fertilizers and rice cultivation methods on the abundance and diversity of phyllosphere microbiome. *J Basic Microbiol.* 58 (2): 172–186.

Venkatachalam, S., Ranjan, K., Prasanna, R. et al. (2016). Diversity and functional traits of culturable microbiome members, including cyanobacteria in the rice phyllosphere. *Plant Biol. (Stuttg.)* 18 (4): 627–637.

Vokou, D., Vareli, K., Zarali, E. et al. (2012). Exploring biodiversity in the bacterial community of the mediterranean phyllosphere and its relationship with airborne bacteria. *Microb. Ecol.* 64 (3): 714–724.

Vorholt, J.A. (2012). Microbial life in the phyllosphere. *Nat. Rev. Microbiol.* 10 (12): 828–840.

Wagner, M., Rudolf, A., Hilde, L. et al. (1993). Probing activated sludge with oligonucleotides specific for Proteobacteria: inadequacy of culture-dependent methods for describing microbial community structure. *Appl. Environ. Microbiol.* 59 (5): 1520–1525.

Whipps, J.M., Hand, P., Pink, D. et al. (2008). Phyllosphere microbiology with special reference to diversity and plant genotype. *J. Appl. Microbiol.* 105 (6): 1744–1755.

Williams, T.R. and Marco, M.L. (2014). Transplantation on lettuce plants grown indoors. *MBio* 5 (4): e01564–e01514.

Williams, T.R., Moyne, A.L., Harris, L.J. et al. (2013). Season, irrigation, leaf age, and *Escherichia coli* inoculation influence the bacterial diversity in the lettuce phyllosphere. *PLoS One* 8 (7): 1–14.

Yang, S., Perna, N.T., Cooksey, D.A. et al. (2004). Genome-wide identification of plant-upregulated genes of *Erwinia chrysanthemi* 3937 using a GFP-based IVET leaf array. *Mol. Plant Microbe Interact.* 17 (9): 999–1008.

Yashiro, E., Spear, R.N., and Mcmanus, P.S. (2011). Culture-dependent and culture-independent assessment of bacteria in the apple phyllosphere. *J. Appl. Microbiol.* 110 (5): 1284–1296.

Zheng, X.Y., Natalie, W.S., Weiqing, Z. et al. (2012). Coronatine promotes *Pseudomonas syringae* virulence in plants by activating a signaling cascade that inhibits salicylic acid accumulation. *Cell Host Microbe* 23 (1): 1–7.

6

Rhizosphere Engineering

An Effective Approach for Sustainable Modern Agriculture

Reema Mishra[1], Tripti Grover[1], Pooja Gulati[2], and Aparajita Mohanty[1]

[1] *Gargi College, University of Delhi, New Delhi, India*
[2] *Maharshi Dayanand University, Rohtak, Haryana, India*

6.1 Introduction

The word "rhizosphere" was coined by the German scientist, Loernz Hiltner, in 1904 (Hiltner 1904). Its origin is from the Greek word "rhiza" that refers to the root and "sphere" meaning field of influence. According to his definition, the rhizosphere is a region of soil around the roots of plants with high microbial colonization and activity, influenced by root exudation. It is the region where positive, negative, or neutral interaction between plants, soil, and microbes takes place. These interactions are very important for the growth and productivity of plants. Rhizosphere consists of three zones, namely innermost endo-rhizosphere (consisting of root cortex and endodermis), middle rhizoplane (including root epidermal cells and mucilage) and outermost ecto-rhizosphere which does not have a definable shape and size.

The microbial communities inhabiting the rhizosphere include fungi, bacteria, viruses, algae, protozoa, arthropods, archaea, etc. (Mendes et al. 2013). The associated microorganisms regulate plant productivity by enhancing nutrient uptake, conferring stress (abiotic and biotic) tolerance/resistance, secreting growth-promoting compounds, and by inducing a systemic immune response (Berg 2009; Babalola 2010; Zamioudis and Pieterse 2012; Mendes et al. 2013). Plants serve as store house of the beneficial plant growth-promoting bacteria, also named as plant growth-promoting rhizobacteria (PGPR). Plants also release nutrient-rich exudates like carbohydrates, amino acids, sugars, hormones, volatile organic compounds (VOCs), enzymes, etc., which are advantageous for the growth of different microorganisms and plant–microbe interactions in the rhizosphere as well (Igiehon and Babalola 2018; Schirawski and Perlin 2018). Thus, the rhizosphere has been regarded as a versatile ecosystem that can be engineered to increase plant productivity. Different components of the rhizosphere have been engineered for sustainable plant yield. One of the strategies includes modifying physiochemical properties of soil, thus making it favorable for plant and microbial growth (Bradford et al. 2008; Beesley et al. 2011;

Phytomicrobiome Interactions and Sustainable Agriculture, First Edition. Edited by Amit Verma, Jitendra Kumar Saini, Abd El-Latif Hesham, and Harikesh Bahadur Singh.

Singh et al. 2012; Shaheen et al. 2014; Dessaux et al. 2016). The other component that has been manipulated is the plant itself. Selection of certain plant species or cultivar with improved root exudate secretion is beneficial for attracting useful microorganisms which in turn can influence plant health (Bakker et al. 2012; Ellouze et al. 2013). Plants have also been engineered to amend root exudation, efflux of ions, root architecture, and soil pH for improving nutrient availability and encouraging the growth of microbes (Ryan et al. 2009; Quiza et al. 2015; Dessaux et al. 2016). Plant-metabolic engineering has been employed for increasing plant productivity by manipulating exudation of semiochemicals (signaling compounds) like strigolactones (SLs), flavonoids, terpenoids (β-caryophyllene) (Zhang et al. 2015b). Rhizobacteria which confers abiotic and biotic stress resistance to the host plants have also been engineered to increase plant productivity (Ryan et al. 2009; Quiza et al. 2015; Dessaux et al. 2016). Another strategy involves the engineering of microbial population to enhance robust plant growth (Dessaux et al. 2016). The components of the rhizosphere that can be engineered require a deep understanding of the molecular mechanisms involved in plant–microbe interactions in the rhizosphere. This chapter reviews the different approaches used to engineer the rhizosphere along with the molecular mechanisms. Further, tools and techniques used to study and implement rhizosphere engineering for improving plant growth and productivity have been discussed.

6.2 Natural Plant–Microbe Interactions in Rhizosphere

For sustainable agriculture, the study of interactions between plants and microbes in nature is essential. A variety of microorganisms inhabit the plant rhizosphere. Different communities of plant-associated microorganisms have been termed as "plant microbiome" or "phytomicrobiome." The association between microorganisms (consisting of viruses, fungi, bacteria, algae, protozoa, and archaea) and plants can be advantageous, harmful, or neutral. It is the positive plant–microbe interactions that can be exploited to improve crop production.

The beneficial rhizosphere microbes collectively named as PGPR have positive effects on plants and regulate plant growth, development, and productivity by alleviating biotic (disease) as well as abiotic (heat, drought, salt, heavy metal) stress. Many nitrogen-fixing bacteria (e.g. *Rhizobium* sp.) and mycorrhizal fungi impact plants by improving nutrient uptake from soil (Berg 2009; Mendes et al. 2013). Microorganisms also influence plant growth by releasing plant growth regulators (auxins, gibberellins), vitamins, siderophores, antibiotics (Diacetylphloroglucinol), volatile compounds, and growth promoters (Babalola 2010). Rhizospheric microorganisms can induce systemic immune response also known as induced systemic resistance (ISR) in different plants and thus, confer resistance against pathogens by microbial antagonism (Berg 2009; Zamioudis and Pieterse 2012).

Plants serve as habitat and nutrient stores for different microorganisms (Schirawski and Perlin 2018). Roots of plants secrete different nutrient-rich exudates that are beneficial for the growth of microbes. These exudates consist of monosaccharides, polysaccharides, amino acids, organic acids, flavanols, fatty acids, lignin, tannins, vitamins, VOCs, and alkaloids which modify the composition of microbial populations and physicochemical properties of soil for facilitating symbiotic interactions (De-la-Pena et al. 2012; Igiehon and

Babalola 2018). The quality and quantity of root exudates are dependent on different factors like plant species or cultivar, plant developmental stage, different types of soil and its pH, environmental conditions like temperature and occurrence of different microbes. These exudates act as environmental cues that regulate microbial genes required for detoxification, cell mobility, chemotaxis, and biofilm formation (Zhang et al. 2015a).

6.3 Molecular Mechanisms in Plant–Microbe Interactions in Rhizosphere

Various molecular mechanisms harmonize the interactions between plants and rhizomicrobiome microorganisms to increase plant productivity (De-la-Pena et al. 2012; Quiza et al. 2015). Identification of genes regulating the plant-PGPR association plays an important role in elucidating the mechanism of interactions as well as in engineering novel microbial strains for improving plant growth and productivity.

The association of the same plant with different PGPRs can alter the expression of different genes. *Arabidopsis thaliana* colonized with PGPR (*Paenibacillus polymyxa)* enhanced the expression of drought stress-induced gene (*ERD15)*, and the hormone-responsive genes for jasmonic acid (JA), salicylic acid (SA) and ethylene (ET) (Timmusk and Wagner 1999). Transcriptome profiling of *A. thaliana* roots inoculated with PGPR (*Pseudomonas fluorescens* strain FPT9601-T5) showed upregulation of signal transduction, auxin-induced, plant metabolism, nodulin like, and stress-responsive genes (Wang et al. 2005). Expression profiling of *A. thaliana* plants exposed to *Pseudomonas putida* (strain MTCC5279) showed upregulation of genes required for amino acid and hormone synthesis, ethylene suppression, Ca^{2+} dependent signaling and activation of abscisic acid (ABA) signaling (Srivastava et al. 2012). An exposure of *Bacillus megaterium* (BP17 strain) to *A. thaliana* plants exhibited better shoot and root length. Expression profiling of infected plants showed upregulation of genes involved in nutrient transport, cell organization, transcription, biotic, and abiotic stress tolerance, and downregulation of hormone- (ET, SA, and JA) responsive genes (Vibhuti et al. 2017). In another report, *A. thaliana* plants inoculated with PGPR mixture showed increased fresh weight, nutrient uptake, plant diameter, chlorophyll content, and increased expression of genes involved in ammonium and nitrate uptake (Calvo et al. 2019).

RNA-sequencing (RNA seq) analysis of sugarcane cv. SP70-1143 plants exposed to nitrogen-fixing bacteria *Gluconacetobacter diazotrophicus* (strain PAL5) showed ABA signaling genes responsible for conferring drought resistance (Vargas et al. 2014). Tobacco plants inoculated with VOC secreting PGPR (*Bacillus subtilis,* strain JS) showed upregulation of genes involved in photosynthetic pathways and downregulation of genes coding for enzymes involved in reactive oxygen species (ROS) scavenging (Kim et al. 2015). Chickpea colonized with different *Streptomyces* species exhibited ISR and alleviated oxidative stress against *Sclerotium rolfsii.* The inoculated plants showed increased phenolic buildup and transcript levels of defense-related genes (Singh and Gaur 2017).

All these reports suggest that the identification and characterization of responsive genes involved in the plant–microbe interaction will be instrumental in explicating the association mechanism and can be exploited for rhizosphere engineering.

6.4 Biochemical Components in Rhizosphere Signaling

The interaction between plants and microorganisms depends mainly on the different signaling molecules secreted by both microorganisms and plants. The interactions may be direct where physical contact is established or indirect where direct contact is not necessary. The interactions facilitated by the secretion of signaling molecules are further classified as plant–microbe, microbe–plant, plant–plant, microbe–microbe, etc. (De-la-Pena et al. 2012).

Arrays of signaling molecules (phytochemicals) released by different parts of the plant are important for the positive, negative, or neutral interactions with microbes. They also regulate gene expression, activity, and growth of microbes (Venturi and Keel 2016). There are only a few classes of phytochemicals known to affect the microbes that have been characterized (Table 6.1). The best signaling network elucidated between plants and microbes is the symbiotic association between nitrogen-fixing rhizobacteria and legumes.

Rhizobacteria also secrete beneficial signaling molecules for establishing mutualistic interaction with plants. This inter-kingdom signaling regulates the expression of signaling genes resulting in improved plant productivity. The most common signaling molecules are extracellular enzymes, antibiotics, organic acids, and microbe-associated molecular patterns (MAMPs) like flagellin, chitin, lipopolysaccharide, and VOCs (Quiza et al. 2015). The phytohormone auxin synthesized by certain PGPRs has also been shown to modulate plant growth and development (Spaepen et al. 2007). The important signaling molecules synthesized by microorganisms and their roles are listed in Table 6.2.

Many bacteria also communicate with each other using the mechanism of quorum sensing (QS) for regulating coordinated gene expression and their population density (Miller and Bassler 2001). The most common QS signals used by different bacterial species (*Ralstonia*, *Pseudomonas*, *Erwinia*) are N-acyl homoserine lactones (AHLs) which govern different features of bacteria like virulence, sporulation, antibiotic production, biofilm formation, and conjugation (Fuqua et al. 2001). QS signaling molecules like diffusible

Table 6.1 Signaling molecules secreted by plants for establishing plant–microbe interaction.

Signaling molecule	Effect on microbial species	References
Flavonoids	Induction of bacterial *NOD* genes coding for lipo-chitooligosaccharides (LCOs) responsible for nodule formation in plants	Limpens et al. (2015)
	Stimulate colonization of plant root, growth of hyphae and germination of arbuscular mycorrhizal fungus (AMF) spore	Mandal et al. (2010)
Strigolactone, 5-deoxy-strigol	Hyphal branching in AMF *Gigaspora margarita*	Akiyama et al. (2005)
Cutin monomers	Stimulate colonization by AMF	Wang et al. (2012b)
Organic acids	Biofilm formation in *Bacillus amyloliquefaciens* SQR9 and *Bacillus subtilis* N11	Zhang et al. (2014)

Table 6.2 Signaling molecules secreted by microbes for establishing plant-microbe interaction.

Signaling molecule	Microbial species	Effect on Plant species	Reference
Small secreted proteins (SSPs)	MiSSP7 protein from *Laccaria bicolor*	JA signaling in *Populus* sp.	Plett et al. (2014)
AHLs	*Pseudomonas putida*, *Serratia liquefaciens*	Induce systemic resistance in tomato	Schuhegger et al. (2006)
	Mesorhizobium tianshanense	Nodulation in *Glycyrrhiza uralensis*	Zheng et al. (2006)
	Sinorhizobium meliloti	Increases nodulation in *Medicago truncatula*	Veliz-Vallejos et al. (2014)
VOCs	*Pseudomonas chlororaphis* O6	Triggers ISR in tobacco against *Erwinia carotovora*	Han et al. (2006)
	Bacillus subtilis, B. amyloliquefaciens	Elicits ISR in *A. thaliana*	Ryu et al. (2004)
	Bacillus B55 strain	Improves sulfur nutrition in *Nicotiana attenuata*	Hofmann (2013)
	Pseudomonas sp., *Burkholderia* sp.	Promotes root development in *A. thaliana*	Bailly et al. (2014)
MAMPs	*Pseudomonas aeruginosa* (rhamnolipids)	Resistance to grapevine against *Botrytis cinerea*	Varnier et al. (2009)
	Pseudomonas fluorescens, Stenotrophomonas maltophilia, Chryseobacterium balustinum	Trigger germination and plant metabolism in *Papaver somniferum*	Bonilla et al. (2014)
Antibiotics	2,4-DAPG from *Pseudomonas fluorescens*	Triggers ISR in *A. thaliana*	Weller et al. (2012)
Lipopeptides	*Bacillus* spp	Triggers ISR in *Phaseolus vulgaris*	Ongena et al. (2007)
	Bacillus amyloliquefaciens sub sp. *plantarum*	Improves plant defense in *Lactuca sativa*	Chowdhury et al. (2015)
Siderophores	*Bacillus subtilis* (strain CTS-G24)	Suppresses disease in *Cicer arietinum*	Patil et al. (2014)

signal factor (DSF), AHLs, dialkylresorcinols, and pyrones secreted by gram-negative bacteria are important for regulating the growth of other microorganisms (Brameyer et al. 2015). Other signaling molecules including antibiotics, pheromones (peptides), VOCs (alcohols, aldehydes, alkenes, terpenes) are also important for modulating different microbial communities present in the rhizosphere (Andersson and Hughes 2014; Monnet et al. 2016).

6.5 Tools and Techniques in Rhizosphere Engineering

Technological advancements have facilitated a deeper understanding of the existing plant–microbial interaction in the rhizosphere and have furthered its appropriate manipulation. Rhizosphere engineering encompasses tools and techniques used for the identification of rhizomicrobiome diversity and for understanding various aspects of plant–microbial interactions in the rhizosphere (Figure 6.1). Over the years, the traditional three-step approach for studies on microbial diversity, involved isolation of microbial DNA, PCR amplification of the conserved regions followed by characterization. Techniques like denaturing gradient gel electrophoresis (DGGE), sequencing, and terminalrestriction fragment length polymorphism (T-RFLP) for characterization have now been taken over by more systematic approaches. Nucleic acid-based methods like stable isotope probing (SIP) and fluorescent in-situ hybridization of mRNA (mRNA-FISH) have helped uncover activities of the microbial community by isotopic labeling of the microbial community and characterizing the tagged cells. Some of the most relevant approaches are discussed further.

6.5.1 Stable Isotope Probing (SIP)

SIP as the name suggests utilizes stable isotopes such as ^{13}C for labeling substrates that are assimilated by the microbes for their growth. This labeling results in the biosynthesis of ^{13}C-labeled biomarkers (DNA or RNA) that differ in density from the unlabeled ones and

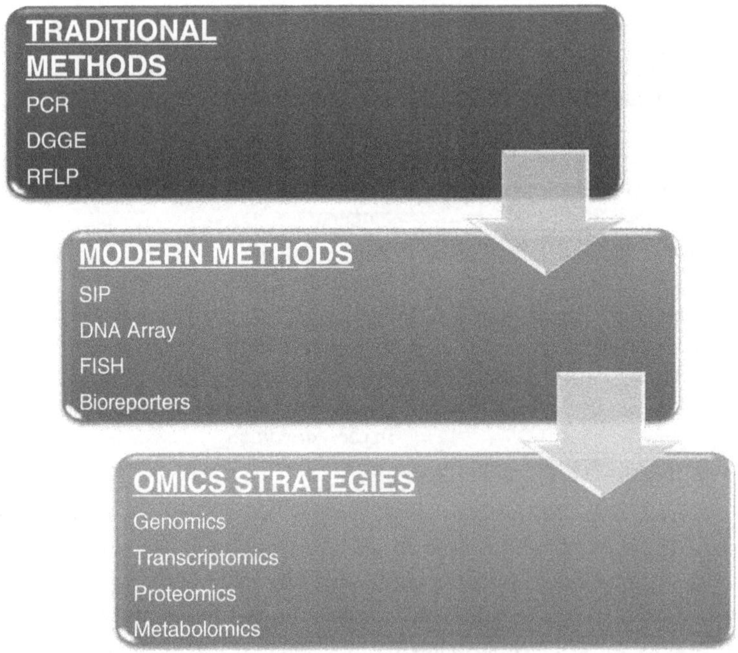

Figure 6.1 Techniques used for rhizosphere engineering.

can be separated by equilibrium density centrifugation. Sequence analysis or DGGE of the DNA or RNA (enriched with ^{13}C) isolated from microorganisms help in the identification of metabolically active microbes in the soil (Dumont and Murrell 2005).

RNA labeling in SIP has been used to assess microbial community in limed and unlimed field plots by delivering short pulses of $^{13}CO_2$. Utilization of the ^{13}C labeled rhizodeposits by the fungi and bacteria in the rhizosphere resulted in RNA enriched pools which were further analyzed by DGGE. The analysis revealed unique phylotypes that could be differentiated as heavy (labeled) and light (unlabeled) fractions depending upon the source of carbon utilization, i.e. labeled $^{13}CO_2$ or unlabeled ^{12}C soil sources. It was observed that the microbial community in the fertile, limed soil was more complex compared to those in unlimed soil (Rangel-Castro et al. 2005).

6.5.2 DNA Arrays

Nucleic acid arrays, specifically DNA arrays, have been used to detect gene expression. This requires an array of oligonucleotide probes that are placed as spots on a surface, which when hybridized with the fluorescently labeled cDNA, enables identification of the genes transcribed in an organism. The fluorescence developed due to the hybridization of the target and probe is an efficient indicator of the gene expression which is analyzed as a heat map. Similarly, functional gene arrays contain oligonucleotides of functional genes (e.g. genes for enzymes involved in nitrogen fixation) from different organisms. A practical implication of this has been to analyze the transcriptionally active genes in the soil by harvesting the soil RNA, converting it to labeled cDNA, and hybridizing it on the array. Hybridization of the labeled soil cDNA with the array indicates the genes present in the soil (Sessitsch et al. 2002). This implies that a careful designing and fabrication of the array is essential for accurate analysis of gene expression.

6.5.3 Fluorescence In Situ Hybridization (FISH)

FISH is a molecular technique that utilizes complementarity between fluorescent probes and specific nucleic acid sequences as the basis for an effective hybridization in the native cell environment. The cells are fixed, made permeable to permit fluorescent probes to enter and hybridize with the target. Visualization of the cells containing fluorescent probes can be performed using epifluorescence microscopy or flow cytometry. The technique has been utilized for decades for characterizing microbial communities (Kobabe et al. 2004). However, advancements have also made it possible to combine FISH with micro-autoradiography or use mRNA-FISH for analyzing gene expression at the level of single cells in a microbial community (Zwirglmaier 2005).

6.5.4 Bioreporters

For analyzing the distribution of microbial populations in the rhizosphere, in-situ microbial bioreporters provide an alternative solution. A bioreporter can be synthesized by fusing a reporter gene with the promoter of interest and inserted into host bacteria. Such bioreporters have been used to test whether the growth-promoting bacteria introduced into the soil can colonize the plant root system. Some bioreporter designs require the promoter to be induced by certain soil compounds while others are constitutively expressed and need not be induced (Boldt et al. 2004).

6.5.5 Genomics

The contribution of genomics in deciphering rhizospheric interactions has been huge with a major application in identifying microbes and their functions using the metagenomics approach (Kakirde et al. 2010). Genomic research has shown that the diverse microbiome in the soil has specific functions in plants. This was also illustrated in a genomic study wherein each of the five *Sinorhizobium* spp. (*Sinorhizobium freddi, Sinorhizobium medicae, Sinorhizobium terange, Sinorhizobium meliloti, and Sinorhizobium saheli*) used a different strategy of interaction with different host plants of the *Medicago* spp. in varied soil conditions (Sugawara et al. 2013). Also, the genes for different functions (e.g. denitrification, synthesis of nod factor) varied within the species as well as between the species (Sugawara et al. 2013).

The metagenomics approach is a step further for understanding the functional potential of the diverse rhizomicrobial community at one go. Recently, the citrus rhizosphere was analyzed using metagenomics approach, highlighting the predominant taxonomic groups, microbial composition of the core rhizosphere (*Agrobacterium, Rhizobium, Paraburkholderia, Pseudomonas,* etc.) and functional traits of the rhizosphere microbiome. The authors reported bacteria as a major component both in the citrus rhizosphere as well as in bulk soil samples collected across six continents (Xu et al. 2018). It is expected that a further understanding of the microbe–microbe and plant–microbe interactions may be achieved through in-situ expression studies using metatranscriptomics, metaproteomics, and metabolomics approaches.

6.5.6 Transcriptomics

Transcriptomics is another omics approach that has enhanced our understanding of the effect of microorganisms on the plant processes and development through changes in plant gene expression (Van Loon 2007). The term transcriptomics is used for single-cultured organisms, whereas a complex microbial community needs a metatranscriptomics approach to be analyzed. Metatranscriptomics requires high-throughput sequencing of the cDNA reverse transcribed from the RNA isolated from the rhizosphere. A combined approach involving plant transcriptional analysis with network analysis can predict genes and metabolic pathways that are impacted during interaction with microbes (Garcia and Kao-Kniffin 2018). This can also pave the way to design future experiments to comprehend the effect of microbes on the plant. Recently, a report demonstrated an interplay of cytosine methylation and alternative splicing mechanisms in regulating plant–fungi interaction while performing transcriptome analysis at different time points of the tomato roots inoculated with the fungi strain *Trichoderma harzianum* (strain T22) (De Palma et al. 2019). Further, it was proposed that similar experiments can provide insight into the underground communications between plant and fungi in a particular environmental niche which may be utilized for plant biofertilization (De Palma et al. 2019).

6.5.7 Proteomics

The proteomics approach has been used to identify proteins exuded into the rhizosphere (Yang et al. 2012). This is important because bacteria such as *B. subtilis* secretes around 300 different proteins in the soil (Tjalsma et al. 2004). Identifying and characterizing those proteins which are crucial for the plant productivity poses a challenge that requires high-end computational tools and analysis.

6.5.8 Metabolomics

Metabolomic strategies can also be employed for identifying the metabolites secreted during plant–microbe interaction. Technologies principally used for metabolite detection include NMR spectroscopy, high-performance liquid chromatography (HPLC), liquid chromatography-mass spectrometry (LC-MS), gas chromatography-mass spectrometry (GC-MS). During the different stages of plant development, the metabolites are known to vary and consequently have varied effects on the microbiota (Chaparro et al. 2013). Annotating these metabolites that drive physiological processes in the plant system (both above and below the ground) will throw light on the complex communication processes between plants and microbes. With technological advancements of MS-based spectroscopy, this is achievable in a short span of time (Mhlongo et al. 2018).

Therefore, integrating the various traditional and modern techniques including omics approaches (White III et al. 2017) will provide a better understanding of the rhizosphere biology and consequently improve overall crop productivity for sustainable agriculture (Figure 6.2).

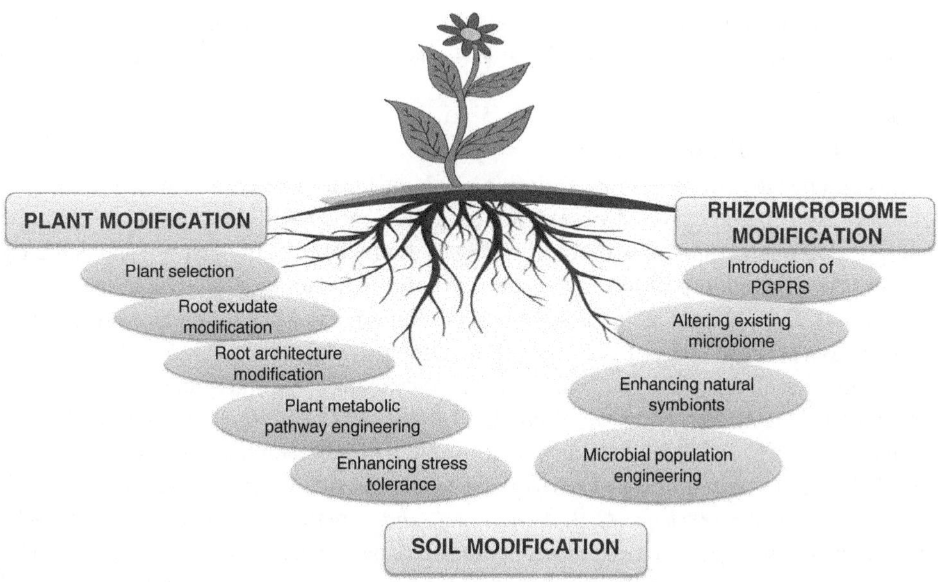

Figure 6.2 Approaches to engineer rhizosphere for sustainable agriculture.

6.6 Rhizosphere Components Amenable to Engineering

Rhizosphere engineering as the name suggests involves manipulating the rhizosphere with a synergistic approach for an efficient plant–microbe interaction (Figure 6.2). This requires an understanding of the interactions or molecular communications occurring in the rhizosphere (Mendes et al. 2013).

The main aim of rhizosphere engineering is to generate stress-resistant and high-yielding crop plants. Thus, engineering of the components of the rhizosphere plays a very important role in enhancing plant productivity and contributing to sustainable modern agriculture. The rhizosphere components amenable to manipulation are discussed under the following sections.

6.6.1 Soil Modification

The first component is the soil which can be modified to alter its physicochemical characteristics for improving plant productivity (Dessaux et al. 2016). Biochar, a kind of porous carbon-rich charcoal increases the fertility of the soil by absorbing inorganic and organic pollutants, improves the water-holding capacity and pH of the soil (Beesley et al. 2011). Other methods make use of coal fly ash (Shaheen et al. 2014), sewage sludge (Singh et al. 2012), and animal manure (Bradford et al. 2008) for improving soil fertility which in turn can modulate plant and microbial communities.

6.6.2 Plant Amendment

The next component that can be engineered is the plant itself. Selecting a plant species or a cultivar with a natural ability to secrete certain compounds for recruiting favorable microorganisms can be used for shaping the microbiome. Further, the physiological features of the plant can be modified to enhance plant–microbe interaction in the rhizosphere.

6.6.2.1 Root Exudate Modification

Plants amend the rhizosphere by secreting exudates from the root. Manipulation of root exudates, soil pH, and H^+ efflux can increase nutrient procurement and can alter the composition of rhizosphere communities (Ryan et al. 2009). Many genes responsible for root exudate secretion have been characterized and they can be exploited to modify the secretion process. Therefore, modification of the genes responsible for root exudations was adopted as a possible mechanism to manipulate the rhizosphere.

The uptake of nutrients and rhizosphere acidification is largely dependent on the efflux of H^+. H^+ efflux is mainly regulated by a family of proteins named H^+-ATPase which makes use of ATP for pumping H^+ through the plasma membrane. Overexpression of H^+-ATPase protein in tobacco (PMA4) led to increased salt tolerance in tobacco plants (Gévaudant et al. 2007). In *A. thaliana*, rice, and tomato, the introduction of *AVP1 pyrophosphatase* gene (*Arabidopsis* vacuolar pyrophosphatase; a type I H^+-pyrophosphatase) resulted in rhizosphere acidification and increased phosphorus (P) nutrition (Yang et al. 2007).

Organic ions are important for a large number of processes taking place in the rhizosphere. Plants can be engineered to synthesize or transport more amount of organic acids (citric acid, malic acid) in the rhizosphere which in turn form complex or solubilize metal ions (Ma et al. 2001; Ryan et al. 2009). However, the two main restrictions for plant development in acidic soil are: (i) aluminum (Al) is toxic to plants and (ii) bioavailability of P is low as it is usually present in an insoluble form (aluminum phosphate). This could be overcome by modifying plants to enhance their ability to grow on soil containing a limited amount of P and a higher concentration of Al. For example, the *mitochondrial citrate synthase* (*mtCS*) gene of *A. thaliana* was used to transform *Daucus carota* which enhanced the release of citrate. This in turn increased the uptake of phosphorus when *D. carota* was grown in medium containing aluminum phosphate (Koyama et al. 1999). In a similar report, ectopic expression of *mtCS* gene of *D. carota* (*DcCS*) in *A. thaliana* showed improved growth in soil containing a limited amount of phosphorus (Koyama et al. 2000). Also, *mtCS*gene of *Arabidopsis* when introduced into *Brassica napus* cv. Westar (canola) conferred increased Al tolerance (Anoop et al. 2003). Further, transgenic canola plants harboring *CS* gene of *Pseudomonas aeruginosa* showed enhanced tolerance to Al toxicity and P deficiency (Wang et al. 2013).

In plants, malate efflux induced by Al is also very important for Al detoxification. In wheat, *aluminum activated malate transporter 1* (*TaALMT1*) gene coding for malate transporter implicated in Al tolerance in plants (Sasaki et al. 2004). Increased Al tolerance was observed in transgenic plants (barley, tobacco) transformed with *TaALMT1* gene (Delhaize et al. 2004; Zhang et al. 2008). In *A. thaliana*, overexpression of a malate transporter gene (*AtALMT1*) conferred Al tolerance (Hoekenga et al. 2006). An Al-induced *BoALMT1* (malate transporter gene) isolated from *Brassica oleracea* when used to transform *A. thaliana,* increased its tolerance to Al stress (Zhang et al. 2018). In alfalfa, the ectopic expression of malate dehydrogenase altered the synthesis of organic acids and led to enhanced Al tolerance (Tesfaye et al. 2001).

6.6.2.2 Root Architecture Modification

Another strategy of rhizosphere engineering w.r.t. plants include modification of root architecture (Dessaux et al. 2016). The root is the most important part of the rhizosphere responsible for the uptake of nutrients. The architecture of root is controlled by the genetic makeup of the plant, environmental factors, and associated microorganisms. These factors have been modified for increasing plant productivity. In rice, different genes (*OsNAC9, DRO1*) regulating root architecture have been shown to confer drought resistance (Redillas et al. 2012). Microorganisms also modify the root architecture by emitting VOCs, inducing the ethylene pathway, and by producing the phytohormone, auxin (Galland et al. 2012).

6.6.2.3 Enhancing Abiotic Stress Tolerance in Plants

Extreme temperature, excess salt, heavy metal toxicity, and water scarcity are the major abiotic stresses that a plant encounters during its lifetime. Rhizosphere engineering to combat abiotic stress in plants can increase productivity dramatically. Of the various approaches, PGPRs for increasing plant fitness during stress conditions has been popularly used. Many PGPRs have been shown to induce abiotic stress tolerance termed as "induced systemic tolerance (IST)" in plants (Yang et al. 2009). PGPRs induce IST in plants by activating stress-responsive genes that produce different phytohormones such as ACC (1-aminocyclopropane-1-carboxylate) deaminase and antioxidative enzymes (Vurukonda et al. 2016). The role of rhizobacteria in engineering drought, salt, temperature, and heavy metal stress resistance in plants has been discussed.

Drought is one of the most adverse threats to the growth, development, and productivity of plants, necessitating engineering for producing drought-tolerant plants. PGPRs can enhance plant tolerance to drought stress by producing phytohormones, antioxidants, VOCs, ACC deaminase, and exopolysaccharides (Waghmode et al. 2019). Wheat plants inoculated with *Azospirillum brasilense* (strain Sp245) showed improved mineral quality and better yield, under water stress conditions (Creus et al. 2004). Tomato and pepper plants exposed to *Achromobacter piechaudii,* strain ARV8 (a PGPR having ACC deaminase activity) showed enhanced tolerance to drought stress (Mayak et al. 2004). Sunflower plants inoculated with *Pseudomonas* sp., strain GAP-P45 displayed increased biomass under drought stress (Sandhya et al. 2009). Similarly, maize plants exposed to five *Pseudomonas* spp. showed higher proline and sugar content, and a decline in the activity of antioxidant enzymes when subjected to drought stress (Sandhya et al. 2010).

Underwater scarcity conditions, higher grain yield was observed when wheat seedlings were inoculated with *Azospirillum lipoferum* (Arzanesh et al. 2011). Plants of cucumber incubated with a mixture of three PGPR namely *Serratia* sp. (strain XY21), *Bacillus cereus* (strain AR156) and *B. subtilis* (strain SM21) displayed darker green leaves and higher proline content under drought stress (Wang et al. 2012a). *Phyllobacterium brassicacearum* (strain STM196) extracted from the oilseed rape rhizosphere was used to inoculate *Arabidopsis* plants. The inoculated plants showed enhanced drought tolerance and better efficacy in water utilization (Bresson et al. 2013). *P. putida* (strain H-2-3) (gibberellin producing rhizobacterium) conferred drought and salt stress tolerance to soybean plants by inducing stress hormones (ABA, SA, JA) and antioxidants (Kang et al. 2014). *Phleum pretense* L. plants inoculated with *B. subtilis,* strain B26 when exposed to drought showed increased biomass, stomatal conductance, photosynthesis, and sugar levels (Gagné-Bourque et al. 2016). A mixture of three ACC deaminase producing beneficial rhizobacteria enhanced drought stress tolerance in *Pisum sativum* and *Vigna mungo* (Saikia et al. 2018). *Sphingomonas* sp. (strain Cra20) enhanced the lateral root growth and changed the rhizosphere microbiome of *Arabidopsis* plants when exposed to drought stress (Luo et al. 2019).

Salinity stress is excess of salt that alters the metabolic activity, impairs photosynthetic apparatus, and causes ion imbalances thereby causing toxicity and a decline in plant productivity. It also triggers the synthesis of ACC deaminase, indole-3-acetic acid (IAA), ET, osmolytes, antioxidants, VOCs, and activation of defense responses (Onaga and Wydra 2016).

P. aeruginosa and *Pseudomonas stutzeri* alleviated salt stress in tomato plants by producing ACC deaminase enzymes, phytohormones, and siderophores (Tank and Saraf 2010). *Arachis hypogaea* plants inoculated with *Brevibacterium casei* (strain JG-08), *Haereroha lobacter* (strain JG-11), and *Brachybacterium saurashtre nse* (strain JG-06) displayed increased biomass, percentage water content and malondialdehyde content under salt stress conditions (Shukla et al. 2012). *Pseudomonas simiae* triggered IST and enhanced plant growth by producing VOCs in soybean exposed to salt stress (Vaishnav et al. 2015).

PGPR, *Dietzia natrono limnaea* conferred salt stress tolerance to wheat plants by inducing expression of SOS pathway, antioxidant enzymes, ion transporter, and ABA signaling related genes (Bharti et al. 2016). *Enterobacter cloacae* and *Bacillus drentensis* enhanced the growth of mung bean under salt stress conditions (Mahmood et al. 2016). *Variovorax paradoxus*, *Micrococcus yunnanensis*, and *Planococcus rifietoensis* rhizobacteria alleviated salt stress effect in *Beta vulgaris* by promoting seed germination, photosynthesis, biomass production, and decreasing ethylene production (Zhou et al. 2017). *Panax ginseng* inoculated with *Paenibacillus yonginensis* (strain DCY84) showed enhanced tolerance to salt stress by inducing osmolyte synthesis, genes responsible for root formation, and ABA biosynthesis genes (Sukweenadhi et al. 2018). Alfalfa plants challenged with *Halomonas* and *Bacillus* rhizobacteria showed better growth under salt stress conditions as compared to uninoculated plants (Kearl et al. 2019).

Temperature stress is another major abiotic stress accountable for crop damage throughout the world. When plants encounter heat stress they display growth retardation, the decline in photosynthesis, stomatal closure, sterility, and reduced yield. Similarly, low temperature (chilling) stress in plants disrupts physiological processes like respiration, photosynthesis, mineral uptake, and the different metabolic pathways. *Triticum aestivum* seeds when exposed to *Azospirillum brasilense* (strain NO40) and *Bacillus amyloliquefaciens* (strain UCMB5113), showed better performance of seedlings under heat stress conditions suggesting that inoculation of seed with PGPR can also be used to alleviate stress (El-Daim et al. 2014). Similarly, *Vitis vinifera* L. cv. Chardonnay inoculated with *Burkholderia phytofirmans* (strain PsJN) PGPR, displayed enhanced root growth, plant biomass, proline levels, and cold tolerance (Barka et al. 2006).

Heavy metals when present in excess, retard growth, and development of plants. Further, heavy metal stress leads to the generation of ROS which causes toxicity thereby reducing yield (Tiwari and Lata 2018). Major studies reporting the effect of PGPR in reducing heavy metal stress in plants of many economically important families have been listed in Table 6.3. The plants where positive effect of PGPR to combat heavy metal stress is observed belong to the families; Fabaceae, Brassicaceae, Poaceae, and Solanaceae.

Therefore, the use of PGPRs for alleviating abiotic stress is a cost-effective, feasible, and environment-friendly approach for increasing crop productivity.

6.6.2.4 Enhancing Biotic Stress Tolerance in Plants

Plants are constantly exposed to different types of biotic stresses. Plant pathogens like bacteria, viruses, insects, nematodes, etc. disturb the normal metabolic functioning of host plants leading to a decline in plant growth and yield. Rhizospheric microorganisms can prevent the growth of plant pathogens using different mechanisms (Lugtenberg and Kamilova 2009). One such mechanism is antibiosis wherein rhizobacteria produce specific and broad-spectrum antibiotics to antagonize the growth of competitive plant pathogens (Raaijmakers and Mazzola 2012). Different species of *Trichoderma* have been implicated in suppressing the plant pathogens thereby enhancing stress tolerance against different pathogens (Druzhinina et al. 2011). *B. subtilis* produces antioxidants, cell wall–degrading enzymes, hormones, siderophores for antagonizing the growth of plant pathogens. It also stimulates plant growth and induces acquired systemic resistance for enhancing plant resistance against pathogens (Hashem et al. 2019). Different rhizobacterial species have

Table 6.3 Rhizobacteria used for alleviation of heavy metal stress in plants.

Rhizobacteria	Plant (Family)	Heavy metal	Stress response	References
Proteobacteria, Firmicutes, Actinobacteria, Bacteroidetes	*Solanum nigrum* (Solanaceae)	Cd	Decline in Cd phytotoxicity and increased dry weight	Luo et al. 2011
Serratia sp. MSMC541	*Lupinus luteus* (Fabaceae)	As, Cd, Pb, Cu, Zn	Phytostabilization	Aafi et al. (2012)
Pseudomonas aeruginosa OSG41	*Cicer arietinum* (Fabaceae)	Cr	Increase in nodule formation, grain yield, dry matter accumulation, and protein	Oves et al. (2013)
Chryseobacteriumhumi, *Ralstoniaeutropha*	*Zea mays* (Poaceae)	Cd	Increase in plant biomass and Cd accumulation in roots	Moreira et al. (2014)
Bacillus licheniformis NCCP-59	*Oryza sativa* (Poaceae)	Ni	Better seed germination and biochemical characteristics	Jamil et al. (2014)
Bacillus sp. SC2b	*Brassica napus* (Brassicaceae) and *Sedum plumbizincicola* (Crassulaceae)	Cd, Pb, Zn	Increase in plant biomass	Ma et al. (2015)
Rhizodermeaveluwensis, Phialocephalafortinii, and Rhizoscyphussp	*Clethra barbinervis Sieb. Et Zucc* (Clethraceae)	Mine soil	Enhanced plant growth, K uptake reduced Pb, Zn, Cu, Ni, and Cd uptake by roots	Yamaji et al. (2016)
Burkholderiacepacia J62, *Microbacteriumoxydans* JYC17, *Pseudomonas thivervalensis* Y1–3-9	*Brassica napus* (Brassicaceae)	Cu	Increase in biomass, Cu uptake, antioxidants	Ren et al. (2019)
Bacillus and *Pseudomonas* consortium	*Spinacea oleracea* (Amaranthaceae)	Cd, Pb, Zn	Increase in plant biomass and decrease in Cd, Zn, and Pb uptake	Shilev et al. (2020)

been shown to secrete VOCs which inhibits fungal mycelial growth (Vespermann et al. 2007). QS molecules like AHLs have also been reported to enhance plant resistance against different plant pathogens (Zarkani et al. 2013).

Rhizobacteria also trigger ISR (phenomenon wherein association of microorganisms with plant roots can make the whole plant body resistant to other pathogenic microorganisms) in plants for enhancing biotic stress tolerance (Pieterse et al. 2014). Different components like flagella, siderophores, lipopeptide, signaling molecules, VOCs, etc., have been shown to elicit ISR in plants (Lugtenberg and Kamilova 2009). *Pseudomonas* sp. CMR12a-induced ISR in rice against *Magnaporthe oryzae* and bean against *Rhizoctonia solani* AG2-2 (Ma et al. 2016). Therefore, it can be concluded that rhizobacteria can stimulate different defense pathways in plants required for the induction of plant resistance. However, there are other mechanisms also by which rhizobacteria confer biotic stress tolerance to plants including predation, parasitism, and competition for food and niches (Lugtenberg and Kamilova 2009).

6.6.2.5 Engineering Metabolic Pathways in Plants

Plant metabolic pathways can be engineered by targeting gene/genes involved in the pathway to enhance plant–microbe interactions. This can be achieved through conventional plant breeding methods also called cultivar selection or through modern genome-editing approaches.

Plant-metabolic engineering may also be utilized to augment yields of plant products by exploiting the production or exudation of signaling molecules into the rhizosphere. Such signaling molecules are generally of low molecular weight, e.g. natural plant products. These molecules are emitted by one organism (e.g. plant) and elicit a response in another (e.g. microbes), thereby mediating interactions (semiochemicals). They facilitate plant–plant, plant–animal, and plant–microbe interaction (Zhang et al. 2015b). Some of these-natural secondary metaboliteslike, SLs, flavonoids, β-caryophyllene, and terpenoids, are discussed. SLs are plant hormones involved in regulating plant growth and development. As signaling molecules in the rhizosphere, they have both positive and negative roles. SLs lead tobranching of the arbuscular mycorrhizal (AM) fungi which support plants in water and nutrient intake through symbiotic interaction (Akiyama et al. 2005). However, the disadvantage is that they may cause yield losses in crops by triggering germination of parasitic weeds, impeding shoot branching, and bud-outgrowth (Umehara et al. 2010). SLs help the plant to sustain unfavorable and limiting soil conditions (phosphorous and nitrogen limiting) by up regulating their synthesis and exudation (Yoneyama et al. 2012). This is required for the interaction of the plant with AM fungi and further symbiosis with *Rhizobium* spp. for inducing root nodulation and hence nitrogen fixation (Akiyama et al. 2005; Foo et al. 2013).

An ATP-binding cassette (ABC) transporter, PDR1, in *Petunia hybrida* was observed to work as an SL exporter regulating the symbiosis between the plant roots and AM fungi. Mutant *P. hybrida* pdr1 with defective SL exudation showed reduced symbiotic interactions (Kretzschmar et al. 2012). Thus, SLs can be considered as important targets for rhizosphere engineering considering their multiple roles in different biological processes.

SLs are biosynthesized through the cascade action of enzymes encoded by *D27*, *CCD7*, *CCD8*, and *MAX1*genes. Cardoso et al. (2014) while working on a major quantitative trait locus (QTL) on chromosome 1 of rice plant identified two homologs of the gene *MAX1* (4-deoxyorobanchol-4-hydroxylase and carlactone oxidase) that are required for SL synthesis. Biochemical characterization of the genes in *Nicotiana benthamiana* by Sainsbury and Lomonossoff (2014) further confirmed that the genes mapped in the QTL encode carlactone oxidase and 4-deoxyorobanchol-4-hydroxylase. Combined experimental approaches like genome mapping using QTLs followed by biochemical characterization may help in understanding SLs synthesis and identify novel genes for effective rhizosphere engineering (Zhang et al. 2015b).

Flavonoids are a structurally diverse class of compounds found in root exudates. They have a polyphenolic structure consisting of a C6–C3–C6 (C15) carbon skeleton. Secreted flavonoids in the soil initiate the symbiosis signaling pathway by inducing expression of bacterial nod genes, stimulating the nodulation factors (nod factors)/lipochitooligosacchar ides (LCOs) (Hassan and Mathesius 2012) and initiating nodule formation in the roots. Interestingly, certain flavonoids also have a role in defense against root pathogens by stimulating spore germination of the AM fungi (Hassan and Mathesius 2012; Mhlongo et al. 2018).

The first product in the biosynthesis of flavonoids is the formation of chalcone, from which other categories of flavonoids are derived, these include flavones, flavonols, flavandiols, anthocyanins, condensed tannins, and isoflavonoids. Genes of the biosynthetic pathway have been used to manipulate flavonoid synthesis with an observable impact in the rhizosphere. Wasson et al. (2006) observed flavonoid deficiency and the absence of root nodules in *Medicago truncatula* as a result of the silencing of the enzymes (chalcone synthase and isoflavone synthase) regulating flavonoid biosynthesis. The ABC transporter of G subfamily (*MtABCG10*) when silenced in the root hairs of *M. truncatula* resulted in decreased exudation of isoflavonoids and increased sensitivity to infection by the root pathogen *Fusarium oxysporum* (Banasiak et al. 2013). According to a recent observation by Xie et al. (2019), the expression of phenolics and flavonoid synthesis genes increased in a symbiotic interaction between an endophytic fungus, *Phomopsis liquidambaris* and the peanut plant with significant improvement in nodulation and nitrogen fixation due to *P. liquidambaris* colonization.

The sesquiterpene, (*E*)-β-caryophyllene [(*E*)-BCP] is a major plant volatile compound that acts as a defense signal against pathogens. In maize plant, (E)-BCP is released by leaves in response to infestation by lepidopteran larvae (*Spodoptera littoralis*), whereas the roots exude it on being damaged by coleopteran larvae (*Diabrotica virgifera virgifera*) (Köllner et al. 2008). Interestingly, in the rhizosphere of transgenic maize, enhanced emission of (E)-BCP can attract nematode, pathogenic to *Diabrotica virgifera virgifera* (Köllner et al. 2008). A practical implication of this was observed in the American maize lines which are highly prone to attack by *D. virgifera* as they do not produce (E)-BCP. However, manipulating the rhizosphere for constitutive expression of (E)-BCP by overexpression of the enzyme (E)-BCP synthase ensured the attraction of the entomopathogenic nematode and therefore reduced loss caused by the root herbivore with a better maize yield (Degenhardt et al. 2009). In the Poaceae family, (E)-BCP emission has been observed to confer resistance against root herbivores, thus enabling grass-roots defense mechanisms (Moore and Johnson 2017).

6.6.3 Engineering Microbial Populations

Microbes are an important component of the rhizosphere because the plant and its associated biome are not considered separate entities but are constituents of a super organism, the holobiont (Dessaux et al. 2016). Manipulating the rhizosphere and the associated microbiome often results in improved plant traits. PGPR are directly or indirectly involved in promoting plant growth, improving fertilization, exhibiting biocontrol activities (production of antibiotics, antifungal, bacteriocins) and triggering plant defense system (Pii et al. 2015). PGPRs have been used successfully to increase the efficiency of root nodulation ability in alfalfa and common bean (Ma et al. 2004; Suárez et al. 2008). Introduction of chitinase gene (from *B. subtilis*) into *Bulkholderia vietnamiensis* PGPR (strain P418), has been utilized in enhancing plant defense against pathogenic fungal infections like tomato gray mold and wheat sheath blight (Zhang et al. 2012). Genetically modified rhizobia have also been effective in improving nodule formation in copper-contaminated soils (Delgadillo et al. 2015).

An alternative strategy of engineering is when the microbes are utilized to exploit the plant hormonal system for better plant growth, development, and yield. Microorganisms are capable of producing phytohormones (auxins, gibberellins, and cytokinins) that mimic the effect of jasmonic acid (JA), crucial for plant defense, plant response to biotic and abiotic stress and in signaling pathway against plant pathogens (Stein et al. 2008; Hause and Schaarschmidt 2009). Biochemical interactions between JA, SA, and ET stimulate plant immune responses like SAR and ISR (Van Loon et al. 2006).

Microbial populations in the rhizosphere can also be engineered and utilized as a biocontrol strategy. In contrast to QS wherein bacteria secrete AHLs as signaling molecules, some bacteria are capable of degrading these signals through quorum quenching (QQ) helping the plant to sustain pathogen attack. Experimental evidence shows the ability to degrade QS signals is inherent to the soil bacteria, *Rhodococcus erythropolis*. This quenching action was visible during the interaction between this bacterium and the virulent plant pathogen *Pectobacterium atrosepticum*. The authors showed *R. erythropolis* worked by limiting/preventing the buildup of the QS signal instead of directly inhibiting the growth of *Pectobacterium*, thereby protecting the plant from harmful effects of macerating (plant cell-wall–degrading enzymes) enzymes produced by the plant pathogen (Kwasiborski et al. 2015).

6.7 Conclusion and Future Prospects

The association between plants and microorganisms is an essential component of sustainable modern agriculture. Rhizosphere demonstrates an attractive zone for beneficial interaction between plant and microbes which ultimately reflect on crop productivity. Therefore, rhizosphere engineering is a niche area to be assessed and manipulated for improved crop production. Different approaches have been used to engineer different components of the rhizosphere (Figure 6.2). The challenges in rhizosphere manipulation are the identification of genes or different exudates and signals responsible for the beneficial plant–microbe interactions. This will help in the genetic modification of plants or microbes for ameliorating abiotic or biotic stress tolerance in plants. The way plants and microorganisms cross talk is very dynamic. A detailed study of these interactions at the molecular level will help in substituting the use of chemical pesticides or fertilizers with genetically modified microorganisms or plants. Of all the constituents of the plant ecosystem, the rhizosphere has been an important hotspot amenable to manipulation for an efficient plant/microbe interaction. This unified plant–microbe holobiont can be engineered at different levels ranging from plants to plant exudates, single microbe, or consortia of microbes, ultimately for producing beneficial plant traits. However, having known the multitude of highly entangled interactions occurring in the soil, deciphering specific mechanisms regulating plant–microbe interactions is a challenging task to be achieved using experimental approaches. Therefore, it is expected that "omics" tools along with next-generation sequencing technology will provide more comprehensive information about the complexity of rhizosphere, the nature of plant–microbe interactions, and signaling in rhizosphere which will ultimately help in devising novel strategies for engineering the rhizosphere. Moreover, it is an environmentally friendly and economical approach for enhancing crop quality and productivity in the present changing scenario of climatic conditions.

Acknowledgment

The authors acknowledge the funding under BTISNET program [Project: BT/BI/25/059/2012 (BIF)] by the Department of Biotechnology (DBT), Govt. of India.

References

Aafi, N.E., Brhada, F., Dary, M. et al. (2012). Rhizostabilization of metals in soils using *Lupinus luteus* inoculated with the metal resistant rhizobacterium *Serratia* sp. MSMC541. *Int. J. Phytorem.* 14: 261–274.

Akiyama, K., Matsuzaki, K.I., and Hayashi, H. (2005). Plant sesquiterpenes induce hyphal branching in arbuscular mycorrhizal fungi. *Nature* 435: 824–827.

Andersson, D.I. and Hughes, D. (2014). Microbiological effects of sublethal levels of antibiotics. *Nat. Rev. Microbiol.* 12: 465–478.

Anoop, V.M., Basu, U., McCammon, M.T. et al. (2003). Modulation of citrate metabolism alters aluminum tolerance in yeast and transgenic canola overexpressing a mitochondrial citrate synthase. *Plant Physiol.* 132: 2205–2217.

Arzanesh, M.H., Alikhani, H.A., Khavazi, K. et al. (2011). Wheat (*Triticum aestivum* L.) growth enhancement by *Azospirillum* sp. under drought stress. *World J. Microbiol.Biotechnol.* 27: 197–205.

Babalola, O.O. (2010). Beneficial bacteria of agricultural importance. *Biotechnol. Lett.* 32: 1559–1570.

Bailly, A., Groenhagen, U., Schulz, S. et al. (2014). The inter-kingdom volatile signal indole promotes root development by interfering with auxin signaling. *Plant J.* 80: 758–771.

Bakker, M.G., Manter, D.K., Sheflin, A.M. et al. (2012). Harnessing the rhizosphere microbiome through plant breeding and agricultural management. *Plant Soil* 360: 1–13.

Banasiak, J., Biała, W., Staszków, A. et al. (2013). A *Medicago truncatula* ABC transporter belonging to subfamily G modulates the level of isoflavonoids. *J. Exp. Bot.* 64: 1005–1015.

Barka, E.A., Nowak, J., and Clément, C. (2006). Enhancement of chilling resistance of inoculated grapevine plantlets with a plant growth-promoting rhizobacterium, *Burkholderia phytofirmans* strain PsJN. *Appl. Environ. Microbiol.* 72: 7246–7252.

Beesley, L., Moreno-Jiménez, E., Gomez-Eyles, J.L. et al. (2011). A review of biochars' potential role in the remediation, revegetation and restoration of contaminated soils. *Environ. Pollut.* 159: 3269–3282.

Berg, G. (2009). Plant-microbe interactions promoting plant growth and health: perspectives for controlled use of microorganisms in agriculture. *Appl. Microbial. Biotechnol.* 84: 11–18.

Bharti, N., Pandey, S.S., Barnawal, D. et al. (2016). Plant growth promoting rhizobacteria *Dietzia natrono limnaea* modulates the expression of stress responsive genes providing protection of wheat from salinity stress. *Sci. Rep.* 6: 34768.

Boldt, T.S., Sørensen, J., Karlson, U. et al. (2004). Combined use of different Gfp reporters for monitoring single-cell activities of a genetically modified PCB degrader in the rhizosphere of alfalfa. *FEMS Microbiol. Ecol.* 48: 139–148.

Bonilla, A., Sarria, A.L.F., and Algar, E. (2014). Microbe associated molecular patterns from rhizosphere bacteria trigger germination and *Papaver somniferum* metabolism under greenhouse conditions. *Plant Physiol. Biochem.* 74: 133–140.

Bradford, S.A., Segal, E., Zheng, W. et al. (2008). Reuse of concentrated animal feeding operation wastewater on agricultural lands. *J. Environ. Qual.* 37: S-97.

Brameyer, S., Bode, H.B., and Heermann, R. (2015). Languages and dialects: bacterial communication beyond homoserine lactones. *Trends Microbiol.* 23: 521–523.

Bresson, J., Varoquaux, F., Bontpart, T. et al. (2013). The PGPR strain *Phyllobacterium brassicacearum* STM 196 induces a reproductive delay and physiological changes that result in improved drought tolerance in *Arabidopsis*. *New Phytol.* 200: 558–569.

Calvo, P., Zebelo, S., McNear, D. et al. (2019). Plant growth-promoting rhizobacteria induce changes in *Arabidopsis thaliana* gene expression of nitrate and ammonium uptake genes. *J. Plant Interact.* 14: 224–231.

Cardoso, C., Zhang, Y., Jamil, M. et al. (2014). Natural variation of rice strigolactone biosynthesis is associated with the deletion of two *MAX1* orthologs. *Proc. Natl. Acad. Sci. U.S.A.* 111: 2379–2384.

Chaparro, J.M., Badri, D.V., Bakker, M.G. et al. (2013). Root exudation of phytochemicals in *Arabidopsis* follows specific patterns that are developmentally programmed and correlate with soil microbial functions. *PloS One* 8: e55731.

Chowdhury, S.P., Uhl, J., Grosch, R. et al. (2015). Cyclic lipopeptides of *Bacillus amyloliquefaciens* subsp. *plantarum* colonizing the lettuce rhizosphere enhances plant defense responses toward the bottom rot pathogen *Rhizoctonia solani*. *Mol. Plant Microbe Interact.* 28: 984–995.

Creus, C.M., Sueldo, R.J., and Barassi, C.A. (2004). Water relations and yield in *Azospirillum*-inoculated wheat exposed to drought in the field. *Can. J. Bot.* 82: 273–281.

De Palma, M., Salzano, M., Villano, C. et al. (2019). Transcriptome reprogramming, epigenetic modifications and alternative splicing orchestrate the tomato root response to the beneficial fungus *Trichoderma harzianum*. *Hortic. Res.* 6: 5.

Degenhardt, J., Hiltpold, I., Köllner, T.G. et al. (2009). Restoring a maize root signal that attracts insect-killing nematodes to control a major pest. *Proc. Natl. Acad. Sci. U.S.A.* 106: 13213–13218.

De-la-Pena, C., Badri, D.V., and Loyola-Vargas, V.M. (2012). Plant root secretions and their interactions with neighbors. In: *Secretions and Exudates in Biological Systems* (eds. J.M. Vivanco and F. Balǔka), 1–26. Berlin, Heidelberg: Springer.

Delgadillo, J., Lafuente, A., Doukkali, B. et al. (2015). Improving legume nodulation and Cu rhizostabilization using a genetically modified rhizobia. *Environ. Technol.* 36: 1237–1245.

Delhaize, E., Ryan, P.R., Hebb, D.M. et al. (2004). Engineering high-level aluminum tolerance in barley with the *ALMT1* gene. *Proc. Natl. Acad. Sci. U.S.A.* 101: 15249–15254.

Dessaux, Y., Grandclément, C., and Faure, D. (2016). Engineering the rhizosphere. *Trends Plant Sci.* 21: 266–278.

Druzhinina, I.S., Seidl-Seiboth, V., Herrera-Estrella, A. et al. (2011). *Trichoderma*: the genomics of opportunistic success. *Nat. Rev. Microbiol.* 9: 749–759.

Dumont, M.G. and Murrell, J.C. (2005). Stable isotope probing—linking microbial identity to function. *Nat. Rev. Microbiol.* 3: 499–504.

El-Daim, I.A.A., Bejai, S., and Meijer, J. (2014). Improved heat stress tolerance of wheat seedlings by bacterial seed treatment. *Plant Soil* 379: 337–350.

Ellouze, W., Hamel, C., Vujanovic, V. et al. (2013). Chickpea genotypes shape the soil microbiome and affect the establishment of the subsequent durum wheat crop in the semiarid North American Great Plains. *Soil Biol. Biochem.* 63: 129–141.

Foo, E., Yoneyama, K., Hugill, C.J. et al. (2013). Strigolactones and the regulation of pea symbioses in response to nitrate and phosphate deficiency. *Mol. Plant* 6: 76–87.

Fuqua, C., Parsek, M.R., and Greenberg, E.P. (2001). Regulation of gene expression by cell-to-cell communication: acyl-homoserine lactone quorum sensing. *Annu. Rev. Genet.* 35: 439–468.

Gagné-Bourque, F., Bertrand, A., Claessens, A. et al. (2016). Alleviation of drought stress and metabolic changes in timothy (*Phleum pratense* L.) colonized with *Bacillus subtilis* B26. *Front. Plant Sci.* 7: 584.

Galland, M., Gamet, L., Varoquaux, F. et al. (2012). The ethylene pathway contributes to root hair elongation induced by the beneficial bacteria *Phyllobacterium brassicacearum* STM196. *Plant Sci.* 190: 74–81.

Garcia, J. and Kao-Kniffin, J. (2018). Microbial group dynamics in plant rhizospheres and their implications on nutrient cycling. *Front. Microbiol.* 9: 1516.

Gévaudant, F., Duby, G., von Stedingk, E. et al. (2007). Expression of a constitutively activated plasma membrane H+-ATPase alters plant development and increases salt tolerance. *Plant Physiol.* 144: 1763–1776.

Han, S.H., Lee, S.J., Moon, J.H. et al. (2006). GacS-dependent production of 2R,3R-butanediol by *Pseudomonas chlororaphis* O6 is a major determinant for eliciting systemic resistance against *Erwinia carotovora* but not against *Pseudomonas syringae* pv. tabaci in tobacco. *Mol. Plant Microbe Interact.* 19: 924–930.

Hashem, A., Tabassum, B., and Abd_Allah, E.F. (2019). *Bacillus subtilis*: a plant-growth promoting rhizobacterium that also impacts biotic stress. *Saudi J. Biol. Sci.* 26: 1291–1297.

Hassan, S. and Mathesius, U. (2012). The role of flavonoids in root–rhizosphere signaling: opportunities and challenges for improving plant-microbe interactions. *J. Exp. Bot.* 63: 3429–3444.

Hause, B. and Schaarschmidt, S. (2009). The role of jasmonates in mutualistic symbioses between plants and soil-born microorganisms. *Phytochemistry* 70: 1589–1599.

Hiltner, L.T. (1904). Uber nevereErfahrungen und Probleme auf dem Gebiet der Boden Bakteriologie und unterbesondererBeurchsichtigung der Grundungung und Broche. *Arbeit. Deut. Landw. Ges. Berlin* 98: 59–78.

Hoekenga, O.A., Maron, L.G., Piñeros, M.A. et al. (2006). AtALMT1, which encodes a malate transporter, is identified as one of several genes critical for aluminum tolerance in *Arabidopsis*. *Proc. Natl. Acad. Sci. U.S.A.* 103: 9738–9743.

Hofmann, N.R. (2013). Volatile organic compounds: a bacterial contribution to plant sulfur nutrition. *Plant Cell* 25: 2381.

Igiehon, N.O. and Babalola, O.O. (2018). Below-ground-above-ground plant-microbial interactions: focusing on soybean, rhizobacteria and mycorrhizal fungi. *Open Microbiol. J.* 12: 261.

Jamil, M., Zeb, S., Anees, M. et al. (2014). Role of *Bacillus licheniformis* in phytoremediation of nickel contaminated soil cultivated with rice. *Int. J. Phytorem.* 16: 554–571.

Kakirde, K.S., Parsley, L.C., and Liles, M.R. (2010). Size does matter: application-driven approaches for soil metagenomics. *Soil Biol. Biochem.* 42: 1911–1923.

Kang, S.M., Radhakrishnan, R., Khan, A.L. et al. (2014). Gibberellin secreting rhizobacterium, *Pseudomonas putida* H-2-3 modulates the hormonal and stress physiology of soybean to improve the plant growth under saline and drought conditions. *Plant Physiol. Biochem.* 84: 115–124.

Kearl, J., McNary, C., Lowman, J.S. et al. (2019). Salt-tolerant halophyte rhizosphere bacteria stimulate growth of alfalfa in salty soil. *Front Microbiol.* 10: 1849.

Kim, J.S., Lee, J., Seo, S.G. et al. (2015). Gene expression profile affected by volatiles of new plant growth promoting rhizobacteria, *Bacillus subtilis* strain JS, in tobacco. *Genes Genom.* 37: 387–397.

Kobabe, S., Wagner, D., and Pfeiffer, E.M. (2004). Characterisation of microbial community composition of a Siberian tundra soil by fluorescence *in situ* hybridisation. *FEMS Microbiol. Ecol.* 50: 13–23.

Köllner, T.G., Held, M., Lenk, C. et al. (2008). A maize (E)-β-caryophyllene synthase implicated in indirect defense responses against herbivores is not expressed in most American maize varieties. *Plant Cell* 20: 482–494.

Koyama, H., Takita, E., Kawamura, A. et al. (1999). Over expression of mitochondrial citrate synthase gene improves the growth of carrot cells in Al-phosphate medium. *Plant Cell Physiol.* 40: 482–488.

Koyama, H., Kawamura, A., Kihara, T. et al. (2000). Overexpression of mitochondrial citrate synthase in *Arabidopsis thaliana* improved growth on a phosphorus-limited soil. *Plant Cell Physiol.* 41: 1030–1037.

Kretzschmar, T., Kohlen, W., Sasse, J. et al. (2012). A petunia ABC protein controls strigolactone-dependent symbiotic signaling and branching. *Nature* 483: 341–344.

Kwasiborski, A., Mondy, S., Chong, T.M. et al. (2015). Transcriptome of the quorum-sensing signal-degrading *Rhodococcus erythropolis* responds differentially to virulent and avirulent *Pectobacterium atrosepticum*. *Heredity* 114: 476–484.

Limpens, E., van Zeijl, A., and Geurts, R. (2015). Lipochitooligosaccharides modulate plant host immunity to enable endosymbioses. *Annu. Rev. Phytopathol.* 53: 311–334.

Lugtenberg, B. and Kamilova, F. (2009). Plant-growth-promoting rhizobacteria. *Annu. Rev. Microbiol.* 63: 541–556.

Luo, S.L., Chen, L., Chen, J.L. et al. (2011). Analysis and characterization of cultivable heavy metal-resistant bacterial endophytes isolated from Cd-hyperaccumulator *Solanum nigrum* L. and their potential use for phytoremediation. *Chemosphere* 85: 1130–1138.

Luo, Y., Wang, F., Zhou, M., and Sheng, H.M. (2019). *Sphingomonas* sp. Cra20 increases plant growth rate and alters rhizosphere microbial community structure of *Arabidopsis thaliana* under drought stress. *Front. Microbiol.* 10: 1221.

Ma, J.F., Ryan, P.R., and Delhaize, E. (2001). Aluminum tolerance in plants and the complexing role of organic acids. *Trends Plant Sci.* 6: 273–278.

Ma, W., Charles, T.C., and Glick, B.R. (2004). Expression of an exogenous 1-aminocyclopropane-1-carboxylate deaminase gene in *Sinorhizobium meliloti* increases its ability to nodulate alfalfa. *Appl. Environ. Microbiol.* 70: 5891–5897.

Ma, Y., Oliveira, R.S., Wu, L. et al. (2015). Inoculation with metal-mobilizing plant-growth-promoting rhizobacterium *Bacillus* sp. SC2b and its role in rhizoremediation. *J. Toxicol. Environ. Health Part A* 78: 931–944.

Ma, Z., Hua, G.K.H., Ongena, M., and Höfte, M. (2016). Role of phenazines and cyclic lipopeptides produced by *Pseudomonas* sp. CMR12a in induced systemic resistance on rice and bean. *Environ. Microbiol. Rep.* 8: 896–904.

Mahmood, S., Daur, I., Al-Solaimani, S.G. et al. (2016). Plant growth promoting rhizobacteria and silicon synergistically enhance salinity tolerance of mung bean. *Front. Plant Sci.* 7: 876.

Mandal, S.M., Chakraborty, D., and Dey, S. (2010). Phenolic acids act as signaling molecules in plant-microbe symbioses. *Plant Signal. Behav.* 5: 359–368.

Mayak, S., Tirosh, T., and Glick, B.R. (2004). Plant growth-promoting bacteria that confer resistance to water stress in tomatoes and peppers. *Plant Sci.* 166: 525–530.

Mendes, R., Garbeva, P., and Raaijmakers, J.M. (2013). The rhizosphere microbiome: significance of plant beneficial, plant pathogenic, and human pathogenic microorganisms. *FEMS Microbiol. Rev.* 37: 634–663.

Mhlongo, M.I., Piater, L.A., Madala, N.E. et al. (2018). The chemistry of plant–microbe interactions in the rhizosphere and the potential for metabolomics to reveal signaling related to defense priming and induced systemic resistance. *Front. Plant Sci.* 9: 112.

Miller, M.B. and Bassler, B.L. (2001). Quorum sensing in bacteria. *Annu. Rev. Microbiol.* 55: 165–199.

Monnet, V., Juillard, V., and Gardan, R. (2016). Peptide conversations in Gram-positive bacteria. *Crit. Rev. Microbiol.* 42: 339–351.

Moore, B.D. and Johnson, S.N. (2017). Get tough, get toxic, or get a bodyguard: identifying candidate traits conferring belowground resistance to herbivores in grasses. *Front. Plant Sci.* 7: 1925.

Moreira, H., Marques, A.P., Franco, A.R. et al. (2014). Phytomanagement of Cd-contaminated soils using maize (*Zea mays* L.) assisted by plant growth-promotingrhizobacteria. *Environ. Sci. Pollut. Res.* 21: 9742–9753.

Onaga, G. and Wydra, K. (2016). Advances in plant tolerance to abiotic stresses. In: *Plant Genomics* (ed. I.Y. Abdurakhmonov), 229–272. Rijeka: Intech.

Ongena, M., Jourdan, E., Adam, A. et al. (2007). Surfactin and fengycin lipopeptides of *Bacillus subtilis* as elicitors of induced systemic resistance in plants. *Environ. Microbiol.* 9: 1084–1090.

Oves, M., Khan, M.S., and Zaidi, A. (2013). Chromium reducing and plant growth promoting novel strain *Pseudomonas aeruginosa* OSG41 enhance chickpea growth in chromium amended soils. *Eur. J. Soil Biol.* 56: 72–83.

Patil, S., Bheemaraddi, M.C., Shivannavar, C.T., and Gaddad, M.S. (2014). Biocontrol activity of siderophore producing *Bacillus subtilis* CTS-G24 against wilt and dry root rot causing fungi in chickpea. *IOSR J. Agric. Vet. Sci.* 7: 63–68.

Pieterse, C.M., Zamioudis, C., Berendsen, R.L. et al. (2014). Induced systemic resistance by beneficial microbes. *Annu. Rev. Phytopathol.* 52: 347–375.

Pii, Y., Mimmo, T., Tomasi, N. et al. (2015). Microbial interactions in the rhizosphere: beneficial influences of plant growth-promoting rhizobacteria on nutrient acquisition process. A review. *Biol. Fertil. Soils* 51: 403–415.

Plett, J.M., Daguerre, Y., Wittulsky, S. et al. (2014). Effector MiSSP7 of the mutualistic fungus *Laccaria bicolor* stabilizes the Populus JAZ6 protein and represses jasmonic acid (JA) responsive genes. *Proc. Natl. Acad. Sci. U.S.A.* 111: 8299–8304.

Quiza, L., St-Arnaud, M., and Yergeau, E. (2015). Harnessing phytomicrobiome signaling for rhizosphere microbiome engineering. *Front. Plant Sci.* 6: 507.

Raaijmakers, J.M. and Mazzola, M. (2012). Diversity and natural functions of antibiotics produced by beneficial and plant pathogenic bacteria. *Annu. Rev. Phytopathol.* 50: 403–424.

Rangel-Castro, J.I., Killham, K., Ostle, N. et al. (2005). Stable isotope probing analysis of the influence of liming on root exudate utilization by soil microorganisms. *Environ. Microbiol.* 7: 828–838.

Redillas, M.C., Jeong, J.S., Kim, Y.S. et al. (2012). The overexpression of *OsNAC9* alters the root architecture of rice plants enhancing drought resistance and grain yield under field conditions. *Plant Biotechnol. J.* 10: 792–805.

Ren, X., Guo, S., Tian, W. et al. (2019). Effects of plant growth-promoting bacteria (PGPB) inoculation on the growth, antioxidant activity, Cu uptake, and bacterial community structure of rape (*Brassica napus* L.) grown in Cu-contaminated agricultural soil. *Front. Microbiol.* 10: 1455.

Ryan, P.R., Dessaux, Y., Thomashow, L.S., and Weller, D.M. (2009). Rhizosphere engineering and management for sustainable agriculture. *Plant Soil* 321: 363–383.

Ryu, C.M., Farag, M.A., Hu, C.H. et al. (2004). Bacterial volatiles induce systemic resistance in *Arabidopsis*. *Plant Physiol.* 134: 1017–1026.

Saikia, J., Sarma, R.K., Dhandia, R. et al. (2018). Alleviation of drought stress in pulse crops with ACC deaminase producing rhizobacteria isolated from acidic soil of Northeast India. *Sci. Rep.* 8: 3560.

Sainsbury, F. and Lomonossoff, G.P. (2014). Transient expressions of synthetic biology in plants. *Curr. Opin. Plant Biol.* 19: 1–7.

Sandhya, V.Z.A.S., Grover, M., Reddy, G., and Venkateswarlu, B. (2009). Alleviation of drought stress effects in sunflower seedlings by the exopolysaccharides producing *Pseudomonas putida* strain GAP-P45. *Biol. Fertil. Soils* 46: 17–26.

Sandhya, V.S.K.Z., Ali, S.Z., Grover, M. et al. (2010). Effect of plant growth promoting *Pseudomonas* spp. on compatible solutes, antioxidant status and plant growth of maize under drought stress. *Plant Growth Regul.* 62: 21–30.

Sasaki, T., Yamamoto, Y., Ezaki, B. et al. (2004). A wheat gene encoding an aluminum-activated malate transporter. *Plant J.* 37: 645–653.

Schirawski, J. and Perlin, M. (2018). Plant–microbe interaction 2017—the good, the bad and the diverse. *Int. J. Mol. Sci.* 19: 1374.

Schuhegger, R., Ihring, A., Gantner, S. et al. (2006). Induction of systemic resistance in tomato by N-acyl-L-homoserine lactone-producing rhizosphere bacteria. *Plant Cell Environ.* 29: 909–918.

Sessitsch, A., Gyamfi, S., Stralis-Pavese, N. et al. (2002). RNA isolation from soil for bacterial community and functional analysis: evaluation of different extraction and soil conservation protocols. *J. Microbiol. Methods* 51: 171–179.

Shaheen, S.M., Hooda, P.S., and Tsadilas, C.D. (2014). Opportunities and challenges in the use of coal fly ash for soil improvements – a review. *J. Environ. Manage.* 145: 249–267.

Shilev, S., Babrikova, I., and Babrikov, T. (2020). Consortium of plant growth-promoting bacteria improves spinach (*Spinacea oleracea* L.) growth under heavy metal stress conditions. *J. Chem. Technol. Biotechnol.* 95: 932–939.

Shukla, P.S., Agarwal, P.K., and Jha, B. (2012). Improved salinity tolerance of *Arachis hypogaea* (L.) by the interaction of halotolerant plant-growth-promoting rhizobacteria. *J. Plant Growth Regul.* 31: 195–206.

Singh, S.P. and Gaur, R. (2017). Endophytic *Streptomyces* spp. underscore induction of defense regulatory genes and confers resistance against *Sclerotium rolfsii* in chickpea. *Biol. Control* 104: 44–56.

Singh, R.P., Singh, P., Ibrahim, M.H., and Hashim, R. (2012). Land application of sewage sludge: physicochemical and microbial response. In: *Reviews of Environmental Contamination and Toxicology* (ed. D. Whitacre), 41–61). New York, NY: Springer.

Spaepen, S., Vanderleyden, J., and Remans, R. (2007). Indole-3-acetic acid in microbial and microorganism-plant signaling. *FEMS Microbiol. Rev.* 31: 425–448.

Srivastava, S., Chaudhry, V., Mishra, A. et al. (2012). Gene expression profiling through microarray analysis in *Arabidopsis thaliana* colonized by *Pseudomonas putida* MTCC5279, a plant growth promoting rhizobacterium. *Plant Signal. Behav.* 7: 235–245.

Stein, E., Molitor, A., Kogel, K.H., and Waller, F. (2008). Systemic resistance in *Arabidopsis* conferred by the mycorrhizal fungus *Piriformospora indica* requires jasmonic acid signaling and the cytoplasmic function of NPR1. *Plant Cell Physiol.* 49: 1747–1751.

Suárez, R., Wong, A., Ramírez, M. et al. (2008). Improvement of drought tolerance and grain yield in common bean by overexpressing trehalose-6-phosphate synthase in rhizobia. *Mol. Plant Microbe Interact.* 21: 958–966.

Sugawara, M., Epstein, B., Badgley, B.D. et al. (2013). Comparative genomics of the core and accessory genomes of 48 *Sinorhizobium* strains comprising five genospecies. *Genome Biol.* 14: R17.

Sukweenadhi, J., Balusamy, S.R., Kim, Y.J. et al. (2018). A growth promoting bacteria, *Paenibacillus yonginensis* DCY84T enhanced salt stress tolerance by activating defense-related systems in *Panax ginseng*. *Front. Plant Sci.* 9: 813.

Tank, N. and Saraf, M. (2010). Salinity-resistant plant growth promoting rhizobacteria ameliorates sodium chloride stress on tomato plants. *J. Plant Interact.* 5: 51–58.

Tesfaye, M., Temple, S.J., Allan, D.L. et al. (2001). Overexpression of malate dehydrogenase in transgenic alfalfa enhances organic acid synthesis and confers tolerance to aluminum. *Plant Physiol.* 127: 1836–1844.

Timmusk, S. and Wagner, E.G.H. (1999). The plant-growth-promoting rhizobacterium *Paenibacillus polymyxa* induces changes in *Arabidopsis thaliana* gene expression: a possible connection between biotic and abiotic stress responses. *Mol. Plant Microbe Interact.* 12: 951–959.

Tiwari, S. and Lata, C. (2018). Heavy metal stress, signaling, and tolerance due to plant-associated microbes: an overview. *Front. Plant Sci.* 9: 452.

Tjalsma, H., Antelmann, H., Jongbloed, J.D. et al. (2004). Proteomics of protein secretion by *Bacillus subtilis*: separating the "secrets" of the secretome. *Microbiol. Mol. Biol. Rev.* 68: 207–233.

Umehara, M., Hanada, A., Magome, H. et al. (2010). Contribution of strigolactones to the inhibition of tiller bud outgrowth under phosphate deficiency in rice. *Plant Cell Physiol.* 51: 1118–1126.

Vaishnav, A., Kumari, S., Jain, S. et al. (2015). Putative bacterial volatile-mediated growth in soybean (*Glycine max* L. Merrill) and expression of induced proteins under salt stress. *J. Appl. Microbiol.* 119: 539–551.

Van Loon, L.C. (2007). Plant responses to plant growth-promoting rhizobacteria. In: *New Perspectives and Approaches in Plant Growth-Promoting Rhizobacteria Research* (eds. P.A.H.M. Bakker, J.M. Raaijmakers, G. Bloemberg, et al.), 243–254. Dordrecht: Springer.

Van Loon, L.C., Rep, M., and Pieterse, C.M. (2006). Significance of inducible defense-related proteins in infected plants. *Annu. Rev. Phytopathol.* 44: 135–162.

Vargas, L., Santa Brigida, A.B., Mota Filho, J.P. et al. (2014). Drought tolerance conferred to sugarcane by association with *Gluconacetobacter diazotrophicus*: a transcriptomic view of hormone pathways. *PLoS One* 9: e114744.

Varnier, A.L., Sanchez, L., Vatsa, P. et al. (2009). Bacterial rhamnolipids are novel MAMPs conferring resistance to *Botrytis cinerea* in grapevine. *Plant Cell Environ.* 32: 178–193.

Veliz-Vallejos, D.F., van Noorden, G.E., Yuan, M., and Mathesius, U. (2014). A *Sinorhizobiummeliloti*-specific N-acyl homoserine lactone quorum-sensing signal increases nodule numbers in *Medicago truncatula* independent of autoregulation. *Front. Plant Sci.* 5: 551.

Venturi, V. and Keel, C. (2016). Signaling in the rhizosphere. *Trends Plant Sci.* 21: 187–198.

Vespermann, A., Kai, M., and Piechulla, B. (2007). Rhizobacterial volatiles affect the growth of fungi and *Arabidopsis thaliana*. *Appl. Environ. Microbiol.* 73: 5639–5641.

Vibhuti, M., Kumar, A., Sheoran, N. et al. (2017). Molecular basis of endophytic *Bacillus megaterium*-induced growth promotion in *Arabidopsis thaliana*: revelation by microarray-based gene expression analysis. *J. Plant Growth Regul.* 36: 118–130.

Vurukonda, S.S.K.P., Vardharajula, S., Shrivastava, M., and SkZ, A. (2016). Enhancement of drought stress tolerance in crops by plant growth promoting rhizobacteria. *Microbiol. Res.* 184: 13–24.

Waghmode, M., Gunjal, A., Patil, N., and Nawani, N. (2019). Role of rhizobacteria in drought tolerance. In: *Plant Growth Promoting Rhizobacteria for Sustainable Stress Management* (eds. R.Z. Sayyed, N.K. Arora and M.S. Reddy), 355–362). Singapore: Springer.

Wang, Y., Ohara, Y., Nakayashiki, H. et al. (2005). Microarray analysis of the gene expression profile induced by the endophytic plant growth-promoting rhizobacteria, *Pseudomonas fluorescens* FPT9601-T5 in *Arabidopsis*. *Mol. Plant Microbe Interact.* 18: 385–396.

Wang, C.J., Yang, W., Wang, C. et al. (2012a). Induction of drought tolerance in cucumber plants by a consortium of three plant growth-promoting rhizobacterium strains. *PLoS One* 7: e52565.

Wang, E., Schornack, S., Marsh, J.F. et al. (2012b). A common signaling process that promotes mycorrhizal and oomycete colonization of plants. *Curr. Biol.* 22: 2242–2246.

Wang, Y., Xu, H., Kou, J. et al. (2013). Dual effects of transgenic *Brassica napus* overexpressing CS gene on tolerances to aluminum toxicity and phosphorus deficiency. *Plant Soil* 362: 231–246.

Wasson, A.P., Pellerone, F.I., and Mathesius, U. (2006). Silencing the flavonoid pathway in *Medicago truncatula* inhibits root nodule formation and prevents auxin transport regulation by rhizobia. *Plant Cell* 18: 1617–1629.

Weller, D.M., Mavrodi, D.V., van Pelt, J.A. et al. (2012). Induced systemic resistance in *Arabidopsis thaliana* against *Pseudomonas syringae*pv. tomato by 2,4-diacetylphloroglucinol-producing *Pseudomonas fluorescens*. *Phytopathology* 102: 403–412.

White, R.A. III, Rivas-Ubach, A., Borkum, M.I. et al. (2017). The state of rhizospheric science in the era of multi-omics: a practical guide to omics technologies. *Rhizosphere* 3: 212–221.

Xie, X.G., Zhang, F.M., Yang, T. et al. (2019). Endophytic fungus drives nodulation and N2fixation attributable to specific root exudates. *mBio* 10: e00728–e00719.

Xu, J., Zhang, Y., Zhang, P. et al. (2018). The structure and function of the global citrus rhizosphere microbiome. *Nat. Commun.* 9: 4894.

Yamaji, K., Watanabe, Y., Masuya, H. et al. (2016). Root fungal endophytes enhance heavy-metal stress tolerance of *Clethra barbinervis* growing naturally at mining sites via growth enhancement, promotion of nutrient uptake and decrease of heavy-metal concentration. *PloS One* 11: e0169089.

Yang, H., Knapp, J., Koirala, P. et al. (2007). Enhanced phosphorus nutrition in monocots and dicots over-expressing a phosphorus-responsive type I H+-pyrophosphatase. *Plant Biotechnol. J.* 5: 735–745.

Yang, J., Kloepper, J.W., and Ryu, C.M. (2009). Rhizosphere bacteria help plants tolerate abiotic stress. *Trends Plant Sci.* 14: 1–4.

Yang, F.E.N., Jensen, J.D., Svensson, B. et al. (2012). Secretomics identifies *Fusarium graminearum* proteins involved in the interaction with barley and wheat. *Mol. Plant Pathol.* 13: 445–453.

Yoneyama, K., Xie, X., Kim, H.I. et al. (2012). How do nitrogen and phosphorus deficiencies affect strigolactone production and exudation? *Planta* 235: 1197–1207.

Zamioudis, C. and Pieterse, C.M. (2012). Modulation of host immunity by beneficial microbes. *Mol. Plant Microbe Interact.* 25: 139–150.

Zarkani, A., Stein, E., Röhrich, C. et al. (2013). Homoserine lactones influence the reaction of plants to rhizobia. *Int. J. Mol.Sci.* 14: 17122–17146.

Zhang, W.H., Ryan, P.R., Sasaki, T. et al. (2008). Characterization of the *TaALMT1* protein as an Al3+−activated anion channel in transformed tobacco (*Nicotiana tabacum* L.) cells. *Plant Cell Physiol.* 49: 1316–1330.

Zhang, X., Huang, Y., Harvey, P.R. et al. (2012). Enhancing plant disease suppression by *Burkholderia vietnamiensis* through chromosomal integration of *Bacillus subtilis* chitinase gene *chi113*. *Biotechnol. Lett.* 34: 287–293.

Zhang, N., Wang, D., Liu, Y. et al. (2014). Effects of different plant root exudates and their organic acid components on chemotaxis, biofilm formation and colonization by beneficial rhizosphere-associated bacterial strains. *Plant Soil* 374: 689–700.

Zhang, N., Yang, D., Wang, D. et al. (2015a). Whole transcriptomic analysis of the plant-beneficial rhizobacterium *Bacillus amyloliquefaciens* SQR9 during enhanced biofilm formation regulated by maize root exudates. *BMC Genomics* 16: 685.

Zhang, Y., Ruyter-Spira, C., and Bouwmeester, H.J. (2015b). Engineering the plant rhizosphere. *Curr. Opin. Plant Biol.* 32: 136–142.

Zhang, L., Wu, X.X., Wang, J. et al. (2018). *BoALMT1*, an Al-induced malate transporter in cabbage, enhances aluminum tolerance in *Arabidopsis thaliana*. *Front. Plant Sci.* 8: 2156.

Zheng, H., Zhong, Z., Lai, X. et al. (2006). A LuxR/LuxI-type quorum-sensing system in a plant bacterium, *Mesorhizobiumtianshanense*, controls symbiotic nodulation. *J. Bacteriol.* 188: 1943–1949.

Zhou, N., Zhao, S., and Tian, C.Y. (2017). Effect of halotolerant rhizobacteria isolated from halophytes on the growth of sugar beet (*Beta vulgaris* L.) under salt stress. *FEMS Microbiol. Lett.* 364: fnx091.

Zwirglmaier, K. (2005). Fluorescence *in situ*hybridisation (FISH)–the next generation. *FEMS Microbiol. Lett.* 246: 151–158.

7

Plant Communication with Associated

Its Components, Composition and Role in Maintaining Plant Homeostasis

Dibyajit Lahiri[1], *Moupriya Nag*[1], *Sayantani Garai*[1], *Bandita Dutta*[2], *and Rina Rani Ray*[2]

[1] *University of Engineering and Management, Kolkata, West Bengal, India*
[2] *Maulana Abul Kalam Azad University of Technology, Kolkata, West Bengal, India*

7.1 Introduction

The association of plants with the microbiome dates back to almost billions of years since their emergence on land (Vorholt 2012; Hardoim et al. 2015; Hurek et al. 2015). The existence of microbes with plant metapopulation is also described as plant microbiota or phytomicrobiome within the endosphere, phyllosphere and the rhizospheric region that has an immense importance for controlling plant health and growth (Vorholt 2012; Hardoim et al. 2015; Brader et al. 2017; Lemanceau et al. 2017). The soil is the consortia of large quantities of metabolically active populations of beneficial microbes which efficiently performs agricultural practices and soil remediation. The microbiota existing within the soil has a wide range of adaptability which ranges its adaptation in various types of the environment having a faster rate of growth and its efficient ability to catabolize various types of natural and xenobiotic chemicals. The majority of these microbial populations are associated with the rhizospheric region of the roots and the quantity of these microbial cells ranges from 109 to 1012 g of soil (Whipps 1990) that is equivalent to the biomass of 500 kg ha^{-1}. The exudations or the rhizospheric deposits from the root into the soil stimulate the microbial growth at the immediate periphery around the root (Brimecombe et al. 2001). This provides the rhizosphere an ideal site for the growth of the microbial species.

Recently, plant microbiome has emerged as a substantial important aspect since it influences plant growth, productivity, and health. It consists of microbial populations that are at constant interaction with that of the plant species. These microbial species include the group of endophytes that settle at the inner parts of the plants (Liu et al. 2017). The group of endophytic bacterial cells tries to settle either extracellularly or intracellularly due to the abundance of large quantities of carbohydrates, proteins, and amino acids. These bacterial cells largely influence plant health, growth, and productivity (Bécard 2017). For example,

Phytomicrobiome Interactions and Sustainable Agriculture, First Edition. Edited by Amit Verma, Jitendra Kumar Saini, Abd El-Latif Hesham, and Harikesh Bahadur Singh.

metabolites such as Fe^{3+} and $PO4^{3-}$ produced by the microbial microcolonies act as essential nutrients for the sustenance of plant species. It has also been found that the process of nitrogen fixation is performed by the group of certain diazotrophic endophytes (Hardoim et al. 2015). The group of endophytic bacteria within the seeds play a pivotal role in the process of germination and enhances the survival rate (Shahzad et al. 2018). The interaction of the microbial species developing at the rhizospheric region with the plants depends upon the type of host and the soil environment in which it is living. It has been seen that the host plant and its developmental stages greatly influence the rhizospheric microbiome (Chaparro et al. 2013; Peiffer and Ley 2013). The phytomicrobiome of the plant is composed of a gene pool of diverse functionality that comprises prokaryotes, viruses, and eukaryotes which remain associated with various plant habitats that comprise a wholesome plant, leaves, flowers, seeds, shoots, and roots (Rout and Southworth 2013). It has been seen that the microbial species develop a consortium containing mixed species of bacterial cells surrounded by extracellular polymeric substances (EPS) that help the cells to adhere to the plant surface known as a biofilm. This EPS is composed of various macro- and micronutrients like carbohydrates, proteins, and nucleic acids that provide nutrition to the developing cells and protects the cells from extremes of conditions like pH, salinity stress, temperature fluctuations, deficiency of nutrients and desiccation. The microbial species thriving within the biofilm exist synergistically or in antagonistic relation by keeping an intimate relationship with plant roots. The biofilm thus formed increases the plant resistances against the environmental stresses and also help in developing tolerance against protozoans and antimicrobial agents. The biofilm also provides consortial metabolism and the mechanism of horizontal gene transfer (HGT). As the biofilm is composed of a high-density group of microorganisms that provides enhanced functioning which single cells are unable to perform like the formation of excreted metabolites or certain groups of exoenzymes that are only effective above a certain limiting value. The phenomenon of biofilm formation is to allow the microbial species to remain at its localized state so as they can bring about its antagonistic or beneficial interactions with that of the plant species. For example, the biofilm formed by the endophytic group of bacterial species can produce plant hormone indole-3-acetic acid, siderophore, and solubilize phosphates (Bastian et al. 1998; Gaiero et al. 2013; Lebeis 2014). The group of these bacterial species also play a predominant role in the process of fixing atmospheric nitrogen to nitrogenous compounds like ammonium and nitrates within the plants (Bhattacharjee et al. 2008; Santi et al. 2013). Different bacterial species like *Rahnella*, *Klebsiella*, *Pantoea*, *Herbaspirillum*, and *Gluconobacter* helps in the growth of plant species at the time of poor nutrient availability (Doty et al. 2009). This review focuses on the phyto-microbiome interactions that lead to the development of the mutualistic relationship between the plant and the microbiomes existing at the root rhizosphere (Figure 7.1).

7.2 Biofilm and Rhizospheric Interactions

We often consider the biofilm to be multi-drug–resistant group of microcolonies embedded within EPS that are mainly responsible for various chronic infections in mammals like urinary tract infection, a surface wound infection, infection upon prosthesis being embedded within body, oral infection, dental plaque infections and many more (Hall-Stoodley et al. 2004). Thus,

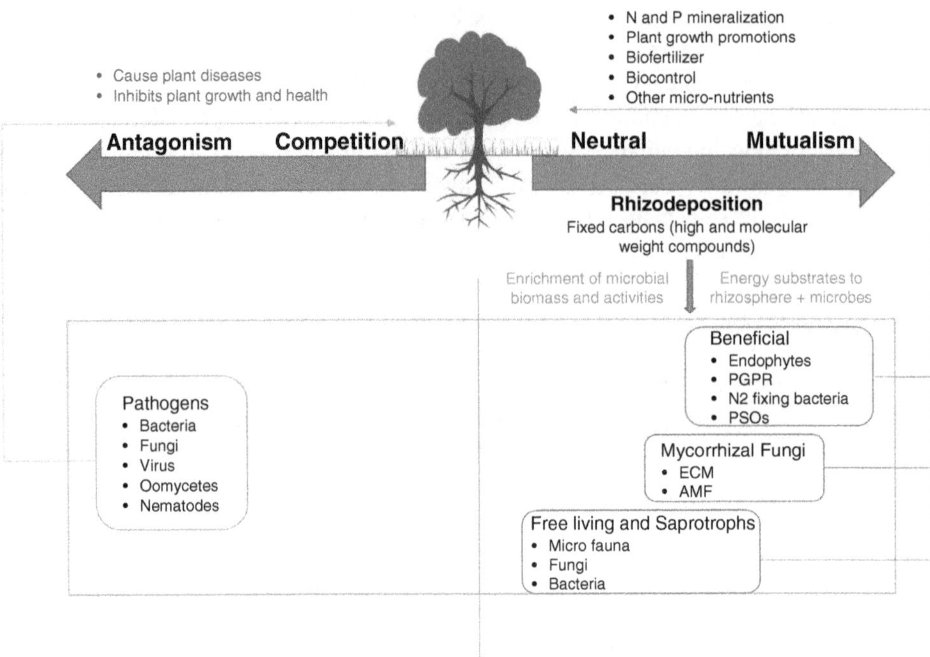

Figure 7.1 Phyto-rhizospheric microbial interaction.

these bacterial species result in the development of biofilm upon the plant surface based on the same mechanism by which they establish within the animal system. In plants, the potent biotic surface which provides a surface for adherence is the root rhizospheric region. Various research works highlight the dynamics and the structure of the biofilm formation taking Pseudomonas species as a model gram-negative organisms and *Staphylococcus aureus* as gram-positive biofilm-forming organisms (Costerton et al. 1995; O'Toole et al. 2000). There exist a large number of multispecies biofilms like the dental plaque biofilm or oral biofilm which forms a consortium having a symbiotic relationship with its host surface. Similarly, various reviews have highlighted the association of biofilm with the surface of the plant and the rhizospheric region (Fujishige et al. 2006; Danhorn and Fuqua 2007).

7.3 Biofilm Formation at the Root Rhizosphere

The functioning of healthy roots is greatly dependent upon the presence of microbial species and on the contrary, root greatly influences the microbial colonies existing at the rhizospheric region of the plants (Merckx et al. 1986). The diffusion of nutrients from the root mucilage through the film upon an abiotic solid surface provides the growth-limiting nutrient for the attachment of the sessile groups of cells (Saunders and Bazin 1973) within the soil column. It was observed that the biomass present within this microbiota comprises 1.6% of photosynthetically convert carbon-14, 15% of this carbon enters into the root rhizosphere and the remaining carbon remains within the soil. It has been observed that 14–20% of the biomass being formed is present at the intermediate zonal region of the root. The bacterial cells adhere to the plant surface with fimbriae/pili and EPS. The interchangeability of proteins and carbohydrates acts important material for the attachment of the cells. It has been observed that bacterial cells belonging to the family Rhizobiaceae which includes *Agrobacterium tumifaciens* produce rhidacedesin and rhizobial which is a small calcium-binding protein that helps them to attach along the root rhizospheric region (Swart et al. 1993; Dardanelli et al. 2003). The rhicadhesin is considered to be the group of Rhizobium adhering proteins which help in the agglutination of the sessile cells and thus helps in the formation of biofilm. The organisms like *Rhizobium leguminosarum* and *Rhizobium etli* produces adhesions that are secreted from the type I secretory system PrsDE that also helps in the mechanism of the formation of the biofilm. They identify the receptors being present upon the pole of the cell having a lectin domain (Ausmees et al. 2001). Rhicadhesin is used by *R. leguminosarum* for biofilm formation at around pH6.5 and with an increase in the acidic conditions prevailing in the soil it uses a type of adhesin glucomannan for the process of biofilm formation. In bacterial species like *Azorhizobium brasilense* cellular aggregation takes place in the presence of polysaccharide-containing extracellular arabinose. Bovine serum albumin conjugates help in carrying agglutination of *A. brasilense* by creating conjugates with sugar residues that remain blocked in the presence of arabinose (Castellanos et al. 1998). This provides evidence that the cells of the organism carry lectin-like proteins that attach to the plant via the arabinose residue. Bacteria like *Xylella fastidiosa* allow the cell-to-cell contact with the help of hemagglutinins. The mutant form of hemagglutinins forms a single layer in the xylem vessel rather than the general type of biofilm usually being shown by the wild types. This mutant type causes the damage of tissues and increases its rate of proliferation within the monolayer (Monier and Lindow 2005). Pseudomonads like *Pseudomonas putida* produces a type of adhesin protein known as large adhesion protein LapA that helps in the attachment of these cells to the abiotic and biotic surfaces (Espinosa-Urgel et al. 2000). The mechanism of the shift of the planktonic to the sessile group of cells and its irreversible attachment by *Pseudomonas fluorescens* requires the assistance of transporter protein like that of the ABC transporter (Figure 7.2).

7.3.1 The Components of Biofilm Matrix

The settlement of plant by the group of endophytic bacteria involves the recognition of the exudates from the root that enhances the movement of the bacterial cells toward the plant, attachment of the bacterial cells to that of the root surface, development of biofilm, penetration of the microcolonies to the root surface, attachment of these cells with the internal parts of the plant body (Kandel et al. 2017). These steps are governed by a set of biomolecules that bring about improvisation at the genetic level both in the case of bacterial cells and the host plant system. This is brought about by a multicomponent approach by taking a multicomponent-relevant mutant that aims in providing a synergistic omic approach (Hardoim and Hardoim 2017). It has been found that the root exudates contain organic components like proteins, amino acids and organic acids which provides nutrition to these group of endophytic bacteria and thus helps in settling the organisms at its proximity (Bulgarelli et al. 2012; Kawasaki et al. 2016) The

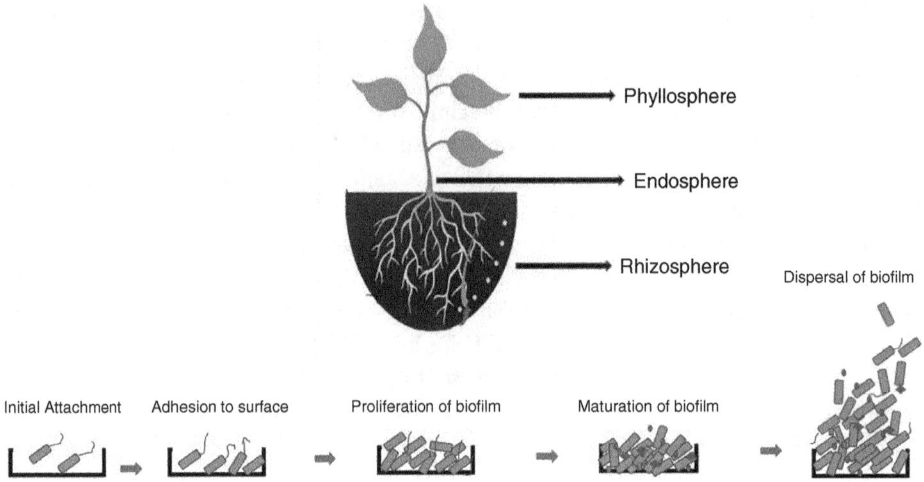

Figure 7.2 Formation of biofilm at the root rhizosphere.

exudates contain chemical metabolites or substances that help the endophytic bacterial cells to bring about communication with that of the host plant and thus help in its settlement. For example, the oxalate being present within the exudates of the wheat plant helps the bacterial strain *Burkholderia phytofirmans* to adhere to its surface. It has also been observed that the bacterial cells develop the biofilm upon the rhizospheric region by the mechanism of quorum sensing (QS) that helps in developing the intimate connection between the cells of the microcolonies.

7.3.2 Bacterial Quorum Sensing

QS is a cell density–dependent cellular communication among bacterial cells that uses signaling molecules that help in cellular colonization and results in the development of biofilm which helps in maintaining a symbiotic relationship. The signaling molecule N-acyl-homoserine lactones (AHLs), autoinducer-2 (AI-2), and 2-heptyl-3-hydroxy-4-quinoline (PQS) help in QS mechanism. *Pseudomonas aeruginosa* contains a lasI gene at the upstream of the quorum-sensing gene, involved in the development of biofilms in wild-type phenotype as well as in undifferentiated mutant phenotype (Davies et al. 1998). Biofilm formation by *Bacillus subtilis* is mediated by other species of the same genus. As this microorganism is gram-positive, the QS mechanism is mediated by AI-2 which is a metabolic product of 4,5-dihydroxy-2,3-pentanedione (DPD) in the following pathway: S-adenosylmethione (SAM) is converted to S-adenosylhomocysteine (SAH). The enzyme Pfscatalyzes SAH to S-ribosyl homocysteine. The luxSgene encodes LuxS which acts on DPD to produce AI-2 and homocysteine (Brito et al. 2013). The LuxS protein is responsible for the synthesis of AI-2, which accumulates extracellularly and then sensed and internalized by the lsr operon in the bacterial cell (Figure 7.3 and Table 7.1).

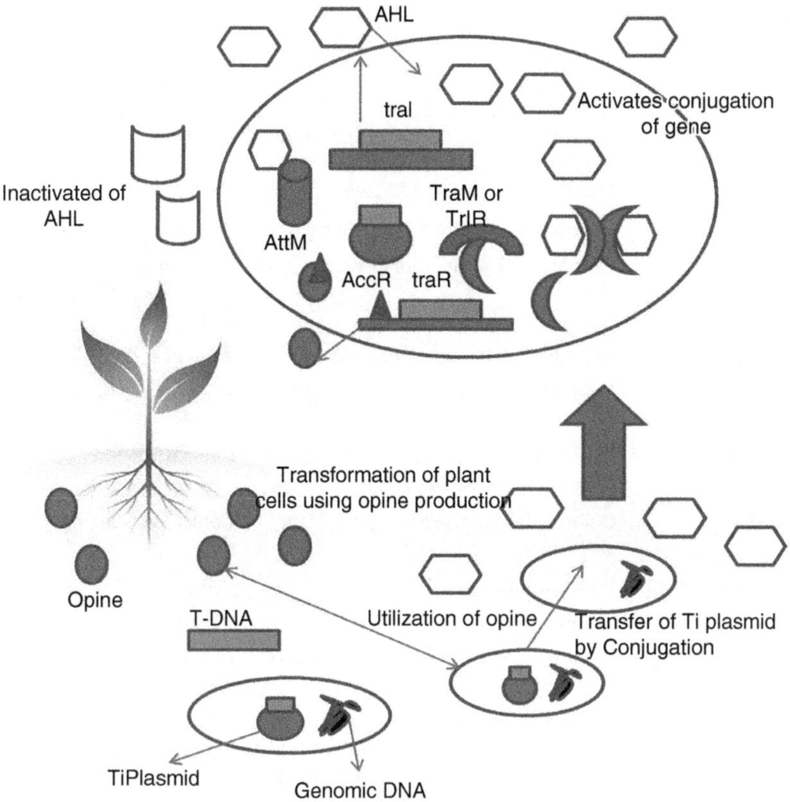

Figure 7.3 Quorum sensing at the root rhizosphere.

Table 7.1 List of microorganism forming biofilm and their mode of attachment.

Organism	Phenotype	Characteristics
Agrobacterium tumefaciens	It results in the development of disease in pea	Focalization
Bacillus polymyxa	Colonization at the root of cucumber	PGPB
Shewanella putrefaciens strain CN-32	Formation of biofilm on mineral surfaces	Metagenomic condition
Pseudomonas chlororaphis	Colonization at the root of the wheat plant	Biological control
Bacillus subtilis	Root colonization of *Arabidopsis thaliana*	Biological control
Burkholderia cepacia and *Burkholderiaceno cepacia*	Develops disease condition in wheat and onion	Develops diseases
Azorhizobium brasilense	Root colonization of wheat	PGPR

(*Continued*)

Table 7.1 (Continued)

Organism	Phenotype	Characteristics
Cyanobacteria spp.	Consortia of biofilm formation in the association of *Azotobacter, Pseudomonas* sp., and *Rhizobium*	Biological control
Enterobacter agglomerans	Colonization at the root of cotton	Biological control
Klebsiella pneumoniae	Colonization at the root of wheat	Helps in the mechanism of nitrogen fixation
Microsphaeropsis spp.	Colonization at the root of onion	Biological control
Paenibacillus lentimorbus	Colonization at the root of chickpea	Tolerance to heavy metals
Paenibacillus polymyxa	Root colonization; significant against crown rot disease in peanut	Biocontrol
Pantoea agglomerans	Colonization at the root of chickpea and wheat	Control the moisture
Azorhizobium caulinodans	Root colonization of rice	PGPR
Pseudomonas aureofaciens	Colonization at the root of wheat	Biocontrol
Acinetobacter calcoaceticus P23	Colonization at the root of duckweed	Helps in bioremediation.
Pseudomonas putida	Colonization within the root of maize and development of competitive colonization with the rhizosphere of *Arabidopsis thaliana*	It helps in bioremediation and develops tolerance against desiccation
Bacillus cereus and *Bacillus pumilus*	Colonization of root in wild group of barley	Significant tolerance against heat, salt and desiccation
Bacillus amyloliquefaciens S499	Colonization at the root within maize, tomato and *Arabidopsis thaliana*	PGPB, biocontrol
Rhizobium alamii	Root colonization of *Arabidopsis thaliana* and rapeseed	Tolerance to heavy metals
Rhizobium leguminosarum bv. *viciae*3841	Root colonization of various legumes	Fixation of nitrogen and tolerance against desiccation
Rhizobium (Sinorhizobium) sp. strain NGR234	Root colonization of numerous leguminous species, competitive colonization in the rhizosphere of cowpea	Nitrogen fixation, PGPR
Pseudomonas aeruginosa PAO1	Disease of the poplar tree	Pathogenesis
Pseudomonas fluorescens	General rhizosphere colonization	Biocontrol
Pseudomonas brassicacearum	Root colonization of *Arabidopsis thaliana*	Biocontrol

7.4 Genetic Features Responsible for Bacterial Cell Adhesion to Plant System

Various biomolecules are associated in these processes which mediate the dynamic changes in the bacterial gene expression in colonized plants (Kandel et al. 2017). Five different stages help the microorganism to colonize in the plant rhizoidal zone.

a) Recognition of exudates from roots thus causes its movement toward the plant.
b) Attachment to the root surface
c) Formation of biofilm
d) Penetrance to the root surface
e) Colonization to the internal parts of a plant

7.4.1 Chemotaxis Motility

The endophytic bacteria can enact to the signals produced by plants during colonization. Chemotaxis followed by attachment to the rhizoplane of the free-living bacteria is the primary stage of colonization which is mediated by methyl-accepting chemotaxis proteins (MCPs). There are transmembrane protein/sensor molecules that surround the bacteria and controls the attachment or away from repellents (Scharf et al. 2016). The che inactivation mutant in *Herbaspirillum seropedicae* SmR1 is one of the examples of plant colonization with the help of MCPs. Sixty-six different genes are encoding MCPs in SmR1, in which nine are involved in root and differentially expressed within the cell. If one of these nine genes get inactivated (e.g. Hsero_3720), it depletes the ability of attachment with roots in mutant type strains as compared to the wild type. The mechanism of MCP protein inactivation possessing the tlp1 gene prevents the terminal electron acceptors like nitrogen and oxygen in various types of the mutant rhizospheric strain like *A. brasilense* Sp7 (Greer-Phillips et al. 2004). The presence of flagella and pili helps microorganisms in their movement toward plant roots. One exception is *Azoarcus* sp. BH72 mutant that lacks flagellum but attaches to the root surface. *Azoarcus* sp. BH72 failed to elicit a plant defense response as their flagella were not recognized by plant but pili are participating in plant colonization at the surface (Buschart et al. 2012). When the pilus-associated pilX gene mutationally gets inactivated, it reduces root colonization of mutant due to impaired twitching of its motility, which also stressed to generate mutants in other genes that code for pilus. Type IV pili *H. seropedicae* SmR1 helps in colonization but mutations in pilus locus decrease the adhesion of cells with one another as well as to the surface and increase motility and colonization efficiency, e.g. rhizosphericstrain *Pseudomonas simiae* WCS417r (Cole et al. 2015).

7.4.2 Substrate Utilization and Transport

Root exudates are the source of nutrients which along with chemotaxis helps microorganisms to colonize in the rhizosphere of the plant. Transporters and enzymes help those bacteria to utilize root exudates, which consist of sugars, aromatic acids, aliphatic acids, amino acids, polysaccharides, sterols, fatty acids, phenolics, plant growth regulators, proteins, enzymes, and secondary metabolites (Badri et al. 2009). Plant root exudates play a role in defense mechanisms as it responds to exogenous stimuli. For example, the presence of oxalate has been noticed in lupin and maize, which is utilized by bacterial cells during oxalate decarboxylase activity (Baetz and Martinoia 2014). Inactivation of oxc gene that encodes oxalate dehydrogenase results in a decrease of colonization of maize and lupin in *B. phytofirmans* PsJN. The enzyme encoded by the AB gene utilizes trehalose and mannitol present in root exudates and converts them in 3-keto derivatives. The thuA and the B genes present in *Rhizobium meliloti* 1021 induces colonization upon the root surface. Disruption of these genes led to the impaired colonization and utilizes trehalose and mannitol (Jensen et al. 2005; Ampomah et al. 2013). In another study, *P. simiae* WCS417r wild type strain was found to utilize the carbohydrate source in root–bacterial interactions. But mutant strains are unable to utilize galacturonate, galactose, inosine or 2-deoxyribose and glucose reduces the colonization. Auxotrophy of some specific amino acids helps the microorganism for their survival in the plant-associated environment. It helps those microbes to respond against the plant-derived compounds and improve colonization and helps the host plant to grow (Cole et al. 2015). Endophytic bacteria are capable of producing polyhydroxyalkanoates (PHA) and polyhydroxy butyrate (PHB), has developed the tolerance in bacteria to various environmental and biotic factors. It was also found in the bacterial cells of *H. seropedicae* SmR1 attached to the root (Pankievicz et al. 2016). A mutant strain of this bacteria unable to produce PHB results in impaired colonization, which was established in a lower number of epiphytic and planktonic cells. Lack of PHB production disturbs the expression of several genes due to the increase of oxidative stress as it alters redox balance (Batista et al. 2018). PHA biosynthesis gene is present in *A. brasilense* FP2, helps in primary wheat–endophyte interaction (Camilios-Neto et al. 2014).

One of the important sites of methanol formation is the plant cell wall, which is a by-product of pectin methylesterase activity. Methanol can be utilized by some endophytic bacteria for their energy source. A mutant strain of facultative methylotrophic bacteria *Methylobacterium extorquens* AM1, is unable to oxidize methanol to formaldehyde and formaldehyde to CO_2 compared to the wild type strain, which can colonize the leaves and roots of *Medicago truncatula* (Sy et al. 2005). An alternative energy source ethanol is not utilized by the mutant Azoarcussp. BH72 and exhibits a reduction of root colonization in rice (Krause et al. 2011). The enzyme alcohol dehydrogenase is regulated by genes being present in *H. seropedicae* SmR1, when it is attached to the root (Alberton et al. 2013). Methanol dehydrogenase was regulated by *Bacillus mycoides* EC18 when it is exposed to root exudates. Nitrogen fixation helps to modify plant environment when diazotrophic endophytic bacteria get accumulated. Nif genes get upregulated when the bacteria are attached to the surface, and encodes the enzyme which is responsible for the nitrogen fixation, and also helps in the growth-promoting mechanism of the plant (Camilios-Neto et al. 2014; Pankievicz et al. 2016).

7.4.3 Lipopolysaccharide and Membrane Proteins

Biofilm is mainly composed of EPS that contains lipopolysaccharide (LPS) (mainly in gram-negative bacteria). Attachment and colonization of bacteria mainly depend on this LPS and membrane protein. There are three components present in LPS which help in bacterial colonization, lipid A, that is usually observed on the outer bacterial membrane followed by the core region and the region of O-antigen comprising five sugar units. Biosynthesis of the O-antigen region, which helps in the early stage of colonization, is regulated by four genes-rfb ABCD (Camilios-Neto et al. 2014), which are upregulated by maize seedlings, apigenin, flavonoids, and naringenin are found at the root exudates. Monosaccharide rhamnose, which is detected as a part O-antigen, biosynthesis is regulated by rfbB and rfbC. Mutation of any of these genes reduces the attachment and endophytic colonization to the root surface 100-fold of maize by *H. seropedicae* SmR1. This also reduces the activity of bacteria in the presence of polymyxin B sulfate, detergent sodium dodecyl sulfate (SDS), and phytohormone SA. The mutation of rfbD gene in *A. brasilense* also results in the similar activities (Jofré et al. 2004; Balsanelli et al. 2010). The biosynthesis of O-antigen is a complex process that is catalyzed by the enzyme O-antigen ligase (encoding gene is upregulated during colonization of maize roots by *H. seropedicae* SmR1) helps in ligation of O-antigen repeat subunits to the lipid A-core. Bacterial LPS can bind to the lectin protein, results in agglutination. Mutation inactivates the O-antigen ligase, failed to interact with lectin protein, resulted in an impaired attachment to the root surface, and decreases colonization efficiency.

Membrane protein of bacteria consists of muropeptide permease, which is necessary for peptidoglycan recycling of the bacterial cell wall. Mutation inactivates gene that encodes permease, alters the LPS biosynthesis in *H. seropedicae* SmR1, reduces endophytic bacterial colonization 10-fold, and increases sensitivity to SDS. *Gluconacetobacter diazotrophicus* PAL5 has a higher amount of lipoprotein Omp16 at the outer membrane, in response to plantlets (Dos Santos et al. 2010). OprF, which is another major membrane protein porin, mainly found in *P. fluorescens* CHA0, mediate the polar solutes pass across the outer membrane and maintain the membrane integrity. Inactivation of the gene encodes this OprF results in decrease attachment and colonization in cucumber and tomato root (Crespo and Valverde 2009)

7.4.4 Plant Cell Wall Modification

Plant cell wall which helps the plant in its defense mechanism consists of cellulose, xylulose, and pectin, can be degraded by the plant cell wall degrading enzyme, secreted by endophytic bacteria. Plant cell wall degradation facilitates the bacterial movement toward the plant root and controls their movement inside the plant parts. Mutation in *Azoarcus* sp. BH72 results in the decreased ability to colonize root as it is devoid of endonuclease activity and losses the ability to spread to the upper part of the plant. The plant cell-wall–degrading enzymes, released from bacteria is regulated by root exudates. When the degrading enzyme from *B. mycoides* EC18 gets exposed to root exudates, it regulates the gene encoding pullulanase, hydrolase, and chitin-binding protein. Bacterial cells are capable of utilizing plant-derived compound, O-glycosyl hydrolase which it regulates by the bacterial gene (Yi et al. 2017). Expansins, a nonenzymatic protein found in the plant cell wall, enable cell wall extension without any actual change and covalent modification of wall polymer. The extension activity of expansin in wild type strain of *B. subtilis* 168, is very low in vitro. The activity of expansin in mutant *B. subtilis* 168, to colonize root with the extension activity gets reduced (Ker et al. 2008), though limited expansin activity is found in bacteria like

Bacillus, *Xanthomonas*, *Xylella*, *Ralstonia*, and *Erwinia*. Expansin is not required for successful colonization but it helps endophytic bacteria to enter into the plant through wounds and natural opening like stomata on young leaves and stems.

7.4.5 Adhesion and Biofilm Formation

Biofilm is a syntrophic consortium of microorganisms that comprise EPS which protects the embedded bacterial cell from a harsh environment. EPS contains water, protein, polysaccharides, extracellular DNA (eDNA), RNA, and ions (Zur et al. 2016). In some species, cellulose is another important component that helps in biofilm formation. The mutation of species along with genetic material of biofilm helps in attachment and colonization of bacteria, e.g. PAL5 exopolysaccharide (EPS) of *G. diazotrophicus* helps in biofilm formation. EPS biosynthesis is activated by an enzyme that is encoded by gumD gene, whose inactivation results in decreased rhizospheric and endophytic colonization of rice root by PAL (Meneses et al. 2011). Mutant *R. leguminosarum* biovar *viciae* 3841, which cannot produce EPS and glucomannan, is deficient in biofilm formation and attachment in the rhizospheric zone. Another example of decreased colonization and biofilm formation is bcsA gene of *Salmonella enterica* (Barak et al. 2005; Williams et al. 2008). There are several genes required for cellulose production. Inactivation of any gene can decrease the biofilm formation and root colonization of the bacteria, e.g. inactivation of wssD gene in *Herbaspirillum rubrisubalbicans* M1 reduces the attachment to the root surface (Monteiro et al. 2012). Absence of colonic acid (a polysaccharide having a repeat unit with D-glucose, l-fucose, d-galactose, and d-glucuronate) biosynthesis reduce the colonization ability of the *S. enterica*. Surface-associated protein LapA in pseudomonads helps in biofilm formation through the regulation of cell hydrophobicity through the cell to cell attachment mechanism. The LapA mutant results in poor impairment resulting in decrease root colonization and poor biofilm formation (Martinez-Gil et al. 2010; Ainelo et al. 2017). Bacteria, which can adhere to the root surface can also colonize the plant interior. Bacterial flagella and pili and proteins like curli and hemagglutinins regulate the bacterial cell adhesion to the various surface, cell aggregation, and biofilm formation (Barnhart and Chapman 2006). In *S. enterica* agfA and agfB plays an important role in biofilm formation. Gene agfA is for encoding the secreted curli subunit, whereas gene agf B is for the nucleator with the exposed surface, around which curli amyloid fibers form. Inactivation of agf B gene affects initial attachment and colonization over the root surface but inactivation of agfA did not affect. Endophytic bacteria *H. seropedicae* SmR1 contains gene Hsero_1294 and fhaB that encode hemagglutinins, which has an important role in pathogenesis and helps in an attachment to the root surface (Taghavi et al. 2010; Pedrosa et al. 2011; Mitter et al. 2013).

7.4.6 Stress Protection

Osmotic pressure, different pH, and oxygen availability helps the colonizing bacteria to adapt to a new environment during the transition from host rhizosphere to endosphere-inducing invasion to overcome plant defense mechanism that involves quick production of reactive nitrogen species (RNS) and reactive oxygen species (ROS) and which get scavenged by enzymes like nitric oxide reductase, glutathione S-transferase, glutathione peroxidase, catalase and secreted by endophytic bacteria (Santoyo et al. 2016). The detoxification of

ROS is enhanced by the superoxide dismutase and glutathione reductase at the early stages of colonization by *G. diazotrophicus* PAL5. The mutant strain of this bacteria reduces colonization efficacy (Alqueres et al. 2013). Stress proteins help in plant defense mechanisms by the upregulation of these genes encode these proteins. For example, upregulation of stress-induced protein, stress-response protein, and general stress protein was observed when *B. mycoides* EC18 wild-type strain get exposed to root exudates (Yi et al. 2017). Phytoalexin, which is produced by plants, helps in the growth and survival of rhizospheric and endophytic bacteria through the activity of the efflux pump. The efflux pump SmeDEF in *Stenotrophomonas maltophilia* D457, upregulated in presence of root exudates, enhances significant impairment and root colonization (Garcia-Leon et al. 2014). ABC is an ATP-binding cassette, a multi-drug transporter present in *H. seropedicae* SmR1 makes the strain resistant against naringenin and quercetin. Inactivation of the gene encoding this transporter decreases the root colonization.

Another ABC transporter putrescine gene gets upregulated by *H. seropedicae* SmR1. This polyamine regulates bacterial response to osmotic stress due to the higher osmolarity in the rhizospheric region of the plant. Another bacterial mutant strain of *Paraburkholderia kururiensis* M130 of NodT family contains the outer membrane efflux transporter. The rhizospheric bacteria involved in the systematic resistance mechanism results spatio-temporal difference in plant defense system results in the production of siderophores, an ion chelator, and promotes growth mechanism (Barak et al. 2005; Badri et al. 2009). Though *Serratia marcescens* have 90–166 siderophore in mutant type, rhizosphere colonization remained the same as the wild type with a reduction of endophytic colonization. ROS uses free iron molecules to produce highly toxic hydroxyl radicals by a Mechanism of Fenton reaction.

The reduction of Siderophore also reduces the bacterial ability to detoxify ROS (Press et al. 2001). 2,3-butanediol, in *B. subtilis* 168, is another molecule which elicits defense against pathogen and protects helpful pathogens from plant root exudates, and also promote growth. Mutant strain decreases the ability to persists in the rhizosphere.

7.4.7 Bacterial Secretion System

Bacterial secretion systems involved in plant defense mechanisms as well as plant–endophyte interactions. The secretion of protein helps the bacteria to get attached to the plant instead of getting eliminated by the plant's immune system (Press et al. 2001). Eight different types of secretion system (type I to the type VII and Sec and Tat) is present in gram-negative bacteria and six different secretion system (Sec, Tat, SecA2, Sortase, Injectosome and type VIII) present in gram-positive bacteria. Type 3 secretion system (T3SS) present in gram-negative bacteria secretes effector protein which is mainly found inner bacterial membrane between periplasmic space and outer membrane. The translocation of this protein manipulates the immune defense response and host metabolism. The efficacy of the secreted effector proteins by T3SS, which is maintained by small cytosolic chaperones helps in the colonization of bacteria. Mutation in *H. rubrisubalbicans* M1 T3SS results in less colonization in endophytes (Thomas et al. 2005; Schmidt et al. 2012). The absence of Tir chaperone proteins in mutant *P. kururiensis* M130 results in the reduction of colonization in the root (Levy et al. 2018). Colonization capacity controlled by type VI secretion system (T6SS), also gets decreased in mutant *Azoarcus sp.*

BH72 compared to the wild type strain. This happens due to the translocation of the effector protein by T6SS eliciting local host defense mechanism that limits host–plant interaction and endophytic colonization (Shidore et al. 2012). The gene-regulating type II and type IV secretion systems get upregulated by *B. phytofirmans* PsJN during colonization in potato (Coutinho et al. 2015).

7.4.8 Transcriptional Regulators and Sensor Proteins

Transcriptional regulators help microorganism to adapt to the environmental changes and regulate migration toward the plant to colonize. The biofilm formation by bacteria gets regulated via the QS mechanism. DSF (diffusible signal factor), a molecule for QS regulates motility of the cell, the formation of biofilm, and efflux of multi-drug in the wild-type bacteria *S. maltophilia* R551-3. Mutant bacteria are devoid of DSF thus unable in colonization and growth promotion (Alavi et al. 2013).

The production of acyl-homoserine lactone by the mutant strain of *P. fluorescens* 2P24 reduces the colonization upon the rhizospheric region. QS mutants (bpI.1 and bpI.2) of *B. phytofirmans* PsJN bacterial controls the colonization on *Arabidopsis thaliana* (Wei and Zhang 2006; Zuniga et al. 2013). The RpoS is a gene regulating biofilm formation and adhesion and also regulates the sigma factor. The mutation of this gene reduces the attachment to the plant root *S. enterica* and biofilm formation. RpoS mutant strain lacks curli and cellulose production. *B. mycoides* EC18, an endophytic strain during responding to root exudates has increased expression of sigma-28 factor, which regulates flagellin, chemotaxis, and motility-related genes (Yi et al. 2017). There is another transcription factor GreA which helps in plant-bacteria interaction, mainly observed in the proteome of *G. diazotrophicus* PAL5. Mutant GreA is unable to adapt to the alteration in hyperosmotic and salt stresses which results in ineffective symbiosis in rhizobia (Dos Santos et al. 2010). The secondary messenger; Cyclic di-GMP regulates bacterial motility and formation of biofilm. C-di-GMP formation and degradation is regulated by a bi-functional enzyme Diguanylate cyclase/phosphodiesterase with a PAS/PAC sensor. Deletion of the gene encoding this enzyme from the genome of *Azoarcus* sp. BH72, reduces root colonization due to the alteration of the c-di-GMP level (Shidore et al. 2012; Valentini and Filloux 2016). Plant-associated bacteria from the Lac-I family are enriched in transcription regulators. The regulators controlled by Lac-I family are involved in carbohydrate metabolism and transport (Levy et al. 2018). The endophytic *B. mycoides* EC18 bacterial strain and soil isolates have the differences in the transcriptomic regulators' expression in response to root exudates. There are five transcriptional regulators including IclR, that facilitate multi-drug resistance and degradation of the aromatic compound and sigma-28 factor, which is upregulated by the endophytic strain. The transcriptional regulator QseB1, that regulates flagellar regulon, get downregulated by the strain *Azoarcus* sp. BH72, in response to root exudates. The initial colonization depends on flagella-based movements and attachment to the surface (Scott 1972).

Two-component systems are regulating respiratory membrane-bound nitrate reductase, comprising of membrane sensor protein NarX and DNA-binding regulator NarL that get upregulated in the presence of plant (Camilios-Neto et al. 2014). The fnr1 and fnr2 genes present in *H. seropedicae* SmR1, encodes frn like protein that regulates intracellular redox potential and expression of other genes depending upon oxygen level change (Pankievicz et al. 2016) (Table 7.2).

Table 7.2 List of endophytic bacterial mutant having genetic alteration and reduction of colonization.

Name of the microorganism	Plant associated	Gene	Mode of action	Function	References
1) **Chemotaxis, Motility**					
A. brasilense Sp7	*Triticum aestivum*	*tlp1*	transducer-like protein 1, chemoreceptor-like protein	Reduced colonization in roots	Greer-Phillips et al. (2004)
		mot3	bacterial flagellar motility	Root adsorption capacity decreases significantly	
Azoarcus sp. BH72	*Oryzasativa ssp. japonica cv.* Nipponbare	*pilX* (azo2916)	type IV fimbrial biogenesis protein PilX	Significant reduction in root colonization	Shidore et al. (2012)
	Oryzae sativa	*pilT* (azo3468)	type IV pilus retraction protein	Endophytic colonization reduces, 50% decrease in surface colonization	
		pilA (azo3355)	major structural component of the pilus structure	Remarkable decrease in surface colonization and endophytic reduction	
		fliC3 (azo2704)	Flagellin the major structural protein	Significant reduction in root colonization	Buschart et al. (2012)
H. seropedicae SmR1	*Zea mays*	(Hsero_3720)	methyl-associated chemotaxis transducer transmembrane protein	It prevents the attachment with the roots by decreasing the chemotaxis toward the plants	
P. fluorescens OE28.3, SBW25, F113, and WCS365	*Solanumlycopersicum*	*cheA*	chemotaxis protein	Decreases 10 to 1000-fold colonization at the root tip	

(Continued)

Table 7.2 (Continued)

Name of the microorganism	Plant associated	Gene	Mode of action	Function	References
2) **Substrate utilization, transport**					
Azoarcus sp. BH72	*O. sativa subsp. indica*cv. IR36	*exaA2* (azo2972)	quino(hemo)protein alcohol dehydrogenase, pyrroloquinolinequinone (PQQ)-dependent (EC 1.1.2.8)	decrease in root colonization	Krause et al. (2011)
		exaA3 (azo2975)	quino(hemo)protein alcohol dehydrogenase, pyrroloquinolinequinone (PQQ)-dependent (EC 1.1.2.8)		
B. phytofirmans PsJN	*Lupinusalbus L.*, cv. Amiga, *Zea mays subsp. mays*, cv. Birko	*oxc*	oxalate decarboxylase	impaired early colonization	
H. seropedicae SmR1	*Z. mays*	*phbC1* (Hsero_2999)	Subunit C of polyhydroxybutyrate	In the early stages of root colonization, it causes eightfold reduction in planktonic and epiphytic cells	
M. extorquens AM1	*M. truncatula*	*mxaF* (MexAM1_ META1p4538)	Alpha subunit of methanol dehydrogenase which oxidizes methanol to formaldehyde.	It helps in persistence at different parts of the plants that decrease the adhering abilities	Sy et al. (2005)
M. extorquens AM1	*M. truncatula*	*mptG* (MexAM1_ META1p1760)	Tetrahydromethanoprotein is synthesized in the presence of beta-ribofuranosylaminobenzene-5′-phosphate synthase.	It helps in increasing the sensitivity to formaldehyde and methanol. It also helps in persistence at different parts of the plants that decreases the adhering abilities	Sy et al. (2005)

R. meliloti 1021	*M. sativa*	*thuA* *thuB*	Utilization of Trehaloserelated protein	Improper colonization at the root	Jensen et al. (2005)
3) **Lipopolysaccharide, membrane proteins**					
A. brasilense	*Z. mays* cv. "Funk's Tronador G422T"	*rfbD* (AMK58_RS28935)	It helps in the synthesis of rhamnose in the presence of dTDP-4-0 dehydrogenase reductase.	It reduces colonization at the root by decreasing the attachment to the root surface	Jofré et al. (2004)
H. seropedicae SmR1	*Z. mays*	*waaL* (Hsero_3570)	Synthesis of LPS in the presence of O-antigen catalase.	Severely weakened colonization	
H. seropedicae SmR1	*Z. mays*	*ampG* (YP_003773531.1)	Muropeptidepermease helps in the mechanism of recycling of the cell wall peptidoglycan	10 times reduction of endophytic growth in roots on first and third day of inoculation	Tadra-Sfeir et al. (2011)
		rfbB (Hsero_4410)	Synthesis of rhamnose in the presence of dTDP-D-glucose 4,6-dehydratase	It reduces 100 times attachment with the root surface	Balsanelli et al. (2010)
		rfbC (Hsero_4411)	Synthesis of rhamnose in the presence of dTDP-4-keto-6-deoxy-D-glucose 3,5-reductase		
P. fluorescens CHA0	*Cucumissativus*var. Poinset 76 and *S. lycopersicumPlatense*	*oprF* (EF592174)	Outer membrane porin F	decrease in adsorption capacity to roots	Crespo and Valverde (2009)

(*Continued*)

Table 7.2 (Continued)

Name of the microorganism	Plant associated	Gene	Mode of action	Function	References
4) **Plant cell wall modifications**					
Azoarcus sp. BH72	*O. sativa subsp. indica*cv. IR36	*eglA* (azo2236)	beta-1,4-glucanase (cellulase) (EC 3.2.1.4)	Reduction in endophytic colonization	
Bacillus amyloliquefaciens TB2	*Brassica rapa subsp. pekinensisandchinensis*	*eglS*	endo-beta-1,4-glucanase	The mechanism of endophytic colonization is reduced	
B. subtilis 168	*Z. mays*	*yoaJ*	Plant cell wall gets loosened in the presence of expansin and also disrupts cell walls by breaking the non-covalent bondings.	Impaired ability to colonize roots	
5) **Adhesion, biofilm formation**					
*H. rubrisubalbicans*M1	*Z. mays*	*wssD* (Hrubri_1119); *bcsZ*	beta-1,4-glucanase (cellulase) (EC 3.2.1.4)	decrease root surface attachment and endophytic colonization	Monteiro et al. (2012)
G. diazotrophicus PAL5	*O. sativa*	*gumD* (YP_001602791)	polysaccharide biosynthesis glycosyltransferase, exopolysaccharide biosynthesis	This causes reduced attachment to the rice root surface by preventing endophytic and root surface colonization	
Pseudomonas putida KT2440	*Z. mays*	*lapA*	surface adhesion protein	It prevents colonization at the root surface	Martinez-Gil et al. (2010)

R. leguminosarum biovarviciae 3841	*Pisumsativum*	*gmsA* (CAK07156.1)	glucomannan production protein	Improper adhesion and biofilm formation upon the root hair cell	Williams et al. (2008)
		pssA (CAK09242.1)	CDP-diacylglycerol-serine O-phosphatidyltransferase, acidic exopolysaccharide biosynthesis	Improper attachment with the root hair cells	
		celA (CAK07141.1); *bcsA*	cellulose synthase catalytic subunit [UDP-forming], cellulose biosynthesis	Improper biofilm formation upon the root	
Salmonella entericserovarEnteritidis	*Medicagosativa*	*agpB*	curli subunit	Reduction in the ability to attach with roots	Barak et al. (2005)
S. enteric	*M. sativa*	*bcsA*	catalytic subunit of cellulose synthase helps in the synthesis of bacterial cellulose	reduced ability to attach and colonize sprouts	Barak et al. (2007)
		yihO	Synthesis of chronic acid by glucuronide transporter, O-antigen capsule		
6) **Stress protection**					
B. subtilis 168	*Capsicum annuum*	*pta, als*	2,3-butanediol biosynthesis	decreased ability to persist in the rhizosphere	
G. diazotrophicus PAL5	*O. sativa subsp. indicacv.* IR42	*sodB* (GDI_2168)	superoxide dismutase [Mn/Fe] (EC 1.15.1.1)	Reduction in the association of root-associated bacterial cells for colonization	Alqueres et al. (2013)
		Gor (GDI_2216)	glutathione reductase (EC 1.8.1.7)		

(*Continued*)

Table 7.2 (Continued)

Name of the microorganism	Plant associated	Gene	Mode of action	Function	References
H. seropedicae SmR1	*Z. mays*	(Hsero_4782)	ABC-type multi-drug transporter	It causes 20-fold reduction in the association endophytic and epiphytic population.	Alqueres et al. (2013)
Stenotrophomonas maltophilia D457	*Brassica napus*cv. Californium	*smeE* (SMD_3657)	RND efflux system, inner membrane transporter, part of the SmeDEF efflux pump involved in quinolone resistance	Impairedcolonization of roots	Garcia-Leon et al. (2014)
S. marcescens 90–166	*Cucumissativus*	*entA* (AAY84_01970)	2,3-dihydro-2,3-dihydroxybenzoate dehydrogenase, enterobactin biosynthesis	significant decrease in root population	Press et al. (2001)
Paraburkholderia kururiensis M130	*O. sativa*var. BALDO	(2553408008)	outer membrane efflux transporter, nodT family	It causes sixfold reduction in the mechanism of colonization with the root.	Levy et al. (2018)
7) **Bacterial secretion systems**					
Azoarcus sp. BH72	*O. sativa ssp. japonica* cv. Nipponbare	*ppkA* (azo3888)	T6SS serine/threonine-protein kinase	Increase in colonization capacity	Shidore et al. (2012)
H. rubrisubalbicans M1	*O. sativa*	*hrcN* (Hrubri_2444)	T3SS ATPase	100 times reduction in endophytic population in roots 9 days after inoculation	Schmidt et al. (2012)
		hrpE (Hrubri_2433)	T3SS structural apparatus		
P. kururiensis M130	*O. sativa*var. BALDO	*cesT* (2553406074)	Tir chaperone protein	Ability to decrease colonization capacity by four to six times.	Levy et al. (2018)

8) **Transcriptional regulators, sensor proteins**					
Azoarcus sp. BH72	*O. sativa ssp. japonica* cv. Nippon bare	(azo1544)	GGDEF/EAL/PAC/PAS-domain-containing protein	Reduces colonization with the root	Shidore et al. (2012)
		(azo2408)	GGDEF domain-containing protein		
B. phytofirmans PsJN	*A. thaliana* Col-0	*bpI.1* (Bphyt_0126)	AHL synthase of chromosome1QS system	Reduces colonization with the root	Zuniga et al. (2013)
		bpI.2 (Bphyt_4275)	AHL synthase of chromosome 2QS system		
S. enterica serovarNewport	*M. sativa*	*rpoS*	Regulation of the biofilm formation in the presence of stationary phase sigma factor.	Reduces colonization of sprouts	Barak et al. (2005)
S. maltophilia R551-3	*B. napus*cv. Californium	*rpfF* (Smal_1830)	Synthesizes quorum-sensing molecule	Reduces colonization with the root efficiency and promotes the growth of plants	
P. fluorescens 2P24	*T. aestivum*	*pcoI*	Synthesis of Acyl-homoserine lactone	Reduces colonization with the root	Wei and Zhang (2006)

7.5 Nutrient Interactions

7.5.1 Release and Activation of Minerals

The liberation or release of organic by-products as metabolites can be a beneficial way of soil nutrient influx or an unfavorable process. The ionic balance of soil and soil pH can be altered by the activation of chelated soil minerals and different metabolic activities of biofilms. Such alterations in the nutrient concentration of the soil profile can affect the biological life processes of microorganisms (Pickett and Bazin 1980; Parke et al. 1986). The permeability of the plant increases thus results in exudation from the root: zinc ions play a discrete role in membrane integrity, which favors the extension of root, and also, postulated ecological implications concerning nutrient mobilization and microbial activity in the region enclosing the rhizosphere. Drastic reductions in soil pH can instigate intake of toxic ions (like aluminum ions which enter the soil solution in low pH) and hence cause retarded or even terminated growth in roots. The microbes can reduce such nonstandard intake of ions into the roots by recycling of the nutrients through their metabolic activities.

7.5.2 Nutrient Recycling

An integral part of the nutrient recycling process occurring in soils is humus dilapidation. Such degradations require specialized enzymes capable of degrading complex plant polymers like cellulases which could be derived from rhizosphere biofilms. As the biofilms interact in the periphery of the plant roots, humus degradation can take place in the soil around the root, facilitating the reabsorption of ions by the same plant. This would increase the efficiency of nutrient recycling. As shown by Cheng and Coleman (1990), living roots have a stimulatory effect on soil organic matter decomposition owing to the higher microbial activity induced by the root itself. Microorganisms also lead to the production of certain extracellular bio-organic substances which are otherwise frequently lost from the soil solution by leaching (Lynch 1990). Thus, they serve as an efficient nutrient recycling and replenishing process.

7.5.3 Nitrogen Dynamics

The nitrogen cycle is an integral part of biological system dynamics, in which dinitrogen fixation and denitrification play a crucial role. Most crops are frequently cited as being highly reliant on the concentration of nitrogen or phosphorus. Nitrogen-fixation capacity would, therefore, have direct influence on crop production. In soil, ammonification and denitrification of inorganic nitrates, are both mediated by the specific rhizospheric bacterial strains. The bacterial populations are more abundant in the periphery of leguminous plant roots than that of non-leguminous crops. Therefore, the presence of a nitrogen-fixing colony, for example, *Rhizobium* species in a biofilm in the rhizosphere can significantly increase the amount of nitrogen available for plant growth, which indirectly leads to further cycling of nutrients to higher trophic levels, resulting in nutrient replenishment through urine and fecal matter. Inoculation with *Azotobacter* and *Bacillus megaterium* intensified the activity of rhizosphere bacteria involved in atmospheric N_2 fixation, nitrification, phosphate mineralization and cellulose decomposition (Samtsevich 1962; Berezova 1963; Mosiashvili et al. 1963; Marendiak 1964). On the other hand, denitrification and the presence of denitrifying microorganisms were observed detrimental as they limited the concentration of essential soil nitrates.

Rhizospheric bacteria, mainly Rhizobium sp. helps to mediate dinitrogen fixation and denitrification which enhances soil fertility. Leguminous plants support the rhizosphere bacterial population compared to the non-leguminous plant (Katznelson 1965). The two nitrogen-fixing bacteria Nitrobacter and Nitrosomonas sp. show a higher cell population where the nitrogenous substrate concentration is high (Ardakani et al. 1973). Plant productivity increases due to the nitrogen fixation as there is a limited amount of nitrogen or phosphorus present in crop plants. The nitrogen-fixing bacteria forms biofilm over the rhizosphere provides nitrogen for plant growth. The two bacteria Azotobacter and *B. megaterium* enhances the rhizosphere bacterial activity like ammonification, anaerobic nitrogen fixation, nitrification, phosphate mineralization, and cellulose decomposition (Samtsevich 1962; Berezova 1963; Mosiashvili et al. 1963; Marendiak 1964). Removal of essential nitrates has a detrimental effect on plants due to denitrification and the presence of denitrifying microorganisms. Biomass containing decomposable residues with high carbon and nitrogen ratios can restrict nitrogen immobilization (Drury et al. 1991).

7.5.4 Ionic Modification

Biofilms utilize some carbon compounds present in the soil to form insoluble complex sugar polymers, thus limiting the soil carbon content. The complex polysaccharides cannot be taken up by other microorganisms that compete with plants for nutrients or may have pathogenic effects. Few biofilms provide prey for predators such as nematodes encouraging soil solution movement and enhance mineral flux around the roots. In a setup containing growing specimens of *Zea mays*, a pattern of endogenous ion currents was observed in the medium around the horizontal meristematic tips and active elongating tissues of the plant (Pilet and Senn 1980) owing to the ion movements in the rhizosphere. Growth-promoting factors (GPF) and growth-inhibiting factors (GPI) was the major cause of enhancement or reduction in the root generated current respectively (Miller et al. 1988). The electrical current is correlated to the process of tip growth (De Silva et al. 1992) and therefore it can be stated that the influence of the biofilm in the electrical current can be incidental to the process of tip growth.

The composition of soluble exudates (water-soluble compounds which diffuse from the plant without the expenditure of metabolic energy) of maize (Kraffczyk et al. 1984) is altered by flux of potassium. Some chemotactic zoospores have been observed to be accumulating near sites of maximum exudation of low-molecular-weight solutes (Dukes and Apple 1961; Katsura and Hosomi 1963; Hickman and Ho 1966; Bimpong and Clerk 1970). Electrotaxis might be a consequence of the mechanism by which zoospores locate regions susceptible to fungal infection (Miller et al. 1988). Normally, an electrical current is generated by emerging fungal hyphae so that positive charge enters the hyphal tip and exits from the rear (Gow 1984; Harold et al. 1985; Miller and Gow 1989a, b). Such directional ion flow induces orientation of germ tubes, hyphae, branches and rhizoids and polarization of sites of germination in several mycelial fungi (De Vries and Wessels 1982;

McGillivray and Gow 1986; Van Laere 1988; Youatt et al. 1988). Thus, the attraction of zoospores can be influenced by the screening or changing of charges on the root surface, and the metabolism of exudates by the biofilm.

7.6 Biotic Interaction

Due to the presence of different microorganisms around the rhizosphere, there are direct biotic interactions including symbiosis, synergy, competition, antagonism, pathogenesis, predation, and mutualism invariably occur over rhizosphere.

7.6.1 Symbiosis

Symbiosis mainly occurs with the help of mycorrhizal fungi and rhizosphere microorganisms (Linderman 1988). The mycorrhizal plant has increased growth due to the presence of free-living bacteria in the rhizosphere and their selective interaction with vesicular-arbuscular mycorrhizal fungi (Azcon 1989). Phosphorus is found in very less content in soil and get utilized by root. Plant symbiosis associated with mycorrhizal infection increases phosphorus uptake (Hartley and Smith 1983). Phosphorus may accumulate partly in the form of polyphosphate in the sheath which is connected tightly with the biofilm when the phosphorus level in the soil is low. Active absorption and passive diffusion through the sheath help in the accumulation of nitrogen and potassium in the sheath.

7.6.2 Synergy

Beneficial rhizobacteria have positive effects on the growth of plants (Burr and Ceasar 1983). Microbial communities present in the rhizosphere influence minerals to produce plant growth hormones and decompose plant toxins that may respectively stimulate or retard plant growth. For example, *Azospirillum* increases the growth of maize root by influencing soil organic matter content (Fallik et al. 1988). Potato tubers rotted in the air may produce Rhishitin that cause 50% reduction in viability of *Erwinia atroseptica* (Matsumoto et al. 1979). Plant age can affect bacterial populations in the rhizosphere and phylloplane and bacterial composition get changed by the change of root exudation patterns during plant growth (Matsumoto et al. 1979). According to Patel (1969) seed inoculated with *Azotobacter* has significantly depressed colonization of emerging roots by bacteria and actinomycetes compared to the uninoculated plant seeds.

7.6.3 Competition

Microorganisms having R selection (uncrowded) and K selection (crowded) maximum specific growth rate minus minimum specific growth rate growth strategies can influences growth efficiencies in batch culture (Dorofeev and Panikov 1989). Root and bacterial biofilm require oxygen that can be obtained from root respiration. This leads to the competition between root and biofilm for the utilization of oxygen, by increasing CO_2

concentration that may affect the plant growth by affecting carbon distribution within the plant and may change the exudation rate in maize (Whipps 1985) by decreasing pH. Ethylene is one of the important plant hormones that stimulates fruit ripening, induces lateral branching in roots, stimulates radial growth of stems and roots, contributes to leaf abscission and other changes characterizing senescence, can aid in breaking dormancy in the buds and seeds, and can help initiate flowering in pineapple. The effect of ethylene gets attributed by the auxin (for example, lateral bud inhibition and root geotropism), as auxin induces ethylene production. Plant hormones have simple organic molecules that can stimulate bacterial growth. Bacteria susceptible to starvation cannot persist in nutrient-poor environments as they fail to compete for organic nutrients. Starvation resistance (e.g. *R. meliloti* and two pseudomonads) is necessary for the bacteria but nutrient-poor environments don't support interspecific competition among bacteria (Sinclair and Alexander 1984). Absence of organic nutrients stimulates the loss of viability in *Streptococcus faecalis, S. aureus, B. subtilis* and *Streptococcus* sp. (Figure 7.4).

7.6.4 Antagonism

Root tissue and biofilm both can produce potential antibiotic and hormone which can alter root community structure. In antagonism microorganisms resistant to the antibiotic and microorganisms that utilize antibiotic as an energy source are mainly favored. The number of microorganisms that are susceptible to the antibiotic has faster growth, get reduced. Hormone production enhances the production of indolyl-3-acetic acid (IAA) with or without added tryptophan by soil organisms (Brown 1972; Scott 1972). Tryptophan and related compounds are present in root exudates that act as precursors for IAA synthesis (Scott 1972).

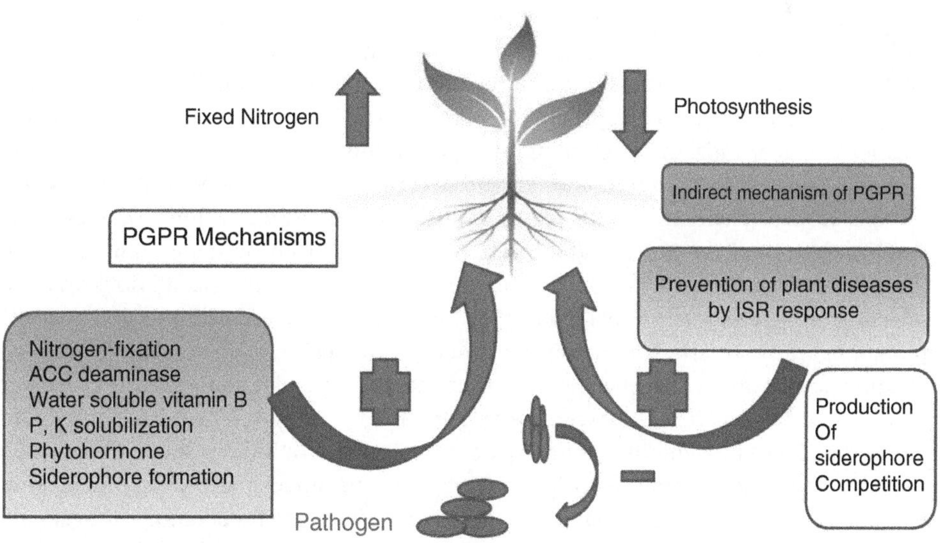

Figure 7.4 PGPR-dependent nitrogen fixation.

7.6.5 Pathogenesis

The exudates of the host plant give the stimulation of the infection by the root pathogens. The physical biofilm barrier leads the pathogenic species to the direct competition for nutrient and indirect competition for space. The amount of the pathogen *Ascochyta imperfecta* in the seeds may decrease the infection of *Medicago lupulina* (Brown and Burlingham 1963). The number of bacteria can be reduced by the interaction between bacteria and actinomycetes. Suppression of damping-off that linked to the microbial activity can be caused by *Pythium ultimum* (Chen et al. 1988; Lumsden et al. 1990). It has been observed that the biofilms which are devoid of pathogenic bacteria cause a reduction of the materials being released for chemotaxis which acts as a substrate for their growth and proliferation. Pathogens that use microorganism for access through root epidermis use biofilm as their infection route. There are simple metabolites present in the plant that helps bacteria to connect to the plant. Metabolites released from damaged tissue attract pathogenic microorganism. Phytoalexins are the chemically heterogeneous group of secondary molecules that contains antimicrobial properties (Lyon and Bayliss 1975). These molecules appear in the infected plant at the site of infection not in healthy tissue (Mansfield et al. 1974; Bell et al. 1984).

7.7 Conclusion and Future Prospects

The rhizosphere is a little soil zone in contact to plant roots and is considered to play an important role in complex interactions between plant soil, plant–microbes, microbe–soil, and microbe–microbe adhered upon plant roots as biofilm. Many studies have been performed in the last few decades to improve nutrient use efficiency in crops for the betterment of its health. Hence considering a sustainable agricultural practice, further research work especially a metabolomics or transcriptomics approach is needed to understand the microbial activities that imbibe better nutrient use efficiency by these plant roots. Here we have discussed the mode of biofilm formation on rhizospheric root surfaces and their symbiotic role in nutrient recycling in plants, identifying new signaling compounds helping in QS and promoting plant growth and health, nitrogen dynamics in locations that are distant from roots and preventing various microbial infections by controlling the microbe communities present in the soil. The present work also sheds light on the various environmental stress conditions (such as water limitation and salt accumulation) and developing adaptation strategies taken up by these plants allowing them to thrive under normally unfavorable conditions. These extreme conditions drive the selection and recruitment of plant-associated microbial colonies known as plant growth promoting rhizobacteria (PGPR) that support plant health improvement in extreme conditions. It is found that the formation of these microbial colonies is dependent on various factors such as soil type, cultivar variety, land history, geo-climatic factors, abiotic stresses, and type of plant. But, their exact mechanism of formation is still not completely understood. Hence, detailed insight is required to understand plant–microbiome interactions, their evolution, and their role in plant ecosystem functioning. Altogether it will help to understand various conditions and mechanisms by which microorganisms exhibit a delicate balance between harmful, beneficial, or neutral effects on phytocompound productions leading to a better knowledge of the plant microbiome.

References

Ainelo, H., Lahesaare, A., Teppo, A. et al. (2017). The promoter region of lap A and its transcriptional regulation by Fis in Pseudomonas putida. *PLoS One* 12 (9): e0185482.

Alberton, D., Müller-Santos, M., and Brusamarello-Santos, L.C. (2013). Comparative proteomics analysis of the rice roots colonized by *Herbaspirillum seropedicae* strain SmR1 reveals induction of the methionine recycling in the plant host. *J. Proteome Res.* 12 (11): 4757–4768.

Alqueres, S., Meneses, C., Rouws, L. et al. (2013). The bacterial superoxide dismutase and glutathione reductase are crucial for endophytic colonization of rice roots by *Gluconacetobacter diazotrophicus* PAL5. *Mol. Plant Microbe Interact.* 26: 937–945.

Ampomah, O.Y., Avetisyan, A., Hansen, E. et al. (2013). The thuEFGKAB operon of rhizobia and *Agrobacterium tumefaciens* code for transport of trehalose, maltitol and isomers of sucrose and their assimilation through the formation of their 3 ketoderivatives. *J. Bacteriol.* 195: 3797–3807.

Ardakani, M.S., Rehbock, J.T., and McLaren, A.D. (1973). Oxidation of nitrite to nitrate in a soil column. *Soil Sci.* 37: 53–56.

Ausmees, N., Jacobsson, K., and Lindberg, M. (2001). A unipolarly located, cell-surface-associated agglutinin, RapA, belongs to a family of Rhizobium-adhering proteins (Rap) in *Rhizobium leguminosarum* bv. trifolii. *Microbiology* 147: 549–559.

Azcon, R. (1989). Selective interaction between free living rhizosphere bacteria and vesicular-arbuscular mycorrhizal fungi. *Soil Biol. Biochem.* 21: 639–644.

Badri, D.V., Weir, T.L., van der Lelie, D., and Vivanco, J.M. (2009). Rhizosphere chemical dialogues: plant-microbe interactions. *Curr. Opin. Biotechnol.* 20: 642–650.

Baetz, U. and Martinoia, E. (2014). Root exudates: The hidden part of plant defense. *Trends Plant Sci.* 19: 90–98.

Balsanelli, E., Serrato, R.V., de Baura, V.A. et al. (2010). *Herbaspirillum seropedicae rfbB* and *rfbC* genes are required formaize colonization. *Environ. Microbiol.* 12: 2233–2244.

Barak, J.D., Gorski, L., Naraghi-Arani, P., and Charkowski, A.O. (2005). *Salmonella enterica* virulence genes are required for bacterial attachment to plant tissue. *Appl. Environ. Microbiol.* 71: 5685–5691.

Barak, J.D., Jahn, C.E., Gibson, D.L., and Charkowski, A.O. (2007). The role of cellulose and O-antigen capsule in the colonization of plants by *Salmonella enterica*. *Mol. Plant Microbe Interact.* 20: 1083–1091.

Barnhart, M.M. and Chapman, M.R. (2006). Curli biogenesis and function. *Annu. Rev. Microbiol.* 60: 131–147.

Bastian, F., Cohen, A., Piccoli, P. et al. (1998). Production of indole-3-acetic acid and gibberellins A1 and A3 by *Acetobacter diazotrophicus* and *Herbaspirillum seropedicae* in chemically-defined culture media. *Plant Growth Regul.* 24: 7–11.

Batista, M.B., Teixeira, C.S., Sfeir, M.Z.T. et al. (2018). PHB biosynthesis counteracts redox stress in *Herbaspirillum seropedicae*. *Front. Microbiol.* 9: 472.

Bécard, G. (2017). Plant communication with associated microorganisms in the rhizosphere. In: *How Plants Communicate with Their Biotic Environment* (ed. G. Bécard), 82. Academic Press.

Bell, J.N., Dixon, R.A., Bailey, J.A. et al. (1984). Differential induction of chalcone synthase mRNA activity at the onset of phytoalexin accumulation in compatible and incompatible plant pathogen interactions. *Proc. Natl. Acad. Sci. U. S. A.* 81: 3384–3388.

Berezova, E.P. (1963). The effectiveness of bacterial fertilizers. *Mikrobiologiya* 32: 358–361.

Bhattacharjee, R.B., Singh, A., and Mukhopadhyay, S.N. (2008). Use of nitrogen-fixing bacteria as biofertiliser for non-legumes: prospects and challenges. *Appl. Microbiol. Biotechnol.* 80: 199–209.

Bimpong, C.E. and Clerk, G.C. (1970). Motility and chemotaxis in zoospores of *Phytophthora palmivora* (Butl.) Butl. *Ann. Bot.* 34: 617–624.

Brader, G., Compant, S., Vescio, K. et al. (2017). Ecology andgenomic insights into plant-pathogenic and plant-nonpathogenic endophytes. *Annu. Rev. Phytopathol.* 55: 61–83.

Brimecombe, M.J., De Leij, F.A., and Lynch, J.M. (2001). The effect of root exudates on rhizosphere microbial populations. In: *The Rhizosphere* (eds. R. Pinton, Z. Varanini and P. Nannipieri), 95–140. New York: Marcel Dekker.

Brito, P.H., Rocha, E.P., Xavier, K.B., and Gordo, I. (2013). Natural genome diversity of AI-2 quorum sensing in *Escherichia coli*: conserved signal production but labile signal reception. *Genome Biol. Evol.* 5: 16–30.

Brown, M.E. (1972). Seed and root bacterization. *Annu. Rev. Phytopathol.* 12: 181–197.

Brown, M.E. and Burlingham, S.K. (1963). *Azotobacter* and plant disease. Annual Report, Rothamsted Experimental Station, 73.

Bulgarelli, D., Rott, M., Schlaeppi, K. et al. (2012). Revealing structure and assembly cues for Arabidopsis root-inhabiting bacterial microbiota. *Nature* 488: 91–95.

Burr, T.J. and Ceasar, A. (1983). Beneficial plant bacteria. *Crit. Rev. Plant Sci.* 2: 1–20.

Buschart, A., Sachs, S., Chen, X. et al. (2012). Flagella mediate endophytic competence rather than act as MAMPS in rice–*Azoarcus* sp. strain BH72 interactions. *Mol. Plant Microbe Interact.* 25: 191–199.

Camilios-Neto, D., Bonato, P., Wassem, R. et al. (2014). Dual RNA-seq transcriptional analysis of wheat roots colonized by *Azospirillum brasilense* reveals up-regulation of nutrient acquisition and cell cycle genes. *BMC Genomics* 15: 378.

Castellanos, T., Ascencio, F., and Bashan, Y. (1998). Cell-surface lectins of *Azospirillum* spp. *Curr. Microbiol.* 36: 241–244.

Chaparro, J.M., Badri, D.V., Bakker, M.G. et al. (2013). Root exudation of phytochemicals in Arabidopsis follows specific patterns that are developmentally programmed and correlate with soil microbial functions. *PLoS One* 8: e55731.

Chen, W., Hoitink, H.A.J., Schmitthenner, A.F., and Tuovinen, O.H. (1988). The role of microbial activity in suppression of damping-off caused by *Pythium ultimum*. *Phytopathology* 78: 314–322.

Cheng, W. and Coleman, D.C. (1990). Effect of living roots on soil organic matter decomposition. *Soil Biol. Biochem.* 22: 781–787.

Cole, B.J., Feltcher, M.E., Waters, R.J. et al. (2015). Genome-wide identification of bacterial plant colonization genes. *PLoS Biol.* 15: e2002860.

Costerton, J.W., Lewandowski, Z., Caldwell, D.E. et al. (1995). Microbial biofilms. *Annu. Rev. Microbiol.* 49: 711–745.

Coutinho, B.G., Licastro, D., Mendonca-Previato, L. et al. (2015). Plant-influenced gene expression in the rice endophyte *Burkholderia kururiensis* M130. *Mol. Plant Microbe Interact.* 28: 10–21.

Crespo, M.C.A. and Valverde, C.A. (2009). A single mutation in the oprF mRNA leader confers strict translational control by the Gac/Rsm system in *Pseudomonas fluorescens* CHA0. *Curr. Microbiol.* 58: 182–188.

Danhorn, T. and Fuqua, C. (2007). Biofilm formation by plant-associated bacteria. *Annu. Rev. Microbiol.* 61: 401–422.

Dardanelli, M., Angelini, J., and Fabra, A. (2003). A calcium-dependent bacterial surface protein is involved in the attachment of rhizobia to peanut roots. *Can. J. Microbiol.* 49: 399–405.

Davies, D.G., Parsek, M.R., Pearson, J.P. et al. (1998). The involvement of cell-to-cell signals in the development of a bacterial biofilm. *Science* 280: 295–298.

De Silva, L.R., Youatt, J., Gooday, G.W., and Gow, N.A.R. (1992). Inwardly directed ionic currents of *Allomyces macrogynus* and other water moulds indicate sites of proton-driven nutrient transport but are incidental to tip growth. *Mycol. Res.* 96: 925–931.

De Vries, S.C. and Wessels, J.G.H. (1982). Polarized outgrowth of hyphae by constant electrical fields during reversion of *Schizophyllum commune* protoplast. *Exp. Mycol.* 6: 95–98.

Dorofeev, A.G. and Panikov, N.S. (1989). Dynamics of the dying off of starving microorganisms as a function of the initial growth rate. *Mikrobiologiya* 60: 814–822.

Dos Santos, M.F., Muniz de Padua, V.L. et al. (2010). Proteome of *Gluconacetobacter diazotrophicus* co-cultivated with sugarcane plantlets. *J. Proteomics* 73: 917–931.

Doty, S.L., Oakley, B., Xin, G. et al. (2009). *Diazotrophic endophytes* of native black cottonwood and willow. *Symbiosis* 47: 23–33.

Drury, C.F., Voroney, R.P., and Beauchamp, E.G. (1991). Availability of NH^{4+}-N to microorganisms and the soil internal N cycle. *Soil Biol. Biochem.* 3: 66–71.

Dukes, P.D. and Apple, J.L. (1961). Chemotaxis of zoospores of *Phytophthora parasitica* var. Nicotianae by plant roots and certain chemical solutions. *Phytopathology* 51: 195–197.

Espinosa-Urgel, M., Salido, A., and Ramos, J.L. (2000). Genetic analysis of functions involved in adhesion of *Pseudomonas putida* to seeds. *J. Bacteriol.* 182: 2363–2369.

Fallik, E., Okon, Y., and Fischer, M. (1988). Growth response of maize roots to *Azospirillum* inoculation., effect of soil organic matter content, number of rhizosphere bacteria and timing of inoculation. *Soil Biol. Biochem.* 20: 45.

Fujishige, N.A., Kapadia, N.N., De Hoff, P.L., and Hirsch, A.M. (2006). Investigations of Rhizobium biofilm formation. *FEMS Microbiol. Ecol.* 56: 195–206.

Gaiero, J.R., McCall, C.A., Thompson, K.A. et al. (2013). Inside the root microbiome: bacterial root endophytes and plant growth promotion. *Am. J. Bot.* 100: 1738–1750.

Garcia-Leon, G., Hernandez, A., Hernando-Amado, S. et al. (2014). A function of SmeDEF, the major quinolone resistance determinant of *Stenotrophomonas maltophilia*, is the colonization of plant roots. *Appl. Environ. Microbiol.* 80: 4559–4565.

Gow, N.A.R. (1984). Transhyphal electrical currents in fungi. *J. Gen. Microbiol.* 130: 3313–3318.

Greer-Phillips, S.E., Stephens, B.B., and Alexandre, G. (2004). An energy taxis transducer promotes root colonization by *Azospirillum brasilense*. *J. Bacteriol.* 186: 6595–6604.

Hall-Stoodley, L., Costerton, J.W., and Stoodley, P. (2004). Bacterial biofilms: from the natural environment to infectious diseases. *Nat. Rev. Microbiol.* 2: 95–108.

Hardoim, P.R. and Hardoim, C.C.P. (2017). Genomic features of mutualistic plant bacteria. In: *Endophytes: Biology and Biotechnology* (ed. D.K. Maheshwari), 99–125. New York: Springer.

Hardoim, P.R., van Overbeek, L.S., Berg, G. et al. (2015). The hidden world within plants: ecological and evolutionary considerations for defining functioning of microbial endophytes. *Microbiol. Mol. Biol. Rev.* 79: 293–320.

Harold, F.M., Schreurs, W.J., Harold, R.L., and Caldwell, J.H. (1985). Electrobiology of fungal hyphae. *Microbiol. Sci.* 2: 363–366.

Hartley, J.L. and Smith, S.E. (1983). *Microbial Symbiosis*. London: Academic Press.

Hickman, C.J. and Ho, H.H. (1966). Behaviour of zoospores in plantpathogenic phycomycetes. *Annu. Rev. Phytopathol.* 4: 195–220.

Jensen, J.B., Ampomah, O.Y., Darrah, R. et al. (2005). Role of trehalose transport and utilization in *Sinorhizobium meliloti*-alfalfa interactions. *Mol. Plant Microbe Interact.* 18: 694–702.

Jofré, E., Lagares, A., and Mori, G. (2004). Disruption of dTDP-rhamnose biosynthesis modifies lipopolysaccharide core, exopolysaccharide production, and root colonization in *Azospirillum brasilense*. *FEMS Microbiol. Lett.* 231: 267–275.

Kandel, S., Joubert, P., and Doty, S. (2017). Bacterial endophyte colonization and distribution within plants. *Microorganisms* 5: 77.

Katsura, K. and Hosomi, T. (1963). Chemotaxis of zoospores for plant roots in relation to infection by *Phytophthora capsici* var. Leonian. *Sci. Rep. Kyoto Prefect. Univ.* 15: 27–30.

Katznelson, H. (1965). Nature and importance of the rhizosphere. In: *Ecology of Soil Borne Pathogens* (eds. K.F. Baker and W.C. William Cowperthwaite Snyder), 187–209. Los Angeles, CA: University of California Press.

Kawasaki, A., Donn, S., Ryan, P.R. et al. (2016). Microbiome and exudates of the root and rhizosphere of *Brachypodium distachyon*, a model for wheat. *PLoS One* 11: e0164533.

Ker, F., Amoroso, A., Herman, R. et al. (2008). Crystal structure and activity of *Bacillus subtilis* YoaJ (EXLX1), a bacterial expansin that promotes root colonization. *Proc. Natl. Acad. Sci. U. S. A.* 105: 16876–16881.

Kraffczyk, I., Trolldenier, G., and Beringer, H. (1984). Soluble root exudates of maize: influence of potassium supply and rhizosphere microorganisms. *Soil Biol. Biochem.* 16: 315–322.

Krause, A., Bischo, B., Miché, L. et al. (2011). Exploring the function of alcohol dehydrogenases during the endophytic life of *Azoarcus* sp. strain BH72. *Mol. Plant Microbe Interact.* 24: 1325–1332.

Lebeis, S.L. (2014). The potential for give and take in plant-microbiome relationships. *Front. Plant Sci.* 5: 287.

Lemanceau, P., Blouin, M., Muller, D., and Moënne-Loccoz, Y. (2017). Let the core microbiota be functional. *Trends Plant Sci.* 22: 583–595.

Levy, A., Gonzalez, I.S., Mittelviefhaus, M. et al. (2018). Genomic features of bacterial adaptation to plants. *Nat. Genet.* 50: 138.

Linderman, R.G. (1988). Mycorrhizal interactions with the rhizosphere microflora: the mycorrhizosphere effect. *Phytopathology* 78: 366–371.

Liu, H., Carvalhais, L.C., Crawford, M. et al. (2017). Inner plant values: diversity, colonization and benefits from endophytic bacteria. *Front. Microbiol.* 8: 2552.

Lumsden, R.D., Carter, J.P., Whipps, J.M., and Lynch, J.M. (1990). Comparison of biomass and viable propagule measurements in the antagonism of *Trichoderma harzianum* against *Pythium ultimum*. *Soil Biol. Biochem.* 22: 187–194.

Lynch, J.M. (1990). Microbial metabolites. In: *The Rhizosphere* (ed. J.M. Lynch), 177–206). Chichester: John Wiley.

Lyon, G.D. and Bayliss, C.E. (1975). The effect of rishitin on *Erwinia carotovoravar. atroseptica* and other bacteria. *Physiol. Plant Pathol.* 6: 177–186.

Mansfield, J.W., Hargreaves, J.A., and Boyle, F.C. (1974). Phytoalexin production by live cells in broad bean leaves infected with *Botrytis cinerea*. *Nature* 252: 316–317.

Marendiak, D. (1964). Contribution to the question of dynamism and quantitative relations of some groups of microorganisms in the rhizosphere of maize. *RostlinnaVyroba* 10: 137–144.

Martinez-Gil, M., Yousef-Coronado, F., and Espinosa-Urgel, M. (2010). LapF, the second largest *Pseudomonas putida* protein, contributes to plant root colonization and determines biofilm architecture. *Mol. Microbiol.* 77: 549–561.

Matsumoto, H., Okada, K., and Takahashi, E. (1979). Excretion products of maize roots from seedlings to seed development stage. *Plant and Soil* 53: 17–26.

McGillivray, A.M. and Gow, N.A.R. (1986). Applied electrical fields polarize the growth of mycelial fungi. *J. Gen. Microbiol.* 113: 2515–2525.

Meneses, C.H., Rouws, L.F., Simões-Araújo, J.L. et al. (2011). Exopolysaccharide production is required for biofilm formation and plant colonization by the nitrogen-fixing endophyte *Gluconacetobacter diazotrophicus*. *Mol. Plant Microbe Interact.* 24: 1448–1458.

Merckx, R., van Ginkel, J.H., Sinnaeve, J., and Cremers, A. (1986). Plant-induced changes in the rhizosphere of maize and wheat. I. Production and turnover of root-derived material in the rhizosphere of maize and wheat. *Plant and Soil* 96: 85–93.

Miller, A.L. and Gow, N.A.R. (1989a). Correlation between root-generated ionic currents, pH, fusicoccin, IAA and the growth of the primary root tip of *Zea mays*. *Plant Physiol.* 89: 1198–1206.

Miller, A.L. and Gow, N.A.R. (1989b). Correlation between profile of ion-current circulation and root development. *Physiol. Plant.* 10: 2–8.

Miller, A.L., Shand, E., and Gow, N.A.R. (1988). Ion currents associated with root tips, emerging laterals and induced wound sites in *Nicotiana tabacum*: spatial relationship proposed between resulting electrical fields and phytophthoran zoospore infection. *Plant Cell Environ.* 11: 21.

Mitter, B., Petric, A., Shin, M.W. et al. (2013). Comparative genome analysis of *Burkholderia phytofirmans* PsJN reveals a wide spectrum of endophytic lifestyles based on interaction strategies with host plants. *Front. Plant Sci.* 4: 120.

Monier, J.M. and Lindow, S.E. (2005). Aggregates of resident bacteria facilitate survival of immigrant bacteria on leaf surfaces. *Microb. Ecol.* 49: 343–352.

Monteiro, R.A., Balsanelli, E., Tuleski, T. et al. (2012). Genomic comparison of the endophyte *Herbaspirillum seropedicae* SmR1 and the phytopathogen *Herbaspirillum rubrisubalbicans* M1 by suppressive subtractive hybridization and partial genome sequencing. *FEMS Microbiol. Ecol.* 80: 441–451.

Mosiashvili, G.I., Topuridze, K.V., and Kiriakova, N.G. (1963). Effectiveness of azotobacterin in vineyard soils. *Mikrobiologiya* 32: 835–837.

O'Toole, G., Kaplan, H.B., and Kolter, R. (2000). Biofilm formation as microbial development. *Annu. Rev. Microbiol.* 54: 49–79.

Pankievicz, V.C., Camilios-Neto, D., Bonato, P. et al. (2016). RNA-seq transcriptional profiling of *Herbaspirillum seropedicae* colonizing wheat (*Triticum aestivum*) roots. *Plant Mol. Biol.* 90: 589–603.

Parke, J.L., Moen, R., Rovira, A.D., and Bowen, G.D. (1986). Soil water flow affects the rhizosphere distribution of a seed-borne biological control agent *Pseudomonas fluorescens*. *Soil Biol. Biochem.* 18: 583–588.

Patel, J.J. (1969). Microorganisms in the rhizosphere of plants inoculated with *Azotobacter chroococcum*. *Plant and Soil* 31: 209–223.

Pedrosa, F.O., Monteiro, R.A., Wassem, R. et al. (2011). Genome of *Herbaspirillum seropedicae* strain SmR1, a specialized diazotrophic endophyte of tropical grasses. *PLoS Genet.* 27: e1002064.

Peiffer, J.A. and Ley, R.E. (2013). Exploring the maize rhizosphere microbiome in the field: a glimpse into a highly complex system. *Commun. Integr. Biol.* 6: e25177.

Pickett, A.M. and Bazin, M.J. (1980). Growth and composition of *Escherichia coli* subjected to squarewave perturbations in nutrient supply: effect of varying amplitudes. *Biotechnol. Bioeng.* 22: 1213–1224.

Pilet, P.E. and Senn, A. (1980). Root growth gradients: a critical analysis. *Z. Pflanzenphysiol.* 99: 121–130.

Press, C., Loper, J., and Kloepper, J. (2001). Role of iron in rhizobacteria-mediated induced systemic resistance of cucumber. *Phytopathology* 91: 593–598.

Reinhold-Hurek, B., Hurek, T., Gillis, M. et al. (1993). Azoarcusgen. nov., Nitrogen-fixing Proteobacteria associated with roots of Kallar grass (*Leptochloa fusca [L.] Kunth*), and description of two species, *Azoarcusindigens* sp. nov. and *Azoarcus communis* sp. nov. *Int. J. Syst. Bacteriol.* 43: 574–584.

Reinhold-Hurek, B., Maes, T., Gemmer, S. et al. (2006). An endoglucanase is involved in infection of rice roots by the not-cellulose-metabolizing endophyte Azoarcus sp. strain BH72. *Mol. Plant Microbe Interact.* 19: 181–188.

Reinhold-Hurek, B., Bünger, W., Burbano, C.S. et al. (2015). Roots shaping their microbiome: global hotspots for microbial activity. *Annu. Rev. Phytopathol.* 53: 403–424.

Rout, M.E. and Southworth, D. (2013). The root microbiome influences scales from molecules to ecosystems: the unseen majority. *Am. J. Bot.* 100: 1689–1691.

Samtsevich, S.A. (1962). Preparation, use and effectiveness of bacterial fertilizers in the Ukrainian SSR. *Mikrobiologiya* 31: 923–933.

Santi, C., Bogusz, D., and Franche, C. (2013). Biological nitrogen fixation in non-legume plants. *Ann. Bot.* 111: 743–767.

Santoyo, G., Moreno-Hagelsieb, G., Orozco-MosquedaMdel, C., and Glick, B.R. (2016). Plant growth-promoting bacterial endophytes. *Microbiol. Res.* 183: 92–99.

Saunders, P.T. and Bazin, M.J. (1973). Attachment of microorganisms in a packed column: metabolite diffusion through the microbial film as a limiting factor. *Appl. Biochem. Biotechnol.* 23: 847–853.

Scharf, B.E., Hynes, M.F., and Alexandre, G.M. (2016). Chemotaxis signaling systems in model beneficial plant-bacteria associations. *Plant Mol. Biol.* 90: 549–559.

Schmidt, M.A., Balsanelli, E., Faoro, H. et al. (2012). The type III secretion system is necessary for the development of a pathogenic and endophytic interaction between *Herbaspirillum rubrisubalbicans* and *Poaceae*. *BMC Microbiol.* 12: 98.

Scott, T.K. (1972). Auxins and roots. *Annu. Rev. Plant Physiol.* 23: 235–258.

Shahzad, R., Khan, A.L., Bilal, S. et al. (2018). What is there in seeds? Vertically transmitted endophytic resources for sustainable improvement in plant growth. *Front. Plant Sci.* 9: 24.

Shidore, T., Dinse, T., Ohrlein, J. et al. (2012). Transcriptomic analysis of responses to exudates reveal genes required for rhizosphere competence of the endophyte *Azoarcus* sp. strain BH72. *Environ. Microbiol.* 14: 2775–2787.

Sinclair, J.L. and Alexander, M. (1984). Role of resistance to starvation in bacterial survival in sewage and lake water. *Appl. Environ. Microbiol.* 48: 410–415.

Swart, S., Smit, G., Lugtenberg, B.J.J., and Kijne, J.W. (1993). Restoration of attachment, virulence and nodulation of *Agrobacterium tumefaciens* chvB mutants by rhicadhesin. *Mol. Microbiol.* 10: 596–605.

Sy, A., Timmers, A.C., Knief, C., and Vorholt, J.A. (2005). Methylotrophic metabolism is advantageous for *Methylobacterium extorquens* during colonization of *Medicago truncatula* under competitive conditions. *Appl. Environ. Microbiol.* 71: 7245–7252.

Tadra-Sfeir, M.Z., Souza, E.M., Faoro, H. et al. (2011). Naringenin regulates expression of genes involved in cell wall synthesis in *Herbaspirillum seropedicae*. *Appl. Environ. Microbiol.* 77: 2180–2183.

Taghavi, S., van der Lelie, D., Homan, A. et al. (2010). Genome sequence of the plant growth promoting endophytic bacterium *Enterobacter* sp. 638. *PLoS Genet.* 6: e1000943.

Thomas, N.A., Deng, W., Puente, J.L. et al. (2005). Finlay, B.B. CesT is a multi-e_ector chaperone and recruitment factor required for the e_cient type III secretion of both LEE and non-LEE-encoded vectors of enteropathogenic *Escherichia coli*. *Mol. Microbiol.* 57: 1762–1779.

Valentini, M. and Filloux, A. (2016). Biofilms and cyclic di-GMP (c-di-GMP) signaling: lessons from *Pseudomonas aeruginosa* and other bacteria. *J. Biol. Chem.* 291: 12547–12555.

Van Laere, A. (1988). Effect of electrical fields on polar growth of *Phycornycesblakesleeanus*. *FEMS Microbiol. Lett.* 49: 111–116.

Vorholt, J.A. (2012). Microbial life in the phyllosphere. *Nat. Rev. Microbiol.* 10: 828–840.

Wei, H.L. and Zhang, L.Q. (2006). Quorum-sensing system influences root colonization and biological control ability in *Pseudomonas fluorescens* 2P24. *Antonie Van Leeuwenhoek.* 89: 267–280.

Whipps, J.M. (1985). Effect of CO_2 concentration on growth, carbon distribution and loss of carbon from the roots of maize. *J. Exp. Bot.* 36: 644–651.

Whipps, J.M. (1990). Carbon economy. In: *The Rhizosphere* (ed. J.M. Lynch), 59–97). New York: Wiley.

Williams, A., Wilkinson, A., Krehenbrink, M. et al. (2008). Glucomannan-mediated attachment of *Rhizobium leguminosarum* to pea root hairs is required for competitive nodule infection. *J. Bacteriol.* 190: 4706–4715.

Yi, Y., de Jong, A., Frenzel, E., and Kuipers, O.P. (2017). Comparative transcriptomics of *Bacillus mycoides* strains in response to potato-root exudates reveals different genetic adaptation of endophytic and soil isolates. *Front. Microbiol.* 8: 1487.

Youatt, J., Gow, N.A.R., and Gooday, G.W. (1988). Bioelectric and biosynthetic aspects of cell polarity in *Allomyces macrogynus*. *Protoplasma* 146: 118–126.

Zuniga, A., Poupin, M.J., Donoso, R. et al. (2013). Quorum sensing and indole-3-acetic acid degradation play a role in colonization and plant growth promotion of Arabidopsis thaliana by *Burkholderia phytofirmans* PsJN. *Mol. Plant Microbe Interact.* 26: 546–553.

Zur, J., Wojcieszynska, D., and Guzik, U. (2016). Metabolic responses of bacterial cells to immobilization. *Molecules* 21: 958.

8

Phytomicrobiome

Synergistic Relationship in Bioremediation of Soil for Sustainable Agriculture

Nimmy Srivastava

Amity Institute of Biotechnology, Amity University, Jharkhand, India

8.1 Introduction

Over the last century rapid urbanization has led to a dramatic elevation in the level of anthropogenic chemical in the environment which has resulted in severe contamination of many area of earth including the land. Currently, India has a total land area of 328 million hectares of which 43.6% is used as agricultural land. Agricultural land soil can be polluted by different sources which includes toxic waste disposal on land, application of fertilizers, animal manure and waste water irrigation. The means of agriculture that are practiced today to meet the growing demand of food consumption has distressed the natural balance of ecosystem. One of the major sources of pollutions of soil are imprudent use of pesticides and chemical fertilizers for insect and pest management which has left the soil contaminated with various organic and inorganic form of pollutants and are also a provenience of water pollution. This increasing level of pollution in the environment has motivated the interest of researchers in developing differing strategies for effective and successful remediation of environment contaminants using any of the chemical, physical or biological processes.

As offsite treatments are extortionate, in-site rehabilitation of contaminated land area has become an attractive tool (Ayoub et al. 2010) which includes processes like Monitored natural attenuation (MNA), bio-stimulation, bio-augmentation, soil vapor extraction, land farming, bioventing, thermal desorption, biopiles, and phytoremediation (Khan et al. 2004; Truu et al. 2015). Use of any single technology has not been found appropriate for all contaminated sites and contaminant types and based on existing site-specific conditions multifarious remediation techniques are often needed (Khan et al. 2004; Truu et al. 2015).

Phytomicrobiome Interactions and Sustainable Agriculture, First Edition. Edited by Amit Verma, Jitendra Kumar Saini, Abd El-Latif Hesham, and Harikesh Bahadur Singh.

8.2 Phytoremediation

The use of remunerative and green technology has become a choice of the present day to combat the side effects caused due to physical and chemical method. A greener and sustainable alternative to physico-chemical method for remediation of contaminated soil is phytoremediation were green plants are utilized for remediation of contaminated land areas.

8.2.1 Process of Phytoremediation

Cunningham and Berti (1993) have delineated phytoremediation as the use of green plants to remove, contain or render harmless environmental contaminants. To accomplish this process successfully specifically selected or genetically engineered green plants are used which are competent of direct uptake of pollutants from the environment (Macek et al. 2000). The process of phytoremediation can be adapted conceivably for both inorganic and organic form of contaminants (Salt et al. 1998).

Remediation and removal of pollutants is a complex process taking inside plants and includes various steps (Schnoor et al. 1995).

1) Uptake of contaminants by plants
2) Translocation to various parts of plants
3) Transfiguration of contaminants from one form to another
4) Its compartmentalization in plant tissues
5) At last transformation to water and CO_2, i.e. mineralization

Physical along with chemical properties of the contaminants like solubilityrate in water, molecular weight, octanol–water partition coefficient and some environmental factors like temperature, pH, organic matter, soil moisture content and different plant attributes like expansion of root system and enzymes present are the main factors which affect the uptake, distribution and transformation of contaminants by plants (Suresh and Ravishankar 2004; Susarla et al. 2002).

8.2.2 Strategies for Phytoremediation

Plants generally adopt any of the below mentioned strategies for removal of contaminants from soil.

a) Phyto immobilization – Organic as well as inorganic contaminants are removed from soil by roots of plants.
b) Phytoextraction/Phytoaccumulation – The contaminants are transported and concentrated in harvestable section of plant
c) Phytovolatilization – In certain cases transpiration to the atmosphere amid stomata present on leave surface may be followed
d) Phytodegradation – Contaminants which are mainly organic in nature may be metabolized in plant tissues in three subsequent steps: (i) transfiguration (ii) conjugation (iii) compartmentalization with help of enzymes, cytochrome P450, glycosyltransferase which results in a depot of contaminants inside various plant tissues like vacuoles, cell wall or they are either excreted from cell.

e) Phyto stabilization – Contaminants which are in mobile state are stabilized by this process
f) Rhizodegradation/Rhizomediation – One of the major dimensions of phytoremediation is a complete plant–microbes endeavor where roots of plants stimulate the microbes present in soil to degrade contaminants, especially organic (Gracia et al. 2018; Pilon-Smits 2005).

When organic or inorganic contaminants are metabolized by plants, sometimes even more toxic by-products can be produced which has been reported in many cases like when trichloroethylene (TCE) is transformed into trichloroethane which is even more lethal from a human health perspective (Burken and Schnoor 1998; Ma and Burken 2003). Plant system do not possess a complete catabolic pathway which can degrade and mineralize contaminants as is often found in microbes (Eapen et al. 2007; Schröder and Collins 2002). Effective and safe remediation of only few contaminants like polychlorinated biphenyls (PCBs), polyaromatic hydrocarbons (PAHs), nitroaromatic and linear halogenated hydrocarbon has been reported by use of plants viz. willow, alfalfa, poplar along with few varieties of grasses due to their extensive root system (Kuiper et al. 2004; Meagher 2000).

8.3 Phytomicrobe Interactions and Rhizomediation

This natural approach takes advantage of an interaction mechanism which exists between plant and microbes leading to the mineralization of contaminants. Organic contaminants can be degraded by microbes present in plant rhizosphere which are stimulated by the release of plant exudates. The root system of plants provides a surface for colonization of microbes (Truu et al. 2015) contributing to the enhanced metabolic activity of microbes known as the rhizosphere effect. The principal scheme of rhizomediation is explained in Figure 8.1.

8.3.1 Principle of Phytomicrobiome Interaction During Rhizomediation

Plant root system are colonized by a subset of a soil-dwelling microorganism, and the exudation of various organic compounds from plant stimulates the activities of microbes (Kuiper et al. 2004) which results in degradation of contaminants. The plant root surface provides a good breeding ground and shelter to microbes for colonization and protects them against biotic as well as abiotic stresses (Kuiper et al. 2004). For its part the rhizospheric microbes act as a plant growth promoter by the fixation of nitrogen, mobilization of nutrients such as phosphorous, decreasing the stress causing hormone level, and plant growth regulators production, which safeguards against plant pathogens and degradation of pollutants. (Chaudhry et al. 2005; Segura et al. 2009). Mutual interaction among plant and microbes which is known popularly as the rhizosphere effect is beneficial for both plants and microbes. Microbes are the main contributor in the process of rhizomediation (Table 8.1).

The root exudates released in the vicinity of roots provide nutrients for growth and a source of energy for microorganism. The main components of root exudates are mainly organic acids, i.e. acetate, oxalate, succinate, lactate, fumarate, malate and citrate which exist in soil as dissociated anions, sugars, and amino acids, and some secondary metabolites like isoprenoids, flavonoids and alkaloids which are listed in Table 8.1 (Jones 1998; Singer et al. 2003; Singh et al. 2004).

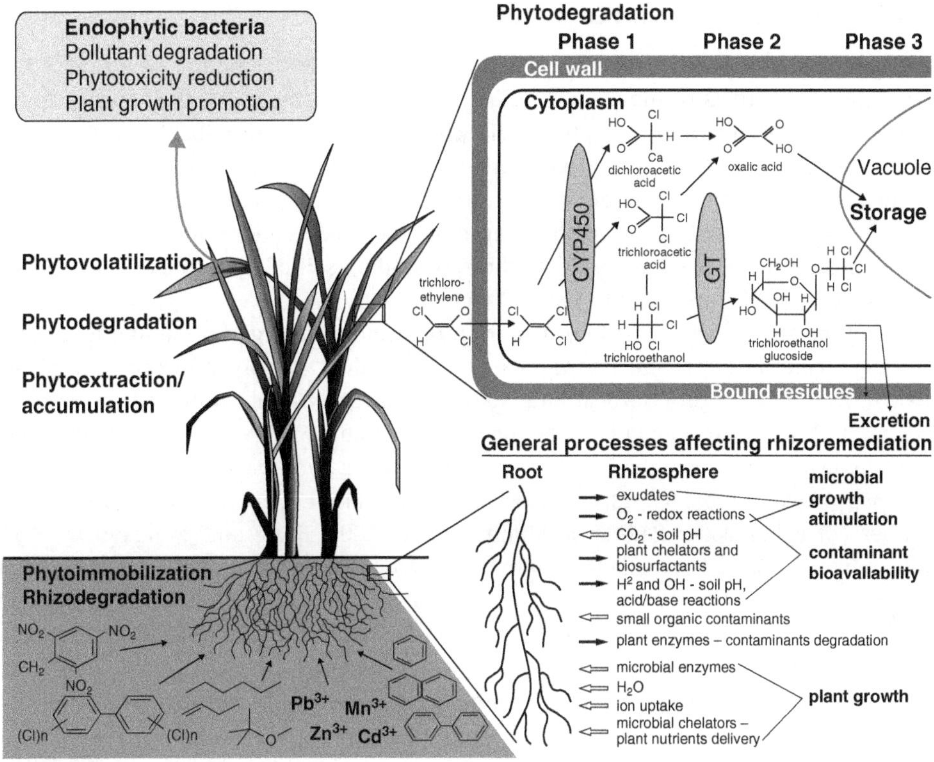

Figure 8.1 Schematic diagram representing degradation of Inorganic and organic components in soil. *Source:* Truu et al. (2015)..

Table 8.1 Composition of root exudates.

Compositions	Substances identified
Organic acid	Acetic, butyric, citric, glutaric, lactic, maleic, malic, malonic, oxalic, propionic, pyruvic, succinic, tartaric, valeric
Amino acid and amide	α-Alanine, β-alanine, arginine, asparagine, aspartic acid, cystine/cysteine, glutamine, glycine, histidine, lysine, methionine, phenylalanine, proline, serine/homoserine
Enzyme	Amylase, invertase, phosphatase, protease, polygalacturonase
Growth factor	*p*-Amino benzoic acid, auxins, biotin, choline, inositol, *n*-methyl nicotinic acid, niacin, pantothenate, pyridoxine, thiamine
Phenolic acid and coumarin	Caffeic acid, cinnamic acid, coumarin, ferulic acid, salycilic acid, syringic acid, vanillic acid
Sugar	Arabinose, fructose, fucose, galactose, glucose, maltose, oligosaccharide, raffinose, rhamnose, ribose, sucrose, xylose
Others	Nucleotide, flavanone, fatty acids, proteins, sterols, lipids, aliphatics, aromatics, carbohydrates

Source: Koo et al. (2005) © 2005, Academic Publishers. Reprinted with permissions of Elsevier.

8.3.2 Removal of Inorganic Contaminants

Inorganic contaminants which mostly include heavy metals and radionuclides are absorbed from the soil by roots of the plants and immobilized. In some cases, they may be transported to plant shoot for phytoextraction (Reichenauer and Germida 2008). Plant possess a highly effective metal uptake system with the help of transporter molecules like zinc-regulated transporter protein and copper transporter protein (Krämer et al. 2007). Plants are also adept of exuding metal-chelating molecules (siderophore, malate, citrate) and some biosurfactants (rhamnolipids) to the surrounding soil of rhizosphere zone (Eapen and D'Souza 2005; Garbisu and Alkorta 2001).

Heavy metals which constitute a large proportion of inorganic pollutants are not biodegradable and after intake from soil by the plants they are only transformed from one oxidation state to other (Garbisu and Alkorta 2001) and are accumulated inside plant tissue. In this context many plants ranging from annual herbs to perennial shrubs along with some trees has been identified and reported having the capacity of hyperaccumulating heavy metals (Table 8.2) especially Ni, Co, Pb, Zn, Mn, Cd, etc., in their above-ground tissues (Meagher 2000; Padmavathiamma and Li 2007; Shah and Nongkynrih 2007; Sheoran et al. 2009). These plants accumulate these heavy metals as defense supports for protection against insect and herbivorous animals (Fattorini et al. 2010; Shah and Nongkynrih 2007). The plants can be harvested from the site after remediation and can be safely disposed (Susarla et al. 2002). The heavy metals like Hg, As, and Se which are considered highly toxic plant absorb them through their root network and then can transport up to the shoot. It has been observed that later these metals are removed from the plant system by transpiration (Padmavathiamma and Li 2007; Pilon-Smits 2005). The volatile form in which Hg and Se are released in the atmosphere from plants are also considered to be toxic so the process of phytoremediation to be used at sites contaminated with Hg and Se compounds are still questionable (Padmavathiamma and Li 2007; Watanabe 1997).

8.3.3 Removal of Organic Pollutants

The organic form of contaminants present in agriculture soil is usually PAHs and some chlorinated solvent. They are usually taken by a plant roots system and immobilized (Gao and Zhu 2004) and at last phytoextracted from the shoot in form of methyl tert-butyl ether-MTBE, TCE, ethyl benzene and xylene (Ma et al. 2004). The plant system is easily able to photo degrade these organic contaminants. The contaminants taken up by the root system are chemically modified by means of reaction like oxidation, reduction and hydrolysis in various water-soluble forms by enzymes such as cytochrome P450 or carboxylesterases. These polar transformed molecules are conjugated with some endogenous molecules, such as sugar or peptides, and with the assistance of enzymes like glycosyltransferases and glutathione S-transferases they are transformed into a less phytotoxic form after being transferred into various compartments of cell-like vacuoles or integrated into the cell wall (Eapen et al. 2007; Ma and Burken 2003; Reichenauer and Germida 2008). The content of root exudates is precise to plant family or species. The Rhizosphere environment is appropriate for the degradation of organic contaminants.

Table 8.2 List of crop plants used in phytoremediation of heavy metals from soil.

Plants	Common name	Heavy metal	References
Allium schoenoprasum L.	Chive	Ni, Co, Cd	Goland-Goldhirsh (2006)
Brassica juncea L.	Mustard	Cd, Cu, Zn, Pb	Belimov et al. (2005), Takeda et al. (2006), Turan and Esringu (2007), Singh and Fulekar (2012)
Brassica napus L.	Canola	Cd, Cu, Zn, Pb	Turan and Esringu (2007), Sheng and Xia (2006), Dell'Amico et al. (2008)
Cajanuscajan L.	Pigeon pea	As, Cd	Garg et al. (2014)
Ciceraeritinum L.	Chick pea	Cd, Pb, Cr, Cu	Dasgupta et al. (2011), Wani et al. (2007), Kambhampati and Vu (2013)
Jatropha curcas L.	Physic nut	Fe, Al, Cu, Mn, Cr, As, Zn, Hg	Jamil et al. (2009), Yadav et al. (2009), Marrugo-Negrete et al. (2015)
Lantana camara L.	Lantana	Pb	Alaribe and Agamuthu (2015)
Lens culinaris	Lentil	Pb	Wani and Khan (2012)
Lepidium sativum L.	Cress	As, Cd, Fe, Pb, Hg	Gunduz et al. (2012), Smolinska and Szczodrowska (2016)
Lactuca sativa L.	Lettuce	Cu, Fe, Mn, Zn, Ni, Cd, Pb, Co, As	Gunduz et al. (2012), Achakzai et al. (2011), Rashid et al. (2014), Quainoo et al. (2015)
Medicago sativa L.	Alfalfa	Cd	Ghnaya et al. (2015)
Oryza sativa L.	Rice	Cu, Cd	Li et al. (2008)
Pisum sativum L.	Pea	Pb, Cu, Zn, Fe, Cd, Ni, As, Cr	Malecka et al. (2008), Wani et al. (2008), Hegedusova et al. (2009), Sharma et al. (2010)
Rapanus sativus L.	Radish	As, Cd, Fe, Pb, Cu	Hatano et al. (2016)
Spinacia oleracea L.	Spinach	Cd, Cu, Fe, Ni, Pb, Zn, Cr	Patel and Subramanian (2006), Salaskar et al. (2011), Pathak et al. (2013), Jahanbakhshi et al. (2014), Abhilash et al. (2016)
Solanum nigrum L.	Black nightshade	Cd	Wei et al. (2006), Wei et al. (2010), Ji et al. (2011)
Sorghum bicolor L.	Sorghum	Cd, Cu, Zn, Fe	Pinto et al. (2004)
Zea mays L.	Corn	Cd, Pb, Zn, Cu	Mojiri (2011)

1) In this instance it has been reported that salicylate, a secondary plant metabolite promotes the microbial degradation of some PAHs, such as naphthalene, fluoranthene, pyrene, chrysene and PCBs (Chen and Aitken 1999; Master and Mohn 2001; Singer et al. 2000). Other secondary metabolite terpenes persuade the microbial degradation of toluene, phenol and TCE (Kim et al. 2002).
2) Plant exudates contain secondary metabolites that are oftentimes structurally analogous to organic contaminants (Singer et al. 2003). This results in an increased presence and activity of microbes which can degrade organic contaminants present in rhizosphere region listed in Table 8.3 (Yergeau et al. 2014).
3) Due to the presence of root exudates, microbial communities present in the rhizosphere region are more active and are more abundant than if found in bulk soil far from the

Table 8.3 Potential microorganisms for biodegradation of different organic pollutants.

Pollutants	Organism	References
Benzene, anthracene, hydrocarbons, PCBs	*Pseudomonas sps.*	Kapley et al. (1999) and Cybulski et al. (2003)
Halogenated hydrocarbons, linear alkylbenzene sulfonates, polycyclic aromatics, PCBs	*Alcaligenes sps.*	Lal and Khanna (1996)
Benzene, hydrocarbons, pentachlorophe-nol, phenoxyacetate, polycyclic aromatic	*Arthrobacter sps.*	Jogdand (1995)
Aromatics, long chain alkanes, phenol, cresol	*Bacillus sps.*	Cybulski et al. (2003)
Halogenated hydrocarbons, phenoxyacetates	*Corynebacterium sps.*	Jogdand (1995)
Aromatics	*Flavobacterium sps.*	Jogdand (1995)
Aromatics	*Azotobacter sps.*	Jogdand (1995)
Naphthalene, biphenyl	*Rhodococcus sps.*	Dean-Ross et al. (2002)
Aromatics, branched hydrocarbons, benzene, cycloparaffins	*Mycobacterium sps.*	Sunggyu (1995)
Hydrocarbons	*Nocardia sps.*	Park et al. (1998)
Aromatics	*Methosinus sps.*	Jogdand (1995)
Hydrocarbons, polycyclic hydrocarbons	*Xanthomonas sps.*	Jogdand (1995) and Ijah (1998)
Phenoxyacetate, halogenated hydrocarbon diazinon	*Streptomyces sps.*	Jogdand (1995)
PCBs, formaldehyde	*Candida tropicalis*	Ijah (1998)
PCBs, polycyclic aromatics, biphenyls	*Cunniughamela elegans*	Jogdand (1995)
PCBs	*P. chrysosporium*	Borazjani et al. (2005)

Source: Chatterjee et al. (2008) © 2008, Elewa BioSciences. Public Domain.

influence of the root (Kent and Triplett 2002; Kowalchuk et al. 2002; Ramos et al. 2000; Smalla et al. 2001)

4) Rhizosphere is a region for horizontal gene transfer (Van Elsas and Bailey 2002) and microorganisms are adapted to this abiotic stress arising due to contamination. Plasmids present in the microorganism help in this process (Sentchilo et al. 2013; Top and Springael 2003)
5) Some researchers have also investigated and reported that root exudates in some cases help to detach organic contaminants from that of organic matter which are present in soil (Gao et al. 2010). Biosurfactants released in some cases escalate the bioavailability and uptake of pollutants (Kuiper et al. 2004; Schwitzguébel et al. 2002; Wenzel 2008).
6) Some root exudates compounds also serve as co-metabolites in some cases where contaminants cannot be utilized by microbes as sole carbon source (Hyman et al. 1995) like degradation of TCE.

8.3.4 Factors Affecting Rhizomediation

Various environmental factors play a decisive role in restricting the outcome of rhizomediation, such as competition offered from some better adapting weed species (Nedunuri et al. 2000). Growth of plants is limited in some cases of heavily contaminant soil, as well as the presence of biotic and abiotic stress (Gerhardt et al. 2009). The process of rhizomediation is found unsuitable for contaminants present in the deeper subsurface layer far away from the reach of root system. In certain cases, many toxic contaminants can bioaccumulate inside plants.

In spite of certain limiting factors, rhizomediation has emerged as an effective means (Azaizeh et al. 2011; Gerhardt et al. 2009; Martin et al. 2014; Reichenauer and Germida 2008) and has been successfully applied at contaminated sites polluted with organic pollutants like petroleum hydrocarbon (Nedunuri et al. 2000; Siciliano et al. 2003), PAHs (Robinson et al. 2003; White et al. 2006), TNT (van Dillewijn et al. 2007), BTEX (Barac et al. 2009) and TCE (Weyens et al. 2009a, b).

8.4 Conclusion and Future Prospects

Rhizomediation is a green technology based on phyto-microbe interaction for biodegradation, transformation and immobilization of toxic pollutant particles present in soil which is utilized for the purpose of agriculture. This technique of utilizing plants and their associated microbe community had been used for processing of many classes of contaminants composing of organic and inorganic elements and has proved beneficial in case of organic contaminants. Despite a few limiting factors but due to its high potentiality to remediate soil, its low cost and it being eco-friendly, it has been accepted by masses. In future, transgenic plants may be developed having a full potential for degradation of xenobiotic pollutants by the introduction of genes isolated from microbes having the potential of degradation. Further research needs to be done to completely understand the plant–microbe interaction mechanism.

References

Abhilash, M.R., Srikantaswamy, S., Shiva Kumar, D. et al. (2016). Phytoremediation of heavy metal industrial contaminated soil by *Spinacia oleracea* L. and *Zea mays* L. *Int. J. Appl. Sci.* 4 (1): 192–199.

Achakzai, A.K.K., Bazai, Z.A., and Kayani, S.A. (2011). Accumulation of heavy metals by lettuce (*Lactuca sativa* L.) irrigated with different levels of wastewater of Quetta City. *Pak. J. Bot.* 43 (6): 2953–2960.

Alaribe, F.O. and Agamuthu, P. (2015). Assessment of phytoremediation potencials of *Lantanacamarain* Pb impacted soil with organic wasted additives. *Ecol. Eng.* 83: 513–520.

Ayoub, K., van Hullebusch, E.D., Cassir, M., and Bermond, A. (2010). Application of advanced oxidation processes for TNT removal: a review. *J. Hazard. Mater.* 178 (1–3): 10–28.

Azaizeh, H., Castro, P.M.L., and Kidd, P. (2011). Biodegradation of organic xenobiotic pollutants in the rhizosphere. In: *Organic Xenobiotics and Plants* (eds. P. Schröder and C.D. Collins), 191–215. Dordrecht: Springer Netherlands.

Barac, T., Weyens, N., Oeyen, L. et al. (2009). Field note: hydraulic containment of a BTEX plume using poplar trees. *Int. J. Phytoremediation* 1 (5): 416–424.

Belimov, A.A., Hontzeas, N., Safranova, V.I. et al. (2005). Cadmium-tolerant plant growth-promoting bacteria associated with the roots of Indian mustard (*Brassica juncea* L. Czern.). *Soil Biol. Biochem.* 37: 241–250.

Borazjani, H., Wiltcher, D., and Diehl, S. (2005). Bioremediation of polychlorinated biphenyl and petroleum contaminated soil. In: *Proceedings of Environmental Science and Technology* (eds. W.G. Lyon, J. Hong and R.K. Reddy), 502–507. New Orleans: American Science Press.

Burken, J.G. and Schnoor, J.L. (1998). Predictive relationships for uptake of organic contaminants by hybrid poplar trees. *Environ. Sci. Technol.* 32: 3379–3385.

Chatterjee, S., Chattopadhyay, P., Roy, S., and Sen, S.K. (2008). Bioremediation: a tool for cleaning polluted environments. *J. Appl. Biosci.* 11: 594–601.

Chaudhry, Q., Blom-Zandstra, M., Gupta, S., and Joner, E.J. (2005). Utilising the synergy between plants and rhizosphere microorganisms to enhance breakdown of organic pollutants in the environment. *Environ. Sci. Pollut. Res.* 12 (1): 34–48.

Chen, S.H. and Aitken, M.D. (1999). Salicylate stimulates the degradation of high molecular weight polycyclic aromatic hydrocarbons by *Pseudomonas saccharophila* p15. *Environ. Sci. Technol.* 33: 435–439.

Cunningham, S.D. and Berti, W.R. (1993). Remediation of contaminated soils with green plants: an overview. *In Vitro Cell. Dev. Biol.* 29: 207–212.

Cybulski, Z., Dzuirla, E., Kaczorek, E., and Olszanowski, A. (2003). The influence of emulsifiers on hydrocar-bon biodegradation by Pseudomonadacea and Bacillacea strains. *Spill Sci. Technol. Bull.* 8: 503–507.

Dasgupta, S., Satvat, P.S., and Mahinrakar, A.B. (2011). Ability of *Cicerarientinum* (L.) for bioremoval of lead and chromium from soil. *Int. J. Technol. Eng. Syst.* 2 (3): 338–341.

Dean-Ross, D., Moody, J., and Cerniglia, C.E. (2002). Utilization of mixtures of polycyclic aromatic hydrocarbons by bacteria isolated from contaminated sediment. *FEMS Microbiol. Ecol.* 41: 1–7.

Dell'Amico, E., Cavalva, L., and Andreoni, V. (2008). Improvement of *Brassica napus* growth under cadmium stress by cadmium-resistant rhizobacteria. *Soil Biol. Biochem.* 40: 74–84.

van Dillewijn, P., Caballero, A., Paz, J.A. et al. (2007). Bioremediation of 2,4,6-trinitrotoluene under field conditions. *Environ. Sci. Technol.* 41 (4): 1378–1383.

Eapen, S. and D'Souza, S.F. (2005). Prospects of genetic engineering of plants for phytoremediation of toxic metals. *Biotechnol. Adv.* 23: 97–114.

Eapen, S., Singh, S., and D'Souza, S.F. (2007). Advances in development of transgenic plants for remediation of xenobiotic pollutants. *Biotechnol. Adv.* 25: 442–451.

Fattorini, D., Notti, A., Nigro, M., and Regoli, F. (2010). Hyperaccumulation of vanadium in the Antarctic polychaete Perkinsianalittoralis as a natural chemical defense against predation. *Environ. Sci. Pollut. Res.* 17: 220–228.

Gao, Y. and Zhu, L. (2004). Plant uptake, accumulation and translocation of phenanthrene and pyrene in soils. *Chemosphere* 55: 1169–1178.

Gao, Y., Ren, L., Ling, W. et al. (2010). Desorption of phenanthrene and pyrene in soils by root exudates. *Bioresour. Technol.* 101: 1159–1165.

Garbisu, C. and Alkorta, I. (2001). Phytoextraction: A cost-effective plant-based technology for the removal of metals from the environment. *Bioresour. Technol.* 77: 229–236.

Garg, N., Singla, P., and Bhandari, P. (2014). Metal uptake, oxidative metabolism, and mycorrhization in pegeon pea and pea under arsenic and cadmium stress. *Turk. J. Agric. For.* 39: 234–250.

Gerhardt, K.E., Huang, X.D., Glick, B.R., and Greenberg, B.M. (2009). Phytoremediation and rhizoremediation of organic soil contaminants: potential and challenges. *Plant Sci.* 176: 20–30.

Ghnaya, T., Mnassri, M., Ghabriche, R. et al. (2015). Nodulation by *Sinorhizobium meliloti* originated from a mining soil alleviates Cd toxicity and increases Cd-phytoextraction in *Medicago sativa* L. *Front. Plant Sci.* 6: 1–10.

Goland-Goldhirsh, A. (2006). Plant tolerance to heavy metals, a risk for food toxicity or a means for food fortification with essential metals: the Allium schoenoprasum model. In: *Soil and Water Pollution Monitoring, Protection and Remediation* (eds. I. Twardowska, H.E. Allen and M.M. Haggblom), 479–486. Amsterdam: Springer.

Gracia, S.C., Pande, P., Seguin, A. et al. (2018). Rhizomediation of petroleum hydrocarbons: a model system for plant microbiome manipulation. *J. Microbial. Biotechnol.* 11: 819–832.

Gunduz, S., Uygur, F.N., and Kahramanoglu, I. (2012). Heavy metal phytoremediation potencials of *Lepidium* sativum L., *Lactuca sativa* L., *Spinacia oleracea* L. and *Raphanus sativus* L. *Herald J. Agric. Food Sci. Res.* 1 (1): 1–5.

Hatano, K., Kanazawa, K., Tomura, K., Yamatsu, T., Tsunoda K., and Kubota K. (2016). Molases melanoidin promotes copper uptake for radish sprouts: the potential for an accelerator of phytoextraction *Environ. Sci. Pollut. Res.* 23: 17656–17663.

Hegedusova, A., Jakabova, S., Vargova, A. et al. (2009). Use of phytoremediation techniques for elimination of lead from polluted soils. *Nova Biotechnol.* 9 (2): 125–132.

Hyman, M.R., Russell, S.A., Ely, R.L. et al. (1995). Inhibition, inactivation, and recovery of ammonia-oxidizing activity in cometabolism of trichloroethylene by *Nitrosomonas europaea*. *Appl. Environ. Microbiol.* 61 (4): 1480–1487.

Ijah, U.J.J. (1998). Studies on relative capabilities of bacterial and yeast isolates from tropical soil in degrading *crude oil*. *Waste Manag.* 18: 293–299.

Jahanbakhshi, S., Rezaei, M.R., and Sayyari-Zahan, M.H. (2014). Optimization of phytoremediation in cd-contaminated soil by using Taguchi method in *Spinacia oleracea*. In: *Proceedings of the International Academy of Ecology and Environmental Sciences*, vol. 4 (ed. W. Zhang), 185–193. Hong Kong: International Academy of Ecology and Evironmental Sciences.

Jamil, S., Abhilash, P.C., Singh, N., and Sharma, P.N. (2009). *Jatropha curcas*: a potential crop for coal fly ash. *J. Hazard. Mater.* 172: 269–275.

Ji, P., Sun, T., Song, Y. et al. (2011). Strategies for enhanching the phytoremediation of cadmium-contaminated agricultural soils by *Solanum nigrum* L. *Environ. Pollut.* 159: 762–768.

Jogdand, S.N. (1995). *Environmental Biotechnology*, 1e, 104–120. Bombay: Himalaya Publishing House.

Jones, D.L. (1998). Organic acids in the rhizosphere – a critical review. *Plant and Soil* 205: 25–44.

Kambhampati, M.S. and Vu, V.T. (2013). EDTA enhanced phytoremediation of copper contaminated soils using chickpea (*Ciceraeritinum* L.). *Bull. Environ. Contam. Toxicol.* 91: 310–313.

Kapley, A., Purohit, H.J., Chhatre, S. et al. (1999). Osmotolerance and hydrocarbon degradation by a genetically engineered microbial consortium. *Bioresour. Technol.* 67: 241–245.

Kent, A.D. and Triplett, E.W. (2002). Microbial communities and their interactions in soil and rhizosphere ecosystems. *Annu. Rev. Microbiol.* 56: 211–236.

Khan, F.I., Husain, T., and Hejazi, R. (2004). An overview and analysis of site remediation technologies. *J. Environ. Manage.* 71: 95–122.

Kim, D., Park, M.J., Koh, S.C. et al. (2002). Three separate pathways for the initial oxidation of limonene, biphenyl, and phenol by *Rhodococcus*sp strain t 104. *J. Microbiol.* 40: 86–89.

Koo, B.J., Adriano, D.C., Bolan, N.S., and Barton, C.D. (2005). Roots exudates and microrganisms. In: *Encyclopedia of Soils in the Environment* (ed. D. Hillel), 421–428. London: Elsevier/Academic.

Kowalchuk, G.A., Buma, D.S., De Boer, W. et al. (2002). Effects of above-ground plant species composition and diversity on the diversity of soil-borne microorganisms. *Int. J. Gen. Mol. Microbiol.* 81: 509–520.

Krämer, U., Talke, I.N., and Hanikenne, M. (2007). Transition metal transport. *FEBS Lett.* 581: 2263–2272.

Kuiper, I., Lagendijk, E.L., Bloemberg, G.V., and Lugtenberg, B.J. (2004). Rhizoremediation: a beneficial plant-microbe interaction. *Mol. Plant Microbe Interact.* 17: 6–15.

Lal, B. and Khanna, S. (1996). Degradation of crude oil by *Acinetobacter calcoaceticus* and *Alcaligenes odorans*. *J. Appl. Bacteriol.* 81: 355–362.

Li, P., Wang, X., Zhang, T. et al. (2008). Effect of several amendments on rice growth and uptake of copper and cadmium from a contaminated soil. *J. Environ. Sci.* 20: 449–455.

Ma, X. and Burken, J.G. (2003). TCE diffusion to the atmosphere in phytoremediation applications. *Environ. Sci. Technol.* 37: 2534–2539.

Ma, X., Richter, A.R., Albers, S., and Burken, J.G. (2004). Phytoremediation of MTBE with hybrid poplar trees. *Int. J. Phytoremediation* 6: 157–167.

Macek, T., Mackova, M., and Kas, J. (2000). Exploitation ofplants for the removal of organics in environmental remediation. *Biotechnol. Adv.* 18: 23–34.

Malecka, A., Piechalak, A., and Morkunas, I. (2008). Accumulation of lead in root cells of *Pisumsativum*. *Acta Physiol. Plant.* 30: 629–637.

Marrugo-Negrete, J., Durango-Hernandez, J., Pinedo-Hernandez, J. et al. (2015). Phytoremediation of mercury-contaminated soils by Jatropha curcas. *Chemosphere* 127: 58–63.

Martin, B.C., George, S.J., Price, C.A. et al. (2014). The role of root exuded low molecular weight organic anions in facilitating petroleum hydrocarbon degradation: current knowledge and future directions. *Sci. Total Environ.* 472: 642–653.

Master, E.R. and Mohn, W.W. (2001). Induction of *bphA*, encoding biphenyl dioxygenase, in two polychlorinated biphenyl-degrading bacteria, psychrotolerant *Pseudomonas* strain cam-1 and mesophilic *Burkholderia* strain lb400. *Appl. Environ. Microbiol.* 67: 2669–2676.

Meagher, R.B. (2000). Phytoremediation of toxic elemental and organic pollutants. *Curr. Opin. Plant Biol.* 3: 153–162.

Mojiri, A. (2011). The potencial of corn (*Zea mays*) for phytoremediation of soil contaminated with cadmium and lead. *J. Biol. Environ. Sci.* 5 (13): 17–22.

Nedunuri, K.V., Govindaraju, R.S., Banks, M.K. et al. (2000). Evaluation of phytoremediation for field-scale degradation of total petroleum hydrocarbons. *J. Environ. Eng.* 126 (6): 483–490.

Padmavathiamma, P.K. and Li, L.Y. (2007). Phytoremediation technology: hyper-accumulation metals in plants. *Water Air Soil Pollut.* 184: 105–126.

Park, A.J., Cha, D.K., and Holsen, T.M. (1998). Enhancing solubilization of sparingly soluble organic com-pounds by biosurfactants produced by *Nocardia erythropolis*. *Water Environ. Res.* 70: 351–355.

Patel, M. and Subramanian, R.B. (2006). Effect of a chelating agent on lead uptake by *Spinaciaolearea*. *Pollut. Res.* 25 (1): 77–79.

Pathak, C., Chopra, A.K., and Srivastava, S. (2013). Accumulation of heavy metals in *Spinaciaoleracea* irrigated with paper mill effluent and sewage. *Environ. Monit. Assess.* 185: 7343–7352.

Pilon-Smits, E. (2005). Phytoremediation. *Annu. Rev. Plant Biol.* 56: 15–39.

Pinto, A.P., Mota, A.M., de Varennes, A., and Pinto, F.C. (2004). Influence of organic matter on uptake of cadmium, zinc, copper and iron by shorghum plant. *Sci. Total Environ.* 326: 239–247.

Quainoo, A.K., Konadu, A., and Kumi, M. (2015). The potential of shea nut shells in phytoremediation of heavy metals in contaminated soil using lettuce (*Lactuca sativa*) as a test crop. *J. Bioremed. Biodegr.* 6 (1): 1–7.

Ramos, C., Molbak, L., and Molin, S. (2000). Bacterial activity in the rhizosphere analyzed at the single-cell level by monitoring ribosome contents and synthesis rates. *Appl. Environ. Microbiol.* 66 (2): 801–809.

Rashid, A., Mahmood, T., Mehmood, F. et al. (2014). Phytoaccumulation, competitive adsorption and evaluation of chelators-metal interaction in lettuce plant. *Environ. Eng. Manag. J.* 13 (10): 2683–2692.

Reichenauer, T.G. and Germida, J.J. (2008). Phytoremediation of organic contaminants in soil and groundwater. *ChemSusChem.* 1: 708–717.

Robinson, S.L., Novak, J.T., Widdowson, M.A. et al. (2003). Field and laboratory evaluation of the impact of tall fescue on polyaromatic hydrocarbon degradation in an aged creosote-contaminated surface soil. *J. Environ. Eng.* 129 (3): 232–240.

Salaskar, D., Shrivastava, M., and Kale, S.P. (2011). Bioremediation potential od spinach (*Spinaciaoleracea* L.) for decontamination of cadmium in soil. *Curr. Sci.* 101 (10): 1359–1363.
Salt, D.E., Smith, R.D., and Raskin, I. (1998). Phytoremediation. *Annu. Rev. Plant. Physiol. Plant. Mol. Biol.* 49: 643–668.
Schnoor, J.L., Licht, L.A., McCutcheon, S.C. et al. (1995). Phytoremediation of organic and nutrient contaminants. *Environ. Sci. Technol.* 29: 318–323.
Schröder, P. and Collins, C. (2002). Conjugating enzymes involved in xenobiotic metabolism of organic xenobiotics in plants. *Int. J. Phytoremediation* 4: 247–265.
Schwitzguébel, J.P., Aubert, S., Grosse, W., and Laturnus, F. (2002). Sulphonated aromatic pollutants: limits of microbial degradability and potential of phytoremediation. *Environ. Sci. Pollut. Res.* 9 (1): 62–72.
Segura, A., Rodríguez-Conde, S., Ramos, C., and Ramos, J.L. (2009). Bacterial responses and interactions with plants during rhizoremediation. *J. Microbial. Biotechnol.* 2 (4): 452–464.
Sentchilo, V., Mayer, A.P., Guy, L. et al. (2013). Community-wide plasmid gene mobilization and selection. *ISME J.* 7: 1173–1186.
Shah, K. and Nongkynrih, J.M. (2007). Metal hyperaccumulation and bioremediation. *Biol. Plant* 51: 618–634.
Sharma, S., Sharma, P., and Mehrotra, P. (2010). Bioaccumulation of heavy metals in *Pisum sativum* L. gowing in fly ash amended soil. *J. Am. Sci.* 6 (6): 43–50.
Sheng, X.F. and Xia, J.J. (2006). Improvement of rape (*Brassica napus*) plant growth and cadmium uptake by cadmium-resistant bacteria. *Chemosphere* 64: 1036–1042.
Sheoran, V., Sheoran, A.S., and Poonia, P. (2009). Phytomining: a review. *Miner. Eng.* 22: 1007–1019.
Siciliano, S.D., Germida, J.J., Banks, K., and Greer, C.W. (2003). Changes in microbial community composition and function during a polyaromatic hydrocarbon phytoremediation field trial. *Appl. Environ. Microbiol.* 69 (1): 483–489.
Singer, A.C., Gilbert, E.S., Luepromchai, E., and Crowley, D.E. (2000). Bioremediation of polychlorinated biphenyl-contaminated soil using carvone and surfactant-grown bacteria. *Appl. Microbiol. Biotechnol.* 54: 838–843.
Singer, A.C., Crowley, D.E., and Thompson, I.P. (2003). Secondary plant metabolites in phytoremediation and biotransformation. *Trends Biotechnol.* 21: 123–130.
Singh, A. and Fulekar, M.H. (2012). Phytoremediation of heavy metals by *Brassica juncea* in aquatic and terrestrial environment. In: *The Plant Family Brassicaceae: Contribution Towards Phytoremediation* (eds. Anjum, N.A. Ahmad, I. Pereira, et al.), 153–169. Amsterdam: Springer Science+Business Media.
Singh, B.K., Millard, P., Whiteley, A.S., and Murrell, J.C. (2004). Unravelling rhizosphere-microbial interactions: opportunities and limitations. *Trends Microbiol.* 12: 386–393.
Smalla, K., Wieland, G., Buchner, A. et al. (2001). Bulk and rhizosphere soil bacterial communities studied by denaturing gradient gel electrophoresis: plant-dependent enrichment and seasonal shifts revealed. *Appl. Environ. Microbiol.* 67: 4742–4751.
Smolinska, B. and Szczodrowska, A. (2016). Antioxidative response of *Lepidium sativum* L. during assisted phytoremediation of Hg contaminated soil. *N. Biotechnol.* 38: 74–83.
Sunggyu, L. (1995). Bioremediation of polycyclic aromatic hydrocarbon-contaminated soil. *J. Clean. Prod.* 3: 255.
Suresh, B. and Ravishankar, G.A. (2004). Phytoremediation – a novel and promising approach for environmental clean-up. *Crit. Rev. Biotechnol.* 24: 97–124.

Susarla, S., Medina, V.F., and McCutcheon, S.C. (2002). Phytoremediation: an ecological solution to organic chemical contamination. *Ecol. Eng.* 18: 647–658.

Takeda, R., Sato, Y., Yoshimura, R., Komemushi, S. et al. (2006). Accumulation of heavy metals by cucumber and Brassica juncea under different cultivation conditions *Proc. Ann. Int. Conf. on Soil Sediments Water Energy (Massachusetts)* 11 (California: The Barkeley Electronic Press): 293–299.

Top, E.M. and Springael, D. (2003). The role of mobile genetic elements in bacterial adaptation to xenobiotic organic compounds. *Curr. Opin. Biotechnol.* 14: 262–269.

Truu, J., Truu, M., Espenberg, M. et al. (2015). Phytoremediation and plant assisted bioremediation in soil and treatment wetlands: a review. *Open Biotechnol. J.* 9 (1): 85–92.

Turan, M. and Esringu, A. (2007). Phytoremediation based on canola (*Brassica napus* L.) and Indian mustard (*Brassica juncea* L.) planted on spiked soil by aliquot amount of Cd, Cu, Pb, and Zn. *Plant Soil Environ.* 53 (1): 7–15.

Van Elsas, J.D., Turner, S., and Bailey, M.J. (2003). Horizontal gene transfer in the phytosphere. *New Phytologist.* 157(3): 525–537.

Wani, P.A. and Khan, M.S. (2012). Bioremediation of lead by a plant growth promoting *Rhizobium* species RL9. *Bacteriology J.* 2 (4): 66–78.

Wani, P.A., Khan, M.S., and Zaidi, A. (2007). Impact of heavy metal toxicity on plant growth, symbiosis, seed yield and nitrogen and metal uptake in chickpea. *Aust. J. Exp. Agric.* 47: 712–720.

Wani, P.A., Khan, M.S., and Zaidi, A. (2008). Effects of heavy metal toxicity on growth, symbiosis, seed yield and metal uptake in pea grown in metal amended soil. *Bull. Environ. Contam. Toxicol.* 81: 152–158.

Watanabe, M.E. (1997). Phytoremediation on the brink of commercialization. *Environ. Sci. Technol.* 31: 182–186.

Wei, S., Zhou, Q., and Koval, P.V. (2006). Flowering stage characteristic of cadmium hyperaccumulator *Solanum nigrum* L. and their significance to phytoremediation. *Sci. Total Environ.* 369: 441–446.

Wei, S., Li, Y., Zhou, Q. et al. (2010). Effect of fertilizer amendments on phytoremediation of Cd-contaminated soil by newly discovered hyperaccumulator *Solanum nigrum* L. *J. Hazard. Mater.* 176: 269–273.

Wenzel, W.W. (2008). Rhizosphere processes and management in plant-assisted bioremediation (phytoremediation) of soils. *Plant Soil* 321 (1–2): 385–408.

Weyens, N., van der Lelie, D., Artois, T. et al. (2009a). Bioaugmentation with engineered endophytic bacteria improves contaminant fate in phytoremediation. *Environ. Sci. Technol.* 43 (24): 9413–9418.

Weyens, N., van der Lelie, D., Taghavi, S., and Vangronsveld, J. (2009b). Phytoremediation: plant-endophyte partnerships take the challenge. *Curr. Opin. Biotechnol.* 20 (2): 248–254.

White, P.M. Jr., Wolf, D.C., Thoma, G.J., and Reynolds, C.M. (2006). Phytoremediation of alkylated polycyclic aromatic hydrocarbons in a crude oil-contaminated soil. *Water Air Soil Pollut.* 169 (1–4): 207–220.

Yadav, S.K., Juwarkar, A.S., Kumar, P. et al. (2009). Bioaccumulation and phyto-translocation of arsenic, chromium, and zinc by *Jatropha curcas* L.: impact of dairy sludge and biofertilizer. *Biosour. Technol.* 100: 4616–4622.

Yergeau, E., Sanschagrin, S., Maynard, C. et al. (2014). Microbial expression profiles in the rhizosphere of willows depend on soil contamination. *ISME J.* 8: 344–358.

9

Rhizospheric Biology

Alternate Tactics for Enhancing Sustainable Agriculture

Kalpana Bhatt[1] *and Pankaj Bhatt*[2]

[1] *Gurukula Kangri Vishwavidhyalaya, Haridwar, Uttarakhand, India*
[2] *Integrative Microbiology Research Centre, South China Agricultural University, Guangzhou, P.R. China*

9.1 Introduction

The soil has a number of miniature life forms, viz., bacteria, fungi, protozoa, and algae. Irrespective of the microscopic life forms, bacteria stand undoubtedly to be the maximum (i.e. ~95%). It has been acknowledged well that soil grasps a massive figure of bacteria (frequently about 10^8 to 10^9cells/g of soil), where culturable bacterial cells tend to be only 1% of the entire cell number so present (Schoenborn et al. 2004). Moreover, bacteria normally are not uniformly dispersed in soil. To be precise, the maximum concentration of bacteria is found to be around the plant roots (rhizosphere) than the other soil areas. Possibly, this may be due to the existence of several nutrients, viz., sugars, organic acids, and amino acids, as well as rest of the molecules from plant root exudates that may perhaps account up to a third of the carbon that is fixed by a plant (Badri et al. 2009; Badri and Vivanco 2009; Bais et al. 2006; Whipps 1990).

Generally, soil microorganisms and plants interact via roots, where the region of their connection is designated as "rhizosphere." The rhizosphere is the vicinity of nearby soil roots explicitly, which is highly influenced via root activity. The relationship of interaction between the plant and its atmosphere is indispensable for water and nutrients acquisition along with favorable connections with soil microorganisms. However, this similar relationship upsurges the susceptibility of plants to a variety of stresses (biotic and abiotic). Several strategies have been manipulated to counter these environmental stresses to minimize the influence of biotic and abiotic stresses, with respect to plant and its rhizosphere. This technique for overcoming environmental stress along with improves plant growth and soil fertility is termed as rhizosphere engineering. Thus, this procedure recommends various methods by which rhizosphere can be modified resulting in enhanced fitness and efficiency of plants. Rhizosphere engineering possibly will lessen our reliance on agrochemicals by substituting with beneficial bacteria, sustainable biostimulants or transgenic plants, etc. This field demands a lot more exploitation, where several new powerful biotechnological and

Phytomicrobiome Interactions and Sustainable Agriculture, First Edition. Edited by Amit Verma, Jitendra Kumar Saini, Abd El-Latif Hesham, and Harikesh Bahadur Singh.

molecular tools will continue in providing complete information regarding the rhizosphere complex chemical and biological interactions, besides safeguarding that these strategies of engineer the rhizosphere tend to be harmless, advantageous in terms of productivity, as well as considerably increasing the agricultural system in a sustainable manner. Henceforth, this chapter provides the various aspects of rhizosphere designing, engineering, and management in line with many emerging novel tactics for sustainable agriculture.

9.2 Engineering the Rhizosphere

Rhizosphere environment (physical and chemical) is the conclusion of several interacting and competing processes depending on the type of soil and water content, root-accompanying microbe configuration and their biological activities and the plant physiology (Pinto et al. 2007). Plant rhizosphere can easily be altered over short phases by means of agronomic administration (Bowen and Rovira 1999). Protracted modifications in the rhizosphere that persevere through the growing sequence could be produced through plant breeding and biotechnology too. Usually, farmers practice chemical-based fertilizers directly onto the surroundings of crop roots, thereby influencing the environment directly or indirectly. On the contrary, the rhizosphere consists of a huge number of microbial populations, effecting the plant and soil atmosphere. So, by the practice of rhizosphere engineering, various methods can be developed for enhancing the rhizosphere microbial populations affecting the plant and soil environment positively.

9.2.1 Rhizosphere and Rhizobia

The term "rhizosphere" was initially coined by Lorenz Hiltner in 1904 and identified as the narrow soil zone that directly surrounds the roots system (Walker et al. 2003). The rhizosphere is broadly categorized into three zones, namely endorhizosphere, rhizoplane, and ectorhizosphere. The root cortex and endodermis portion, where microbial populations followed by mineral ions exist in the apoplastic spaces among the cells, is termed as endorhizosphere. Likewise, the intermediate region, subsequent to root epidermal cells and mucilage, is termed as rhizoplane, whereas the outmost region extending as of rhizoplane out into the bulk soil as ectorhizosphere. Considering the rhizosphere as gradient of chemical, physical, and biological assets alongside roots is more relevant than considering it as the definable size or shape region. Plant metabolism intensely influences the rhizosphere via liberating carbon dioxide (CO_2) and secreting photosynthate as collection of root exudates (primarily from the ectorhizosphere and rhizoplane). Principally, the rhizosphere is central for ecosystem amenities, for instance, carbon and water cycling, nutrient accumulating, crop production, uptakeas well as storage of carbon (Adl 2016). Worldwide climatic variation, with an increase in temperatures, disorderly weather patterns due to an increase in atmospheric carbon dioxide (CO_2) levels affecting the rhizosphere ecosystem, thus ecological system function, via the diversity of direct as well as indirect means.

The rhizosphere comprises a number of microbes, viz., bacteria, fungi, algae, and protozoa, occurring together in the rhizosphere, bacteria being copious among them all. The term "rhizobacteria" is used to indicate an accumulation of rhizosphere bacteria proficient in colonizing the root environment (Kloepper et al. 1991). These bacteria interact and correlate with plants in a codependent way, where a plant requires nitrogen for their development; however, bacteria liberates that nitrogen thereby assisting in fixing atmospheric nitrogen henceforth sustaining the earth's atmosphere composition. However, rhizosphere complexity has been unraveled, so various attempts possibly could be created favorable for plant growth and promotion via alteration in plant–microorganism interactions. One of the foremost means by which plants alter the rhizosphere is by liberating organic and inorganic substances from their roots (rhizodeposition), referred to as root exudates. These exudates significantly augment nutrient acquisition, besides aiding in avoiding mineral stresses as well as encouraging the growth of beneficial microorganisms.The microorganisms are one of the important constituents of the rhizosphere, where these rhizospheric microbial populations affect plant and soil environment interactions. So, there is substantial attention in evolving procedures for boosting the exponentiation of such advantageous microbial populations (e.g., rhizobia and mycorrhiza) assisting in uptake of nutrients, directly promoting plant development as well as suppressing plant pathogens.

Rhizobia is a group of root and stem nodulating bacteria, which are gram-negative, short rods, aerobic and motile (Jordan 1984) and possess the ability to fix atmospheric nitrogen in symbiotic association with leguminous plants. This is a taxonomically diverse group of organisms, placed in 13 genera and have more than about 80 species belonging to both α-, β-proteobacteria, and γ-proteobacteria; namely, *Azorhizobium*, *Bradyrhizobium*, *Burkholderia*, *Cupriavidus*, *Devosia*, *Ensifer*, *Herbaspirillum*, *Mesorhizobium*, *Methylobacterium*, *Ochrobactrum*, *Phyllobacterium*, *Rhizobium* and *Shinella* (Weir 2008). However, the association among plants (leguminous) and rhizobia is quite a multifaceted interaction. Rhizobia are categorized based on their capability to nodulate various members of the leguminaceae family. Rhizobial-legume association begins with two free-living organisms, whereas it concludes with an intimate cellular coexistence. Diversity of cultivable microorganisms of *Davidia involucrate* in rhizosphere soil was recently studied, and it was stated that *Bacillus* is a dominant rhizobacteria that possibly has the potential in the growth promotion of host plants. The taxonomy of rhizobia has considerably differed over the past 20 years, with the original genus *Rhizobium*, a member of the alpha-Proteobacteria, presently distributed into numerous genera. These soil bacteria distinguish specific plants and aggravate the growth of root-nodules, thus invading the plant tissue (Liu et al. 2016; Salamone et al. 1996). Ultimately transferring them into the host cell, it surrounds itself with plant membrane, besides arranging a nutrient exchange in which bacteria supply ammonia to the plant by fixing a gaseous nitrogen. Such bacteria that inhabit this specialized niche are said to be "root nodulating bacteria." Rhizospheric bacteria acquire various nutrients excreted from the roots, viz., amino acids, organic acids, enzymes, and complex carbohydrates (Anderson et al. 1993; Kuiper et al. 2004). In response, bacteria modify nutrient into minerals accessible for plants. Additionally, the tips of root deliver a stable-state redox condition along with a structural surface for microbial colonization.

Microorganisms existing in the rhizosphere are metabolically stimulated by plant exudate and organisms capable of metabolizing the contaminants (Table 9.1). Researchers have exploited this synergetic association for rhizoremediation. Successful rhizoremediation depends on various aspects, viz. primary and secondary metabolites, colonization, existence, and environmental interactions with other microorganisms. Perhaps, this novel application is referred to as rhizoremediation or phytoremediation. The rhizospheric bacteria are accountable for contamination eradication, whereas the roots for providing nutrients used by microbes for propagation (Bais et al. 2008; Shukla et al. 2010) demonstrated the process of rhizoremediation to highlight the roles of the root exudates as well as the rhizosphere competent microorganisms. Microorganisms have the potential to remediate the toxic contaminant from the soil by producing various enzymes (Bhatt 2019; Bhatt et al. 2019a, 2019b, 2019c, 2019d). The microbial enzymes acting on the different chemical bonds and help to degradation and transformation of these compounds (Bhatt et al. 2020a, 2020b, 2020c, 2020d) Similar to this, the microbial population in rhizosphere tends to have higher activity than the microbial population residing in bulk soil.

Moreover, the microbial population in the rhizosphere employs noteworthy effects on plant growth and development by providing an ample amount of soluble nutrients essential for plants' uptake. Rhizobacteria also play a vital role in fixing $N_2$and producing various types of plant growth hormones, which directly boosts plant growth (Patten and Glick 1996). Rhizosphere biology field has profoundly affected the plant–microbe interactions, wherein, the rhizosphere is likely to be considered as the "hot spot" for microorganism diversity in addition to their colonizing activity (Bolton et al. 1993).

9.2.2 Root Exudates: Chemical Nature and Types

One of the central ways by which plants establish rhizosphere modifications is by root exudation. The root exudates, along with phytohormones, assist rhizosphere interactions by aiding as a source of energy the microbes as well as act as chemical repellents and attractants (Bais et al. 2001).They assist by initiating physiological and biological interactions among the soil microbiome and plant roots via inducing soil assets (chemical and physical),and soil microbial population, impeding competing plant species growth, assisting unfavorable symbioses (along with nitrogen-fixing bacteria, epiphytes, and mycorrhizal fungi, etc.), avoiding the proliferation of pathogenic microorganism (bacterial and fungal), and insect attacks (Nardi et al. 2000). Since the root exudates are the principal source of nutrients for microorganisms, consequently, the suitability of these exudates to microbes is indispensable for effective root colonization. The microbes are not restricted by the carbon quantity as well as by energy sources in the rhizosphere; microorganisms figure in the rhizosphere up to 19–32 folds greater than the other soil (root-free) (Bodelier et al. 1997). Soil microorganisms are chemotactically fascinated to plant compounds, viz. root exudates, rhizodeposition, volatile organic carbon, followed by multiplication in the same environment, which is rich in carbon (Rovira 1969). The major amount of carbon loss generally occurs at the tips of the roots, but a substantial amount is misplaced from root hair and elongation region as diffusible substrates. An additional exudation hot spot is the root dividing points. These complex rhizospheric microorganisms grow resulting in increasing population density, but also in a community structure distinct from that in the bulk soil (Bowen and Rovira 1991). The enormous associations between the two are based on root exudates that mediate their association in different ways (Weir et al. 2004; Broeckling et al. 2008).

Plant roots vigorously emerging discharge various organic compounds into the rhizosphere, organic compounds, for instance, sloughed-off cells, lysates plus exudates, into the rhizosphere (Lynch and Whipps 1990; Bowen and Rovira 1991). The plant–microbe interaction

Table 9.1 Various compounds in root exudates of different plant species.

Compounds	Types
Amino acids	a-Alanine, a-Aminoadipic, arginine, b-alanine, c-Aminobutyric, asparagine, aspartate, glutamate, glycine, histidine, homoserine, isoleucine, leucine, lysine, methionine, ornithine, phenylalanine, proline, serine, threonine, tryptophan, valine acid
Organic acids	Acetic acid, aconitic acid, aldonic acid, butyric acid, citric acid, erythronic acid, fumaric acid, formic acid, glutaric acid, glycolic acid, lactic acid, malic acid, malonic acid, oxalic acid, piscidic acid, pyruvic acid, succinic acid, tetronic acid, valeric acid
Sugars	Arabinose, glucose, fructose, galactose, maltose, oligosaccharides, ribose, rhamnose, raffinose, xylose
Vitamins	Biotin, thiamine, pantothenate, riboflavin, niacin
Purines/nucleosides	Adenine, cytidine, guanine, uridine
Enzymes	Amylase, acid/alkaline-phosphatase, invertase, protease
Inorganic ions and gaseous molecules	H^+,H_2, HCO_3, OH, CO_2

Source: Tailor and Joshi 2014.

nature can directly be determined by factors released by plants, viz., amount, and composition of organic materials (Griffiths et al. 1999; Jaeger et al. 1999). The root exudates can be characterized as, viz., amino acids, enzymes, nucleotides, organic acids, sugars, antimicrobial compounds, and flavonoids (Kirk et al. 2005; Baetz and Martinoia 2014). In general, carbohydrates components discharged in the root exudates are arabinose, galactose, glucose, fructose, mannitol, pentose, rhamnose, raffinose, ribose, sucrose, and xylose. The dominance of glucose in most of the root exudates has been reported by Toal et al. (2000). All twenty amino acids, as well as 1-hydroxyproline, homoserine, mugineic acid, and aminobutyric acid, are included in the root exudates. Presence of acetic, aspartic, caffeic, citric, gallic, glutamic, isocitric, succinic, malic, salicylic, shikimic, oxalic, and protocatacheuic acids has been reported by Alexa et al. (2004). Different derivatives of lignin, cyanide, enzymes, fatty acids, hormones, proteins, sugar derivatives, cations, and mucilages are the major constituent of root exudates of different plant species that specifically attract different types of microorganism (Bais et al. 2006). Various types of root exudates are represented in Table 9.1. It is imperative to know that the engineering plant influencing the rhizosphere may perhaps be a multifaceted procedure because of the deprivation of engineered compound in the soil, insignificant exudation rate to effect rhizosphere, inadequate information about root exudates configuration, as well as fluctuations of the exudate discharging period and levels with plant growth and exterior stimuli.

9.2.3 Factors Affecting Root Exudate

Various environmental factors, viz., abiotic and biotic, soil physio-chemical properties (organic matter content, nutrient availability, pH, and texture), and soil microbials also can affect the root exudates' accessibility. Commencement of climatic change and global warming, hindrance in harvesting crops, eventually due to both extreme heat and cold conditions. However, to clarify temperature effects on root exudates, researchers cultivated host plants (strawberry) at 5–10 °C paralleled with other plants grown at 20–30 °C. As a result, further amino acids in exudates were observed in plants that are grown at low soil temperature, evidently affected *Rhizoctonia fragariae* pathogenicity (Selvakumar et al. 2008). Similar to temperature, soil pH, and nutrients (carbon, nitrogen, and phosphate) status recorded adversely affect the liberation of plant root exudates. With respect to this, a noticeable nutrient grading effect on the amides and amino acid exudation *Pinus radiate* seedlings roots has been demonstrated well (Ali et al. 2010). The consequences specified amplification of amides/amino acids in exudates from plants lacking phosphate. Since the reduction in cereals, harvesting often has been faced globally as a consequence of flood and drought. Numerous studies have revealed the adverse effect of soil moisture in the root exudates liberation. The provisionally wilting of plants augmented amino acid discharge from roots, which perhaps associated with the pathogen frequency in the rhizosphere (Fatima et al. 2009). Since soil microbes play a central role in the growth of plants and in plant exudates too. They do influence exudation by disturbing the absorptivity of root cells and root metabolism. Microbes captivate certain complexes in root exudates and then expel the further one. They can produce secondary metabolites, affecting signaling and metabolism in plants, considered as "plant secondary genome," providing plant hosts with microbe consequent compounds (Kuz'mina and Melent'ev 2003).

9.3 Engineering Soil Microbial Populations and Plant–Microbe Interactions

9.3.1 Microorganisms in Soil

As root exudates directly influence the soil's chemical and physical properties, it also supports huge microbial populations that safeguard the roots against pathogens through various metabolic activity. Various soil microbes, viz. *Pseudomonas*, *Enterobacter*, and *Arthobacter*, etc., play an imperative role in several characteristics of the native ecosystem, viz. soil fecundity (throughout growth and development of plants), promoting plant growth activities and biogeochemistry of organic and inorganic compounds cycling through an entire natural ecosystem (Molin and Molin 1997; Trevors 1998). Soil rhizospheric region is inhabited by a large number of plant growth-promoting rhizobacteria (PGPR) having an advantageous influence on the plants. PGPR are free-living strains capable of colonizing the roots, thereby stimulating the plant growth. These microorganism affects plant growth by direct ways, i.e. influencing plant growth hormones levels and aiding in nutrient acquisition, or indirect ways via inhibiting several pathogens effects on the advancement of plant growth as biocontrol agents form. Practicing soil microbes capable of fixing atmospheric nitrogen, solubilizing phosphorous, as well as stimulating plant growth by means of synthesizing growth-promoting constituents has gained importance (Hoflich et al. 1994; Thomashow et al. 1997). Subsequently, the efficiency and sustainability of agronomic systems intensely depend upon the functional processes of microbial communities in soil (Doran and Zeiss 2000). It is likewise well recognized that plants have a substantial impact on the variety, spatial distribution, and copiousness of soil microbes through the rhizosphere (Wall and Virginia 1999). Since there are thousands of microbial strains well recognized for their valuable properties, so engineering an ecological artificial microbial community exemplifies a substantial task.

For two strain co-cultures, around six biological interaction exists (Grosskopf and Soyer 2014), viz. cooperation: in which both the strains are mutually benefitted from each other, competition: where both strains compete for attaining advantage (e.g., substrate competition), predation: where predator benefits unlike prey, no interaction: where no influence with no shared substrates or no predator–prey relations, as well as no competition takes place, commensalism: where one strain gets benefits, without affecting another one, and amensalism: where one strain is negatively affected, unlike the other one. For instance, a *Bacillus* sp. may be modified to cover the machinery of nitrogen-fixation, e.g., NifH from *Paenibacillus* (Kim and Timmusk 2013), aiming maximum phytohormones concentrations (Arkhipova et al. 2005), adding several paths from other *Bacillus* spp. to rheostat pathogens (Köberl et al. 2013, 2015). Likewise, *Rhizobium*, *Bradyrhizobium,* and *Pseudomonas* genera amplifying nitrogen fixation along with co-inoculation with engineered strains. Similarly, a consortium of three strains along with an engineered *Bacillus* sp. and additional natural or engineered nitrogen fixers, viz. *Pseudomonas*, *Rhizobium* either, or both *Bradyrhizobium* possibly will offer a number of benefits to the natural rhizosphere community. Bacteria motile in nature are chemotactically driven by host plant root exudates to roots and by solid attachment to the plant permits more solid linkage leading improved effects to the plant (Bashan et al. 2014).

9.3.2 Soil Modification: Altering Microbial Populations

Soil alteration or its microbial populations is a frequently unintended magnitude of various humanoid activities. For example, impurities in the soil may well extremely disturb the soil configuration along with microflora associated with the plant (Colores et al. 2000; Siciliano et al. 2001; Lynch 2002). Similarly, recurrent crop cultivation will lead to the development of disease-suppressive soils. Advancement of innovative practices in microbiology, along with microbial ecology, delivered various chances that alter the soil microbiota in a way corresponding to the selective "rhizosphere engineering" so occurring in the environment. As the microorganisms residing in the roots of the leguminous plant, they get equally influenced by root exudates. Numerous cases have been involved with respect to microbial engineering associated with amendments to the soil (chemical and biological). Among the particular bacteria, PGPR (detergent-adapted) *Pseudomonas* strain lasted well in modified rhizosphere soils as compared to the control one. The alterations have been documented well in the courtesy of plant growth. Previously, a strain of *Pseudomonas fluorescens* was engineered to employ salicylic acid as a carbon source, where *P. fluorescensn* growth was subsequently improved in salicylate-modified bulk and rhizosphere soils (Colbert et al. 1993a, 1993b).

9.4 Plant Growth-Promoting Rhizobacteria: Mechanisms, Potential, and Usages

Adverse environmental impacts, eradication of natural enemy populations, improved resistance, plus negative health effects of chemical pesticides have encouraged researchers to recognize and explore substitute methods for pest control. So, one such tactic involves the treatments of crops with PGPR, whereby means of PGPR plants tolerate abiotic stresses, viz. drought or high salinity soils stress, thereby enhancing uptake of various nutrients (nitrate and phosphorus) and minerals (Yang et al. 2008). Though exact mechanisms by which PGPR accomplish the same have not yet been recognized, PGPR has remarkably been shown to deliver resistance in contradiction of several phytopathogens, viz. bacteria, fungi, viruses, nematodes, and insect-transmitted diseases (Kloepper et al. 1999; Zehnder et al. 2001). Presently, the PGPR in formulation form is widely retailed for seed, for soil application, and drip irrigation, respectively. Among all PGPR strains, the *Bacillus* genus is predominantly practiced owing to their spore-forming assets by which they can store and dispatched for a long duration, making them suitable as a commercial product fulfilling a range of conditions. PGPR aside, enhancing plant health by providing resistance is an application that can be favorable in terms of alternative methods for supervising sustainable agriculture. PGPR integration in agriculture management practices has several advantages, of which various reports have already established, yet a number of potential aspects require exploration.

Furthermore, PGPR effects on the treatment of growth and disease resistance of plant growth have also been extensively studied. Likewise, various evidence related to the role of PGPR in plant–insect interactions was also supported in later studies. Although evaluation of PGPR strain (*Bacillus subtilis*)is well recognized in enhancing plant tolerance toward salinity stress, researchers also illustrated the treatment of PGPR in tomato plants to lower whitefly populations in greenhouse environments settings (Hanafi et al. 2007). However, resistance mechanism of this was not recognized but few reports displayed PGPR treatment on tomato plants in the reduction of adult whitefly emergence, whereas re-established suppression of whitefly population is usually attained by jasmonic acid (JA)–pathway plant responses (Valenzuela-Soto et al. 2010). PGPR enhances the growth promotion of plants in a sustainable manner. Although the accomplishment and effectiveness of PGPR as potential inoculants for agronomic crops are influenced by numerous aspects, these beneficial PGPR stimulate plant growth at several phases of host plant life cycle via diverse mechanisms (Figure 9.1; Table 9.2).

9.4.1 Direct Mechanisms

9.4.1.1 Biological N_2 Fixation

Biological N_2 fixation (BNF), a biological procedure converting atmospheric nitrogen into a form that is readily taken up by the plant, thus contributing as an alternative approach for sustainable agriculture (Bohlool et al. 1992; Parr et al. 1992; Buyer and Kaufman 1996). This nitrogen fixation process is facilitated solitary by bacteria in nature, thereby establishing a mutual relationship between plant and bacterium (Table 9.3). Rhizobia in legumes live in roots nodule, where inside of these nodules, symbiotic rhizobia forms (bacteroids) fixes the atmospheric nitrogen and NH^{4+} produce is transforming to organic nitrogen. Organic nitrogen so produced in the root nodules is further translocated to plant. Legumes nitrogen fixation may range from 25 to 75 lbs. of nitrogen per acre per year in a natural bionetwork plus some hundred pounds in a cropping system. Principally, it fixes up to 250 lbs. of nitrogen per acre and is not usually fertilized. Legumes (perennial and forage), viz. sweet clover, alfalfa, vetches, and true clovers, fix about 250–500 lbs. of nitrogen per acre. The substitution of it with N fertilizer would cost about $7 to $10 billion per annum

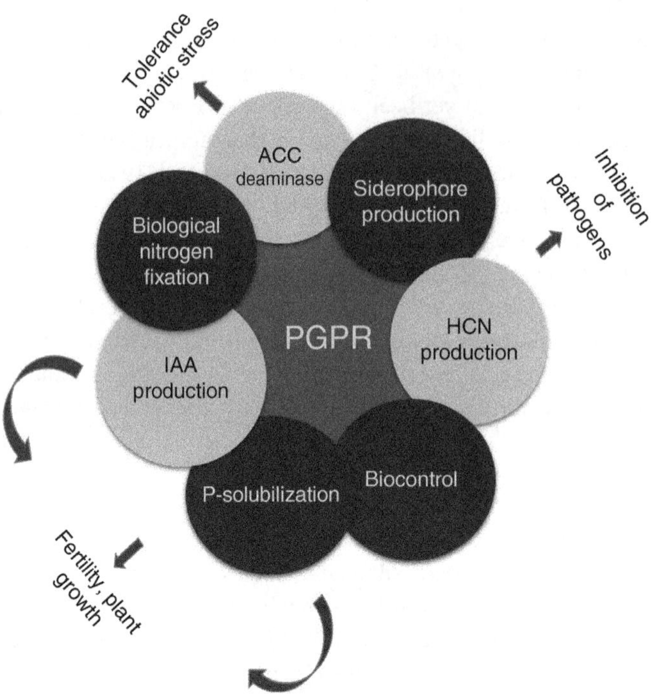

Figure 9.1 Overview of PGPR mechanisms. *Sources:* Pankaj et al. 2013; Bhatt and Maheshwari 2019.

Table 9.2 Mechanisms of PGPR.

Term	Definition	Mechanism	References
Biofertilizer	An inoculant containing live microbes which, on application to plant seeds or surface or to the soil, colonizes rhizosphere or to an integral part of the plant thereby promoting plant growth via enhancing the availability of nutrients supply to the host plant.	Uptake of insoluble forms of minerals (P), biological nitrogen fixation.	Vessey (2003), Fuentes-Ramírez and Caballero-Mellado (2006)
Phytostimulator	Microbes possessing the ability of production or improvement of plant growth regulators, viz. gibberellic acid, indole acetic acid, cytokinins, and ethylene.	Phytohormones production (auxins, gibberellins, and cytokinins) Decrement in ethylene concentration	Lugtenberg et al. (2001)
Biopesticide or biocontrol agent	Microbes promoting plant growth via controlling various phytopathogenic agents, principally for antibiotics production, and antifungal metabolites.	Antibiotics production (HCN, siderophore, antifungal metabolites). Enzyme production capable of degrading the cellular wall of the fungi.	Vessey (2003), Chandler et al. (2008)

Source: Nandal and Hooda 2013.

Table 9.3 Nitrogen-fixing bacteria with their relationship to host plants.

PGPR	Relationship with host	Host crops	References
Azospirillium sp.	Rhizospheric	Wheat, maize, kallar grass and rice	Dénarié et al. (1996), Malik et al. (1997), Boddey et al. (1986)
Azoarcus sp.	Endophytic	Sorghum, rice	Hurek et al. (2002), Stein et al. (1997), Egener et al. (1999)
*Azotobacter*sp.	Rhizospheric	Wheat, maize	Pandey et al. (1998), Mrkovacki and Milic (2001)
Bacillus polymyxa	Rhizospheric	Wheat	Omar et al. (1996)
Burkholderia sp.	Endophytic	Rice	Baldani et al. (2001)
Cyanobacteria	Rhizospheric	Wheat, rice	Obreht et al. (1993), Hashem (2001)
Gluconacetobacter diazotrophicus	Endophytic	Sorghum, sugarcane	Isopi et al. (1995); Sevilla et al. (2001)
Herbaspirillum sp.	Endophytic	Rice, sorghum, Sugarcane	Pimentel et al. (1991), James et al. (1997), James et al. (2002)

Source: Tailor and Joshi 2014.

(Graham and Vance 2000). Some of the legumes are capable of enriching soil nitrogen. Irrespective of this, common beans fix less nitrogen as per their needs, i.e. less than 50 lbs. per acre, hence considered as poor fixers. While, legumes, viz. peanuts, soybeans, cowpeas, etc., are considered to be good nitrogen fixers, as they fix all of their nitrogen needs other than that absorbed from the soil. Since, N_2 fertilizer is unreachable to farmers, leaving them reliant on legumes N_2 fixation or other organisms fixing nitrogen. Since, the plant, viz. *Mimosa* sp., and *Leucaena* sp., produce "mimosine" as an amino acid, which tends to be lethal for several bacteria (Hammond 1995). In context to this, the nitrogen-fixing Rhizobiaceae residing nodules on these plants have attained mimosine degradation capacity, specific characteristics persuade only by root colonizing Rhizobiaceae (Soedarjo et al. 1994). The mimosine "engineering" illustration found to be beneficial for both the plant and symbionts (Soedarjo and Borthakur 1996; Fox and Borthakur 2001; Bhatt and Barh 2018).

9.4.1.2 Phosphate Solubilization

Phosphorous (P) is one of the main macronutrients required by plants but is present in available form because of its reactivity with aluminum, iron, and calcium; resulting in phosphorous precipitation thereby making it inaccessible for plants. Some rhizobacteria can transform the unavailable form of phosphorous into orthophosphate which is an attainable form to the plants (Saharan and Nehra 2011). Presently it is quite captivated that legumes per use foremost role in phosphorous (P) sustainability, by possessing various mechanisms for P solubilizing, resulting in improved P acquisition, predominantly in low

P containing soils. Legumes combination into rotational or intercropping systems offers a cheap and low-cost substitute for incorporating P fertilizer for upgraded soil productiveness (Uhde-Stone et al. 2003).

9.4.1.3 Zinc Solubilization

Alike, another micronutrient, Zinc (Zn) is also an essential required for proper functioning of plants. Zn inadequacy negatively influences plant growth and productiveness. Usually, Zn is in ample amount but is primarily associated with hydrous Fe^{3+} and Al^{3+}oxides, making it insoluble, thus hindering its deliverance. Although, various PGPB have been recognized well, aiding in its solubilization and mobilization, thereby increasing the uptake by plants. The efficiency of PGPB is through the interaction with plant roots, involving various processes including mobilization, solubilization, and biofortification, making them readily available from soil to plants. Moreover, this bacteria inoculation does help in improving the growth, quality, and productivity of various host plant (Bhatt and Maheshwari 2020). Since zinc solubilization directly depends on various factors, viz. temperature, pH, concentration, nutrients, and sugars, consequently, alteration in any the factors perhaps may lead to accomplishing its maximum solubilization.

9.4.1.4 Siderophore Production

In parallel to P, iron (Fe) is also a crucial nutrient present in the soil but is insoluble and in a limited amount. Generally, all microorganisms are recognized as hitherto, except certain lactobacilli and fundamentally necessitated iron (Neilands 1995). The iron occurs as Fe^{3+} in an aerobic environment, which likely forms hydroxides and oxyhydroxides in insoluble forms, thus making it unavailable for plants and microorganisms both (Rajkumar et al. 2010). Plant growth promotory rhizobacteria produce compounds called siderophore, chelating ferric iron (Fe^{3+}) which is further easily taken up by the roots. Where siderophore have high association constants for complexing iron and forms stable complexes with various heavy metals, viz. Al, Cu, In, Cd, Pb, Zn and Gaalong, with radionuclides (Np and U); they are of environmental concern, (Kiss and Farkas 1998; Neubauer et al. 2000). Henceforth, these rhizobacteria siderophore aid in alleviating stresses imposed on plants because of high soil levels of heavy metals. Various environmental factors, viz. pH, levels of iron, other nutrients (carbon, nitrogen, and phosphorus) supplies, presence of other trace elements correspondingly modify siderophore production (Duffy and Défago 1999).

9.4.1.5 Production of Phytohormones

These rhizobacteria also produce several hormones, viz. auxins, gibberellins, and cytokinins, which are prerequisites for the augmentation of plant growth, referred to as phytohormones. Among all, IAA (indole-3-acetic acid), is the important phytohormones required for the promotion of lateral root formation and elongation of shoot tips, besides increasing lateral root formation leading ability for nutrient uptake by the plant (Loon 2007). Usually, IAA secreted by rhizobacteria affects various plant developmental processes since the endogenous pool of plants. Apart from IAA, gibberellins and cytokinins are also involved in stimulating shoot development and hence increasing plant height. Likewise, ethylene phytohormone significantly influences growth and development of plant by several aspects; besides this it also inhibits elongation of the root, whereas it stimulates root hair formation (Negi et al. 2008). Several rhizobacteria synthesize phytohormones which positively affect root growth; however, rhizosphere engineering through hormone alterations focuses mainly on the degradation of ethylene, produced by plants in contact with various stresses (flooding, drought, salt, metals existence, pathogens).

Abiotic stresses including drought, salinity, and high temperatures are the principal root causes damaging natural vegetation, decrement in crop productivity, consequently, reducing capacity for CO_2 uptake. Among the abiotic stresses, drought stress rigorously hinders the photosynthesis process along with the development of root (Verslues 2017). Similarly, salinity stress causes ion toxicity as causing disadvantageous effects on plant survival and development (Negrao et al. 2017). Survivability of plants under abiotic stress circumstances necessitates widespread physiological variations. In context to this, phytohormones play critical parts in the procedure of root development along with regulating root morphological responses to abiotic stress, where abscisic acid (ABA) in responses to abiotic stresses specifically in drought stress has been categorized well. Xiong et al. (2006) reported well that during drought stress, ABA inhibits the primary root elongation. Although, plants may respond well to abiotic stress tolerance by various processes, the interaction of plants with beneficial microbial population residing in the rhizosphere provides additional support for boosting resistance against several abiotic stresses, along with providing strong protection (Farrar et al. 2014). Henceforth, these plant hormones are also significant w.r.t. favorable plant–microbe interactions. Hence, plant phytohormones should be considered as important aspects while dealing with rhizosphere engineering.

9.4.1.6 ACC (1-Aminocyclopropane-1-Carboxylate) Deaminase Activity

As ethylene is indispensable for plant growth and survivability (Khalid et al. 2006). It is one of the phytohormone produced almost by entire plants, as well as produced by various processes (biotic and abiotic) in soils, along with convincing diverse physiological alterations in plants. Since ethylene is well recognized as "stress hormone" also (Saleem et al. 2007), wherein certain stress, viz. drought, salinity, heavy metals, waterlogging, and pathogenicity, thereby affecting plants growth in negatively. The increase in ethylene levels due to salinity and drought stresses the inhibition of root growth and various plant physiological paths also gets affected. However, PGPR possesses an enzyme, 1-aminocyclopropane-1-carboxylate (ACC) deaminase, catalyzing ACC cleavage, the instant precursor of ethylene, which decreases the ethylene concentrations thereby enhancing plant growth, by reducing and inducing drought stress in plants respectively (Nadeem et al. 2007; Zahir et al. 2008). The principal regulatory role of ethylene in survivability and development of the plant, as well as the capability of PGPR to modify the ethylene stress levels, making it a more fascinating aim for rhizosphere engineering, principally when it comes to incomplete plant growth as a result of abiotic stress.

9.4.2 Indirect Mechanisms

Microbial application in controlling diseases is an eco-friendly tactic. The foremost indirect mechanism is by acting as biocontrol agents (Glick 2012). Generally, nutrients struggle, niche elimination, production of antifungal metabolites, and induced systemic resistance

(ISR) are some of the topmost approaches of biocontrol activity in PGPR (Lugtenberg and Kamilova 2009). Several rhizobacteria have been documented well for producing antifungal metabolites, viz. HCN, pyrrolnitrin, phenazines, pyoluteorin, 2,4-diacetylphloroglucinol, tensin, and viscosinamide (Bhattacharyya and Jha 2012). The interface of rhizobacteria with the plant roots can affect plant resistance against pathogenic agents (bacteria, fungi, and viruses), the phenomenon thus is referred to as called ISR (Lugtenberg and Kamilova 2009).

9.5 Plant–Microbe Interaction

Plant–microbe interaction is quite a complex phenomenon, occurring via different processes, viz. symbiotic, endophytic. The plant–microbe interaction covers a range of spectrum associating various relationships from mutualistic, commensalism and parasitic, besides representing primeval co-evolution (Heckman et al. 2001). This multifaceted plant–microbe association is important for improved plant health and survivability (Berendsen et al. 2012). In the context of rhizobia, its interaction with legumes is host-specific. Based on which interaction rhizobia is classified as broad and narrow host range rhizobia. For instance, *Rhizobium leguminosarum* bv. *viciae* nodulates *Lathyrus*, *Vicia, Lens,* and *Pisum* whereas *Rhizobium leguminosarum* bv. *Trifolii* nodulates *Trifolium* spp. On the other hand, *Rhizobium* sp. strain NGR 234 falls under broad host range rhizobia thereby nodulating around 35 legumes (VanRhijn and Vanderleyden 1995). Henceforth, the host specificity degree differs enormously among rhizobia. Likewise, the variation of Nod factors amount and the structure are imperative for defining host specificity in rhizobia–legume interaction (Dénarié et al. 1996). On the other hand, plant genotype extremely affects the number and arrangement of native microbes and the population dynamics of acquainting with rhizobacteria. This happens for the reason that rhizosphere microbes are reliant on the substrates released from the plant roots for their growth and development since rhizo deposition is principally in the plant's genetic control. Several studies indicating this concept have been mentioned in the previous works (Miller et al. 1989; Negi et al. 2014). Likewise, studies representing employing chromosomal substitution lines of wheat revealed exquisitely influence plant genotype on microbes of the rhizosphere (Neal et al. 1970, 1973; Atkinson et al. 1975).To enhance plant performance, modification of plant rhizospheric microorganisms tends to be an innovative research prospect in the field of agriculture and medicine as well. Such novel and unexploited methodology benefits plant–microbe association to engineer microbial role. The objective of microbe selection is to recover the host routine through non-natural selection on the microbe, whereas host routine may comprise various attribute those are specifically important in terms of biologically, medically, and economically.

Furthermore, microbe PGPR mainly interacts with the plant by root colonization thus creating a favorable environment for growth and development of plants. Besides, extensive application of chemical fertilizers has been practiced on agricultural land (Wartiainen et al. 2008) with an intuition of achieving high crop yield. However, extreme usage of these chemical components causes unanticipated adverse effects on the environment (Shenoy et al. 2001; Adesemoye et al. 2009) along with degrading the plant product quality.

9.6 Biofertilizers and its Applications

Chemical-based fertilizers have had a remarkable impact on the food production viewpoint in the past and today, making it a crucial part of modern agriculture. These fertilizers assure food production for a continually increasing population. Nevertheless, adverse impact on environment degradation along with human health poses a foremost limit to their extreme practice. Nitrogen fertilizer input efficacy is one of the lowest, contributing directly to environmental pollution. Nitrate in surface and groundwater is also the main threat to the ozone layer stability from various gaseous oxides of nitrogen, posing major concern from a health and environmental viewpoint (Sunitha et al. 2012). Hence, nitrogen fertilizers besides being expensive also damage the environment by contaminating water with nitrates, soil acidification, and greenhouse-gas emissions, whereas continuous and overuse practice of nitrogen fertilizers decline the crop quality and yields. Moreover, plants where ammonium-based fertilizers are often practiced and have a tendency of acidifying rhizosphere. However, the nitrate-based fertilizer effects more in an alkaline rhizosphere. Variations in pH can modify soil roots profile, besides influencing growth along with the configuration of the microbial population. On the contrary, microorganisms residing in nature have been exploited for biofertilizers and bio-pesticides development for supporting plant growth and controlling pests and weeds, thus increasing agricultural productivity. This design of the fertilization system has been planned by nature itself. Generally, rhizobium (PGPR) are used for biofertilizers preparations. Where these bacteria aid in accumulating the essential nutrients from the soil and utilizing waste by-products as their food. Scientists have utilized these assets for developing the biofertilizers. PGPRs as bioinoculant/biofertilizers has gear much more important as they are capable of promoting plant growth, increasing yield, enhancing soil profile via accumulating organic material in huge amounts as well as valuable nutrients too, avoiding soil erosion and acidification, nutrients leaching and improving soil water-holding capacity. These beneficial inoculants work all together of each other with multiple mechanisms, including biological nitrogen fixation, production of phytohormones and siderophore, phosphate solubilization, ACC deaminase, and antifungal activity, etc., and are responsible directly and directly enhancing plant vegetative and reproductive growth (Bashan and Holguin 1997). Henceforth, these environmental and production considerations dictate the need for potential biological alternatives augmenting essential nutrients like nitrogen, thereby replacing nitrogen fertilizers, which must be practiced.

9.7 Plant Genetic Engineering

All the approaches defined above are either achieved naturally occurring processes or by amending exogenous soil for altering the rhizosphere microorganism community. Several other alternatives also exist which describe engineering the plant for obtaining various compounds that can modify the rhizosphere microflora or their properties.

The opine model: Several reports reveal transgenic plants to liberate xenobiotic complexes in the rhizosphere. Xenobiotic complexes can be well defined as those compounds that do not exist naturally in a certain environment. Regarding this, previously studied molecule "opines" are targeted in the current chapter, which can be defined as the amino acid derivatives with low molecular weight occurring in tumors induced by *Agrobacterium tumefaciens* (Dessaux et al. 1998). Plants offer these molecules that are further transformed with one to three biosynthesis genes (Savka et al. 1996). Previously studies reveal bacteria capable of colonizing the rhizosphere of *Lotus corniculatus* (host plant), which was further engineered to synthesize opines, thereby matched with those colonizing near-isogenic wild-type plants (Oger et al. 1997). Several other studies by Murphy et al. (1987) have also attempted engineering plant–bacteria interaction centered by synthesis by plants and bacterial biodegradation of a compound responsible for nitrogen-fixing nodules of legumes (rhizoplane: 3-O-methyl-scyllo-inosamine) by them respectively, besides using *Agrobacterium* opine system for the selection of precise interaction between plant and bacteria. Moreover, the involvement of rhizopine genes for biodegradation and synthesis have been isolated from *Sinorhizobium meliloti* (Murphy et al. 1993; Saint et al. 1993). However, this methodology is worth following since rhizopine promotes the growth of advantageous bacteria capable of nitrogen fixation and rhizopine-degrading strains are preferred in arhizopine-rich atmosphere (Gordon et al. 1996; Pankaj et al. 2016a, 2016b). All these study reports reveal rhizosphere engineering to be an attractive practice not only in terms of opine production but also in ecological engineering.

9.8 Conclusion and Future Prospects

All the reports demonstrate well that plant root exudates significantly influence microbial populations around the rhizosphere, by providing a protective remark that configuration of rhizosphere microbial population gets influenced by root exudation. Various reports to date described rhizosphere engineering to be suitable for crop species and varieties selection, by incorporation of microbial population, and by altering the genetic composition of plants and biological activities of the microbial population. Besides, PGPR is continuously practiced in various research as well as in agricultural sectors. Since PGPR-partaking manifold activities concerning growth promotion of plant by establishing different mechanisms, demonstrating bioremediating potentials by decontaminating pollutants, viz. chemical fertilizers (pesticides) and heavy metals and regulatory a variety of plant pathogens as biopesticides, reveals enormous outcomes in diverse crop studies. So, by modification in any of the characters directly influences plant and microbe interaction by the implementation of rhizosphere engineering. Further, discoveries in these fields will give depth knowledge of various diseases threatening plants and animals as well, against which a large-scale treatment will be established. It is also domineering that researchers carry on their study so that this becomes friendlier and safe to people in context to sustainable and eco-friendly agricultural practices.

Acknowledgments

Authors all acknowledge the researcher for this topic, and their research articles are cited in the manuscript.

References

Adl, S. (2016). Rhizosphere, food security, and climate change: a critical role for plant-soil research. *Rhizosphere* 1: 1–3.

Alexa, N.S., Haig, T., and James, E. (2004). Evaluation of putative allelochemicals in rice root exudates for their role in the suppression of arrowhead root growth. *J. Chem. Ecol.* 30 (8): 1663–1678.

Ali, B., Sabri, A.N., and Hasnain, S. (2010). Rhizobacterial potential to alter auxin content and growth of *Vigna radiata* (L.). *World J. Microbiol. Biotechnol.* 26 (8): 1379–1384.

Anderson, T.A., Guthrie, E.A., and Walton, B.T. (1993). Bioremediation in the rhizosphere: plant roots and associated microbes clean contaminated soil. *Environ. Sci. Technol.* 27 (13): 2630–2636.

Arkhipova, T.N., Veselov, S.U., Melentiev, A.I. et al. (2005). Ability of bacterium *Bacillus subtilis* to produce cytokinins and to influence the growth and endogenous hormone content of lettuce plants. *Plant Soil* 272: 201–209.

Atkinson, T.G., Neal, J.L. Jr., and Larson, R.I. (1975). Genetic control of the rhizosphere microflora of wheat. In: *Biology and Control of Soil-Borne Plant Pathogens* (ed. G.W. Bruehl), 116–122. Saint-Paul: American Phytopathological Society.

Badri, D.V. and Vivanco, J.M. (2009). Regulation and function of root exudates. *Plant Cell Environ.* 32: 666–681.

Badri, D.V., Weir, T.L., van der Lelie, D., and Vivanco, J.M. (2009). Rhizosphere chemical dialogues: plant-microbe interactions. *Curr. Opin. Biotechnol.* 20 (6): 642–650.

Baetz, U. and Martinoia, E. (2014). Root exudates: the hidden part of plant defense. *Trends Plant Sci.* 19 (2): 90–98.

Bais, H.P., Loyola-Vargas, V.M., Flores, H.E. et al. (2001). Invited review: rootspecific metabolism: the biology and biochemistry of underground organs. *In Vitro Cell. Dev. Biol. Plant* 37: 730–741.

Bais, H.P., Weir, T.L., Perry, L.G. et al. (2006). The role of root exudates in rhizosphere interactions with plants and other organisms. *Annu. Rev. Plant Biol.* 57: 233–266.

Bais, H.P., Broeckling, C.D., and Vivanco, J.M. (2008). Root exudates modulate plant-microbe interactions in the rhizosphere in secondary metabolites in soil ecology. *Soil Biol.* 14: 241–252.

Baldani, V.L.D., Baldani, J.I., and Dobereiner, J. (2001). Inoculation of rice plants with the endophytic diazotrophs *Herbaspirillum seropedicae* and *Burkholderia* spp. *Biol. Fertil. Soils* 30: 485–491.

Bashan, Y. and Holguin, G. (1997). *Azospirillum*-plant relationships: environmental and physiological advances (1990–1996). *Can. J. Microbiol.* 43: 103–121.

Bashan, Y., de Bashan, L.E., Prabhu, S.R. et al. (2014). Advances in plant growth promoting bacterial inoculant technology: formulations and practical perspectives. *Plant and Soil* 378: 1–33.

Berendsen, R.L., Pieterse, C.M., and Bakker, P.A. (2012). The rhizosphere microbiome and plant health. *Trends Plant Sci.* 17: 478–486.

Bhatt, P. (2019). *Smart Bioremediation Technologies: Microbial Enzymes.* Amsterdam: Elsevier Science.

Bhatt, P. and Barh, A. (2018). Bioinformatic tools to study the soil microorganisms: an *in silico* approach for sustainable agriculture. In: *In Silico Approach for Sustainable Agriculture* (eds. D. Choudhary, M. Kumar, R. Prasad and V. Kumar), 169–182. Singapore: Springer.

Bhatt, K. and Maheshwari, D.K. (2019). Decoding multifarious role of cow dung bacteria in mobilization of zinc fractions along with growth promotion of *C. annuum* L. *Sci. Rep.* 9 (1): 1–10.

Bhatt, K. and Maheshwari, D.K. (2020). Zinc solubilizing bacteria (*Bacillus megaterium*) with multifarious plant growth promoting activities alleviate growth in *Capsicum annuum* L. *3 Biotech.* 10 (2): 36.

Bhatt, P., Pal, K., Bhandari, G., and Barh, A. (2019a). Modeling of the methyl halidebiodegradation in bacteria and its effect on environmental systems. *Pest. Biochem. Physiol.* 158: 88–100.

Bhatt, P., Gangola, S., Chaudhary, P. et al. (2019b). Pesticide induced up-regulation of esterase and aldehyde dehydrogenase in indigenous *Bacillus* spp. *Biorem. J.* 23: 42–52.

Bhatt, P., Huang, Y., Zhan, H., and Chen, S. (2019c). Insight into microbial applications for the biodegradation of pyrethroid insecticides. *Front. Microbiol.* 10: 1778.

Bhatt, P., Pathak, R., and Bhatt, K. (2019d). System biology, simulation and network analysis of enzymes in waste removal from environment. In: Smart bioremediation technologies: Microbial enzymes series, pp. 347–358.

Bhatt, P., Bhatt, K., Huang, Y. et al. (2020a). Esterase is a powerful tool for the biodegradation of pyrethroid insecticides. *Chemosphere* 244: 125507.

Bhatt, P., Huang, Y., Zhang, W. et al. (2020b). Enhanced cypermethrin degradation kinetics and metabolic pathway in *Bacillus thuringiensis* strain SG4. *Microorganisms* 8: 223.

Bhatt, P., Huang, Y., Rene, E.R. et al. (2020c). Mechanism of allethrin biodegradation by a newly isolated *Sphingomonas trueperi* strain CW3 from wastewater sludge. *Bioresour. Technol.* 305: 123074.

Bhatt, P., Zhang, W., Lin, Z. et al. (2020d). Biodegradation of the allethrin by a novel fungus *Fusarium proliferatum* strain CF2 isolated from contaminated sites. *Microorganisms* 8: 593.

Bhattacharyya, P.N. and Jha, D.K. (2012). Plant growth-promoting rhizobacteria (PGPR): emergence in agriculture. *World J. Microbiol. Biotechnol.* 28: 1327–1350.

Boddey, R.M., Baldani, V.L.D., Baldani, J.I. et al. (1986). Effect of inoculation of *Azospirillum* spp. on nitrogen accumulation by field grown wheat. *Plant Soil* 95: 109–121.

Bodelier, P.L.E., Wijhuizen, A.G., Blom, C.W.P.M. et al. (1997). Effects of photoperiod on growth of and denitrification by *Pseudomonas chlororaphis*in the root zone of *Glyceria maxima*, studied in thegnotobiotics microcosm. *Plant Soil* 190: 91–103.

Bohlool, B.B., Ladha, J.K., Garrity, D.P. et al. (1992). Biological nitrogen fixation for sustainable agriculture: a perspective. *Plant Soil* 141: 1–11.

Bolton, B., Fredrickson, J.K., and Elliott, F.E. (1993). Microbial ecology of the rhizosphere. In: *Soil Microbial Ecology: Applications in Agricultural and Environmental Management* (ed. F.B. Metting Jr.), 27–63. New York: Marcel Dekker, Inc.

Bowen, G.D. and Rovira, A.D. (1991). The rhizosphere, the hidden half of the hidden half. In: *Plant Roots – the Hidden Half* (eds. Y. Waisel, A. Eshel and U. Kafkafi), 641–649. New York: Marcel Dekker.

Bowen, G.D. and Rovira, A.D. (1999). The rhizosphere and its management to improve plant growth. *Adv. Agron.* 66: 1–102.

Broeckling, C.D., Broz, A.K., Bergelson, J. et al. (2008). Root exudates regulate soil fungal community composition and diversity. *Appl. Environ. Microbiol.* 74: 738–744.

Buyer, J.S. and Kaufman, D.D. (1996). Microbial diversity in the rhizosphere of corn grown under conventional and low-input systems. *Appl. Soil Ecol.* 5: 21–27.

Chandler, D., Davidson, G., Grant, W.P. et al. (2008). Microbial biopesticides for integrated crop management: an assessment of environmental and regulatory sustainability. *Trends Food Sci. Technol.* 19: 275–283.

Colbert, S.F., Hendson, M., Ferri, M. et al. (1993a). Enhanced growth and activity of a biocontrol bacterium genetically engineered to utilize salicylate. *Appl. Environ. Microbiol.* 59: 2071–2076.

Colbert, S.F., Schroth, M.N., Weinhold, A.R. et al. (1993b). Enhancement of population densities of *Pseudomonas putida* ppg7 in agricultural ecosystems by selective feeding with the carbon source salicylate. *Appl. Environ. Microbiol.* 59: 2064–2070.

Colores, G.M., Macur, R.E., Ward, D.M. et al. (2000). Molecular analysis of surfactant-driven microbial population shifts in hydrocarbon-contaminated soil. *Appl. Environ. Microbiol.* 66: 2959–2964.

De Salamone, I.E.G., Dobereiner, J., Uquiaga, S., and Boddy, R.M. (1996). Biological nitrogen fixation in Azospirillum strain maize genotype associations as evaluated by 15N isotope dilution technique. *Biol. Fertil. Soils* 23: 249–256.

Dénarié, J., Debelle, F., and Prome, J.C. (1996). *Rhizobium lipochitooligosaccharide* nodulation factor: signaling molecules mediating & morphogenesis. *Annu. Rev. Biochem.* 65: 503–535.

Dessaux, Y., Petit, A., Farrand, S.K., and Murphy, P.M. (1998). Opines and opine-like molecules in plant-Rhizobiaceae interactions. In: *The Rhizobiaceae* (eds. A. Kondorosi, H. Spaink and P. Hooykaas), 173–197. Dordrecht: Kluwer Academic.

Doran, J.W. and Zeiss, M.R. (2000). Soil health and sustainability: managing the biotic component of soil quality. *Appl. Soil Ecol.* 15: 3–11.

Duffy, B.K. and Défago, G. (1999). Environmental factors modulating antibiotic and siderophore biosynthesis by *Pseudomonas fluorescens* biocontrol strains. *Appl. Environ. Microbiol.* 65: 2429–2438.

Egener, T., Hurek, T., Reinhold, A., and Hurek, B. (1999). Endophytic expression of nif genes of *Azoarcus* sp. strain BH72 in rice roots. *Mol. Plant Microbe Interact.* 12: 813–819.

Farrar, K., Bryant, D., and Cope-Selby, N. (2014). Understanding and engineering beneficial plant-microbe interactions: plant growth promotion in energy crops. *Plant Biotechnol. J.* 12: 1193–1206.

Fatima, Z., Saleemi, M., Zia, M. et al. (2009). Antifungal activity of plant growth-promoting rhizobacteria isolates against *Rhizoctonia solani* in wheat. *Afr. J. Biotechnol.* 8 (2): 18–21.

Fox, P.M. and Borthakur, D. (2001). Selection of several classes of mimosine-degradation-defective Tn3Hogus-insertion mutants of *Rhizobium* sp. strain TAL1145 on the basis of mimosine-inducible GUS activity. *Can. J. Microbiol.* 47: 488–494.

Fuentes-Ramírez, L.E., and Caballero-Mellado, J. (2006). Bacterial biofertilizers. In: Siddiqui, Z.A. (ed.), *PGPR: Biocontrol and Biofertilization* (pp. 143–172). Netherlands: Springer.

Glick, B.R. (1995). The enhancement of plant growth by freeliving bacteria. *Can. J. Microbiol.* 41: 109–117.

Glick, B.R. (2012). Plant growth-promoting bacteria: mechanisms and applications. *Scientifica* 2012: 1–15.

Gordon, D.M., Ryder, M.H., Heinrich, K. et al. (1996). An experimental test of the rhizopine concept in *Rhizobium meliloti*. *Appl. Environ. Microbiol.* 62: 3991–3996.

Graham, P.H. and Vance, C.P. (2000). Nitrogen fixation in perspective: an overview of research and extension needs. *Field Crop Res* 65: 93–106.

Griffiths, B.S., Ritz, K., Ebblewhite, N. et al. (1999). Soil microbial community structure: effects of substrate loading rates. *Soil Biol. Biochem.* 31: 145–153.

Grosskopf, T. and Soyer, O.S. (2014). Synthetic microbial communities. *Curr. Opin. Microbiol.* 18: 72–77.

Hammond, A.C. (1995). *Leucaena toxicosis* and its control in ruminants. *J. Anim. Sci.* 73: 1487–1492.

Hanafi, A., Traore, M., Schnitzler, W.H. et al. (2007). Induced resistance of tomato to whiteflies and Phytium with the PGPR *Bacillus subtilis* in a soilless crop grown under greenhouseconditions. In: *Proceedings of VIIIth IS on Protected Cultivation in Mild Winter Climates*, vol. 747 (eds. A. Hanafi and W.H. Schnitzler), 315–322. Acta Horticulturae.

Hashem, M.A. (2001). Problems and prospects of cyanobacterial biofertilizer for rice cultivation. *Aust. J. Plant Physiol.* 28: 881–888.

Heckman, D.S., Geiser, D.M., Eidell, B.R. et al. (2001). Molecular evidence for the early colonization of land by fungi and plants. *Science* 293: 29–30.

Höflich, G., Wiehe, W., and Kühn, G. (1994). Plant growth stimulation by inoculation with symbiotic and associative rhizosphere microorganisms. *Experientia* 50 (10): 897–905.

Hurek, T., Handley, L.L., Reinhold-Hurek, B. et al. (2002). *Azoarcus*grass endophytes contribute fixed nitrogen to the plant in an unculturable state. *Mol. Plant Microbe Interact.* 15: 233–242.

Isopi, R., Fabbri, P., Del Gallo, M. et al. (1995). Dual inoculation of Sorghum bicolor (L.) Moench ssp. bicolor with vesicular arbuscular mycorrhizas and *Acetobacter diazotrophicus*. *Symbiosis* 18: 43–55.

Jaeger, C.H., Lindow, S.E., Miller, W. et al. (1999). Mapping of sugar and amino acid availability in soil around roots with bacterial sensors of sucrose and tryptophan. *Appl. Environ. Microbiol.* 65: 2685–2690.

James, E.K., Olivares, F.L., Baldani, J.I. et al. (1997). *Herbaspirillum*, an endophytic diazotroph colonizing vascular tissue in leaves of *Sorghum bicolor* L. Moench. *J. Exp. Bot.* 48: 785–797.

James, E.K., Gyaneshwar, P., Mathan, N. et al. (2002). Infection and colonization of rice seedlings by the plant growth-promoting bacterium *Herbaspirillum seropedicae* Z67. *Mol. Plant Microbe Interact.* 15: 894–906.

Jordan, D.C. (1984). Family II Rhizobiaceae. *Bergey's Manual Syst. Bacteriol.* 1: 232–242.

Khalid, A., Akhtar, M.J., Mahmood, M.H. et al. (2006). Effect of substrate-dependent microbial ethylene production on plant growth. *Microbiology* 75: 231–236.

Kim, S.B. and Timmusk, S. (2013). A simplified method for gene knockout and direct screening of recombinant clones for application in *Paenibacilluspolymyxa*. *PLoS One* 8 (6): e68092.

Kirk, J.L., Klironomos, J.N., Lee, H. et al. (2005). The effects of perennial ryegrass and alfalfa on microbial abundance and diversity in petroleumcontaminated soil. *Environ. Pollut.* 133 (3): 455–465.

Kiss, T. and Farkas, E. (1998). Metal-binding ability of desferrioxamine B. *J. Inclusion Phenom. Mol. Recognit. Chem.* 32: 385–403.

Kloepper, J.W., Zablotowick, R.M., Tipping, E.M. et al. (1991). Plant growth promotion mediated by bacterial rhizosphere colonizers. In: *The Rhizosphere and Plant Growth* (eds.

D.L. Keister and P.B. Cregan), 315–326. Dordrecht, Netherlands: Kluwer Academic Publishers.

Kloepper, J.W., Rodriguez-Ubana, R., Zehnder, G.W. et al. (1999). Plant root-bacterial interactions in biological control of soilborne diseases and potential extension to systemic and foliar diseases. *Aust. Plant Pathol.* 28: 21–26.

Köberl, M., Ramadan, E.M., Adam, M. et al. (2013). *Bacillus* and *Streptomyces* were selected as broad-spectrum antagonists against soilborne pathogens from arid areas in Egypt. *FEMS Microbiol. Lett.* 342: 168–178.

Köberl, M., White, R.A. III, Erschen, S. et al. (2015). Draft genome sequence of Streptomyces sp. strain Wb2n-11, a desert isolate with broad-spectrum antagonism against soilborne phytopathogens. *Genome Announc.* 3 (4).

Kuiper, I., Lagendijk, E.L., Bloemberg, G.V. et al. (2004). Rhizoremediation: a beneficial plant-microbe interaction. *Mol. Plant Microbe Interact.* 17: 6–15.

Kuz'mina, L.Y. and Melent'ev, A. (2003). The effect of seed bacterization by *Bacillus* Cohn bacteria on their colonization of the spring wheat rhizosphere. *Microbiologica* 72 (2): 230–235.

Liu, W., Wang, J., Yin, D.X. et al. (2016). Genetic diversity and structure of the threatened species *Sinopodophyllum hexadrum* (Royle) Ying. *Genet. Mol. Res.* 15 (2).

Lugtenberg, B. and Kamilova, F. (2009). Plant-growth-promoting rhizobacteria. *Annu. Rev. Microbiol.* 63: 541–556.

Lugtenberg, B.J.J., Dekkers, L., and Bloemberg, G.V. (2001). Molecular determinants of rhizosphere colonization by Pseudomonas. *Annu. Rev. Phytopathol.* 39: 461–490.

Lynch, J.M. (2002). Resilience of the rhizosphere to anthropogenic disturbance. *Biodegradation* 13: 21–27.

Lynch, J.M. and Whipps, J.M. (1990). Substrate flow in rhizosphere. *Plant Soil* 129: 1–10.

Malik, K.A., Bilal, R., Mehnaz, S. et al. (1997). Association of nitrogen fixing, plant growth promoting rhizobacteria (PGPR) with kallar grass and rice. *Plant Soil* 194: 37–44.

Miller, H.J., Henken, G., and van Veen, J.A. (1989). Variation and composition of bacterial populations in the rhizosphere of maize, wheat, and grass cultivars. *Can. J. Microbiol.* 35: 656–660.

Molin, J. and Molin, S. (1997). CASE: complex adaptive systems ecology. In: *Advances Microbial Ecology*, vol. 15 (ed. J.G. Jones), 27–79.

Mrkovacki, N. and Milic, V. (2001). Use of *Azotobacterchroococcum* as potentially useful in agricultural application. *Ann. Microbiol.* 51: 145–158.

Murphy, P.J., Heycke, N., Banfalvi, Z. et al. (1987). Genes for the catabolism and synthesis of an opine-like compound in *Rhizobium meliloti* are closely linked and on the Sym plasmid. *Proc. Natl. Acad. Sci. U. S. A.* 84: 493–497.

Murphy, P.J., Trenz, S.P., Grzemski, W. et al. (1993). *Rhizobium meliloti*rhizopinemos locus is a mosaic structure facilitating its symbiotic regulation. *J. Bacteriol.* 175: 5193–5204.

Nadeem, S.M., Zahir, Z.A., Naveed, M. et al. (2007). Preliminary investigations on inducing salt tolerance in maize through inoculation with rhizobacteria containing ACC deaminase activity. *Can. J. Microbiol.* 53: 1141–1149.

Nandal, M. and Hooda, R. (2013). Plant growth promoting rhizobacteria: A review article. Int. J. Cur. Res. 5(12):3863–3871. *Nat. Prod. Res.* 7: 1–5.

Nardi, S., Concheri, G., Pizzeghello, D. et al. (2000). Soil organic matter mobilization by root exudates. *Chemosphere* 41: 653–658.

Neal, J.L. Jr., Atkinson, T.G., and Larson, R.I. (1970). Changes in the rhizosphere microflora of spring wheat induced by disomic substitution of a chromosome. *Can. J. Microbiol.* 16: 153–158.

Neal, J.L. Jr., Larson, R.I., and Atkinson, T.G. (1973). Changes in rhizosphere populations of selected physiological groups of bacteria related to substitution of specific pairs of chromosomes in spring wheat. *Plant Soil* 39: 209–212.

Negi, S., Ivanchenko, M.G., and Muday, G.K. (2008). Ethylene regulates lateral root formation and auxin transport in *Arabidopsis thaliana*. *Plant J.* 55: 175–187.

Negi, G., Pankaj, S.A., and Sharma, A. (2014). In situ biodegradation of endosulfan, imidacloprid, and carbendazim using indigenous bacterial cultures of agriculture fields of Uttarakhand, India. *Int. J. Bioeng. Life Sci.* 8: 973–981.

Negrao, S., Schmockel, S.M., and Tester, M. (2017). Evaluating physiological responses of plants to salinity stress. *Ann. Bot.* 119: 1–11.

Neilands, J.B. (1995). Siderophores: structure and function of microbial iron transport compounds. *J. Biol. Chem.* 270: 26723–26726.

Neubauer, U., Furrer, G., Kayser, A., and Schulin, R. (2000). Siderophores, NTA, and citrate: potential soil amendments to enhance heavy metal mobility in phytoremediation. *Int. J. Phytoremediation* 2: 353–368.

Obreht, Z., Kerby, N.W., Gantar, M. et al. (1993). Effects of root associated N2 fixing cyanobacteria on the growth and nitrogen content of wheat (*Triticum vulgare* L.) seedlings.*Biol. Fertil. Soils* 15: 68–72.

Oger, P., Petit, A., and Dessaux, Y. (1997). Transgenic plants producing opines alter their biological environment. *Nat. Biotechnol.* 15: 369–372.

Omar, M.N.A., Mahrous, N.M., and Hamouda, A.M. (1996). Evaluating the efficiency of inoculating some diazatrophs on yield and protein content of 3 wheat cultivars under graded levels of nitrogen fertilization. *Ann. Agric. Sci.* 41: 579–590.

Pandey, A., Sharma, E., and Palni, L.M.S. (1998). Influence of bacterial inoculation on maize in upland farming systems of the Sikkim Himalaya. *Soil Biol. Biochem.* 30: 379–384.

Pankaj, N., Tapan, K., Negi, L., and Panwar, A. (2013). Isolation an characterization of rhizobial isolates from the rhizospheric soil of an endangered plant *Meizotropispellita*. *Asian J. Microbiol. Biotechnol. Environ. Sci.* 16 (2): 65–70.

Pankaj, G.N., Gangola, S., Khati, P. et al. (2016a). Differential expression and characterization of cypermethrin-degrading potential proteins in *Bacillus thuringiensis* strain, SG4. *3 Biotech* 6: 225.

Pankaj, G.N., Sharma, A., Gangola, S. et al. (2016b). Novel pathway of cypermethrin biodegradation in a *Bacillus* sp. strain SG2 isolated from cypermethrin-contaminated agriculture field. *3 Biotech* 6: 45.

Parr, J., Papendick, R., Hornick, S., and Meyer, R. (1992). Soil quality: attributes and relationship to alternative and sustainable agriculture. *Am. J. Altern. Agric.* 7: 5–11.

Patten, C.L. and Glick, B.R. (1996). Bacterial biosynthesis of indole-3-acetic acid. *Can. J. Microbiol.* 42: 207–220.

Pimentel, J.P., Olivares, F., Pitard, R.M. et al. (1991). Dinitrogen fixation and infection of grass leaves by *Pseudomonas rubrisubalbicans* and *Herbaspirillum seropedicae*. *Dev. Plant Soil Sci.* 48: 225–229.

Pinto, R., Varanini, Z., and Nannipieri, P. (2007). *The Rhizosphere: Biochemistry and Organic Substances at the Soil-Plant Interface*. Boca Raton: CRC.

Rajkumar, M., Ae, N., Prasad, M.N.V., and Freitas, H. (2010). Potential of siderophore-producing bacteria for improving heavy metal phytoextraction. *Trends Biotechnol.* 28: 142–149.

Rovira, A.D. (1969). Plant root exudates. *Bot. Rev.* 35 (1): 35–57.

Saharan, B.S. and Nehra, V. (2011). Plant growth promoting rhizobacteria: A critical review. *Life Sci. Med. Res.* 1: 21–31.

Saint, C.P., Wexler, M., Murphy, P.J. et al. (1993). Characterization of genes for synthesis and catabolism of a new rhizopine induced in nodules by *Rhizobium meliloti* Rm220–3: extension of the rhizopine concept. *J. Bacteriol.* 175: 5205–5215.

Saleem, M., Arshad, M., Hussain, S., and Bhatti, A.S. (2007). Perspective of plant growth promoting rhizobacteria (PGPR) containing ACC deaminase in stress agriculture. *J. Ind. Microbiol. Biotechnol.* 34: 635–648.

Savka, M.A., Black, R.C., Binns, A.N. et al. (1996). Translocation and exudation of tumor metabolites in crown galled plants. *Mol. Plant Microbe Interact.* 9: 310–313.

Schoenborn, L., Yates, P.S., Grinton, B.E. et al. (2004). Liquid serial dilution is inferior to solid media for isolation of cultures representative of the phylum-level diversity of soil bacteria. *Appl. Environ. Microbiol.* 7: 4363–4366.

Selvakumar, G., Kundu, S., Gupta, A.D. et al. (2008). Isolation and characterization of nonrhizobial plant growth promoting bacteria from nodules of Kudzu (*Pueraria thunbergiana*) and their effect on wheat seedling growth. *Curr. Microbial.* 56 (2): 134–139.

Sevilla, M., Burris, R.H., Gunapala, N. et al. (2001). Comparison of benefit to sugarcane plant growth and 15N2 incorporation following inoculation of sterile plants with *Acetobacter diazotrophicus* wild type and Nif – mutant strains. *Mol. Plant Microbe Interact.* 14: 358–366.

Shenoy, V.V., Kalagudi, G.M., and Gurudatta, B.V. (2001). Towards nitrogen autotrophic rice. *Curr. Sci.* 81: 451–457.

Shukla, K.P., Singh, N.K., and Sharma, S. (2010). Bioremediation: developments, current practices and perspectives. *Gen. Eng. Biotechnol. J.* 3: 1–20.

Siciliano, S.D., Fortin, N., Mihoc, A. et al. (2001). Selection of specific endophytic bacterial genotypes by plants in response to soil contamination. *Appl. Environ. Microbiol.* 67: 2469–2475.

Soedarjo, M. and Borthakur, D. (1996). Mimosine produced by the tree legume Leucaena provides growth advantages to some rhizobium strains that utilize it as a source of carbon and nitrogen. *Plant Soil* 186: 87–92.

Soedarjo, M., Hemscheidt, T.K., and Borthakur, D. (1994). Mimosine, a toxin present in leguminous trees (*Leucaena* spp.), induces a mimosine-degrading enzyme activity in some *Rhizobium* strains. *Appl. Environ. Microbiol.* 60: 4268–4272.

Stein, T., Hayen, N., and Fendrik, I. (1997). Contribution of BNF by *Azoarcus* sp. BH72 in Sorghum vulgare. *Soil Biol. Biochem.* 29: 969–971.

Sunitha, V., Muralidhara, R., and Reddy, M. (2012). Groundwater contamination from agro-chemicals n irrigated environment: field trials. *Adv. Appl. Sci. Res.* 3 (5): 3382–3386.

Tailor, A.J. and Joshi, B.H. (2014). Harnessing plant growth promoting rhizobacteria beyond nature: a review. *J. Plant Nutr.* 37: 1534–1571.

Thomashow, L.S., Bonsall, R.F., and Weller, D.M. (1997). *Antibiotic Production by Soil and Rhizosphere Microbes In Situ. Manual of Environmental Microbiology*, 493–499. Washington, DC: American Society for Microbiology.

Toal, M.E., Yeomans, C., Killham, K. et al. (2000). A review of rhizosphere carbon flow modelling. *Plant Soil* 222: 263–281.

Trevors, J.T. (1998). Molecular evolution in bacteria: cell division. *Rev. Microbiol.* 29: 237–245.

Uhde-Stone, C., Zinn, K.E., Yanez, M.R. et al. (2003). Nylon filter arrays reveal differential gene expression in proteoid roots of white lupin in response to phosphorus deficiency. *Plant Physiol.* 131: 1064–1079.

Valenzuela-Soto, J.H., Estrada-Herna'ndez, M.G., Ibarra-Laclette, E. et al. (2010). Inoculation of tomato plants (*Solanum lycopersicum*) with growth-promoting *Bacillus subtilis* retards whitefly *Bemisiatabaci* development. *Planta* 231: 397–410.

Van Loon, L.C. (2007). Plant response to plant growth promoting Rhizobacteria. *Eur. J. Plant Pathol.* 119 (3): 243–254.

VanRhijn, P. and Vanderleyden, J. (1995). The *Rhizobium* – plant symbiosis. *Microbiol. Rev.* 59 (1): 124–142.

Verslues, P.E. (2017). Time to grow: factors that control plant growth during mild to moderate drought stress. *Plant Cell Environ.* 40: 177–179.

Vessey, J.K. (2003). Plant growth promoting rhizobacteria as biofertilizers. *Plant Soil* 255: 571–586.

Walker, T.S., Bais, H.P., Grotewold, E., and Vivanco, J.M. (2003). Root exudation and rhizosphere biology. *Plant Physiol.* 132: 44–51.

Wall, D.H. and Virginia, R.A. (1999). Controls on soil biodiversity: insights from extreme environments. *Appl. Soil Ecol.* 13: 137–150.

Wartiainen, I., Eriksson, T., Zheng, W. et al. (2008). Variation in the active diazotrophic community in rice paddy-nifH PCR-DGGE analysis of rhizosphere and bulk soil. *Appl. Soil Ecol.* 39: 65–75.

Weir, B.S. (2008). The current taxonomy of rhizobia. New Zealand rhizobia website. Last updated: 16 Nov 2008.

Weir, T.L., Park, S., and Vivanco, J.M. (2004). Biochemical and physiological mechanisms mediated by allelochemicals. *Curr. Opin. Plant Biol.* 7: 472–479.

Whipps, J.M. (1990). Carbon utilization. In: *The Rhizosphere* (ed. J.M. Lynch), 59–97). Chichester, UK: Wiley.

Xiong, L., Wang, R.G., Mao, G. et al. (2006). Identification of drought tolerance determinants by genetic analysis of root response to drought stress and abscisic acid. *Plant Physiol.* 142: 1065–1074.

Yang, J., Kloepper, J.W., and Ryu, C. (2008). Rhizobacteria help plants tolerate abiotic stress. *Trends Plant Sci.* 14: 1–4.

Zahir, Z.A., Munir, A., Asghar, H.N. et al. (2008). Effectiveness of rhizobacteria containing ACC-deaminase for growth promotion of pea (*Pisum sativum*) under drought conditions. *J. Microbiol. Biotechnol.* 18: 958–963.

Zehnder, G.W., Murphy, J.F., Sikora, E.J. et al. (2001). Application of rhizobacteria for induced resistance. *Eur. J. Plant Pathol.* 107: 39–50.

10

Application of Inorganic Amendments to Improve Soil Fertility

Sunita Chauhan[1] *and Shweta Kulshreshtha*[2]

[1] *Amity Institute of Biotechnology, Amity University, Kant, Rajasthan, India*
[2] *Department of Biotechnology, Kumarappa National Handmade Paper Institute, Jaipur, Rajasthan, India*

10.1 Introduction

Maintaining proper soil health and fertility is the foremost requirement for sustainable agriculture to feed the rapidly growing populations all over the world as well as for safeguarding the ecosystem services. Soil health is described as the constant ability of soil to endure growth of crops or animals over the long term which is possible through efficient recycling of waste and maintenance of soil fertility through nutrient supplementation and water. The soil health is maintained by several different factors, microbes, and their interactions. There are various factors involved in assessing the soil quality which is known as the "soil quality index." The soil quality and fertility may be improved by soil amendments. Soil amendments are the substances that are mixed with the topsoil to promote plant growth and maintain their health. They are broadly categorized into two groups: inorganic and organic amendments. Precise diagnosis of soil health is the foremost challenge to choose appropriate amendment(s).

In this chapter, the use of inorganic amendments in improving soil fertility is focused on. Inorganic amendments are either mined or are manmade materials. Various types of inorganic soil amendments can be used for improving soil quality. The inorganic amendments can play a very effective role if applied on a need basis after proper soil tests. By considering the *Bhoochetna* initiative of Karnataka, it has been explained as a holistic approach that can be utilized based on the need for amendments. The amendment leads to the increase in the bioavailable nutrients and therefore, supports the growth of specific bio-fortified cultivars to get higher crop yields. Besides, inorganic amendments can also be useful in addressing the concerns of alkaline or saline soils and the reclamation of disturbed soil areas, viz. heavily mined lands.

Phytomicrobiome Interactions and Sustainable Agriculture, First Edition. Edited by Amit Verma, Jitendra Kumar Saini, Abd El-Latif Hesham, and Harikesh Bahadur Singh.

10.2 Impact of Bhoochetna Movement in Southern India

The government of Karnataka started a novel project under Rashtriya Krishi Vikas Yojana (RKVY) called "Bhoochetana" to improve the livelihoods of dryland farmers in the state by increasing the agricultural productivity of rain-fed agriculture. The Bhoochetna project was initiated in 2009 with the prime focus of the revival of soil fertility status until the period 2013–2014. As a part of this project, the soil test was recommended before the supplementation of micronutrients. Moreover, inorganic amendments are provided to the farmers at 50% subsidy rates. The awareness among the farmers is created through meetings, posters, advertising posters and wall writings, and mass media.

Impact assessment of Bhoochetana has been reported by Dhanalakshmi et al. (2017). Exhaustive efforts and improved practices during the five years of the Bhoochetna movement in the Raichur district of Karnataka resulted in a significant improvement of soil health of the IP adopted plots. A survey was conducted in six villages to find out the impact of the Bhoochetna movement. Soil organic carbon percentage increased and reached the medium level (0.5–0.71%). This was high as compared to the percentage of soil organic carbon before the implementation of improved practices. Besides soil organic carbon, other macro and micronutrients increased due to improved practices. Crop yield of major crops like sunflower (11%), cotton (19%), pigeon pea (13%), sorghum (11%), and groundnut (17%) was vastly improved.

Out of six villages, soil organic carbon of five villages under IP was found to increase to medium (0.5–0.71%) levels as compared to the low (0.26–0.43%) levels observed under FP (farmer practices). Soil quality under IP improved in terms of macro and micronutrients like N, P, K, S, B, and Zn. Increase in crop yield in all major crops like cotton (19%), groundnut (17%), pigeon pea (13%), sorghum (11%), and sunflower (11%) was observed under improved practices.

10.3 Sustainable Agriculture

The major concern of almost all countries over the world is satisfying the need for continuously growing populations sustainably. Sustainable agriculture focused on meeting the need of the present generation without compromising the productive potential for future generations. Therefore, the maintenance and recovery of soil health are important. The maintenance and recovery of soil include practices that are economically and environmentally sustainable.

In contrast to the traditional agricultural practices, the advent of the technologies of "green revolution," the dependency of agriculture on external inputs increased. The application of these inputs without soil testing, not only polluted the soil but also depleted the soil quality. Besides, it affected the health of human beings. Thus, the introduction of the green revolution has threatened long-term soil health, human health, and agricultural sustainability (Biswas et al. 2014). Maintenance of soil fertility and soil health is the primary requirement to achieve sustainable crop production.

10.3.1 Healthy Soil and Soil Quality

As per the ancient Indian practices, farmers used to treat soil as a living entity, which is regarded as one of the important components of the five elements of life including air, water, sky, and fire, collectively regarded as "Panchabhuta." In Vedas, the soil was considered as "Mother" and the human beings as her "son", i.e. "matabhumiputraaham-prathvaya" *(Atharveda 12.1.12.)*. In the modern context, according to Doran (2002), the soil is considered as the mother of agricultural activity.

Soil is the zone in which organisms, land, air, and water interact, and therefore, it creates a dynamic interface. In the soil, lithosphere (rock), atmosphere (air), hydrosphere (water), and biosphere (living things) interact with one another and influence the soil characteristics and quality. The soil characteristics determine the way of soil function toward the establishment of an ecosystem, i.e. natural or man-made.

Healthy soil has been defined as "the continued capacity of soil to function as a vital living system, within an ecosystem and land-use boundaries to sustain biological productivity, promote the quality of air and water environments; and maintain the plant, animal, and human health" (Doran and Safley 1997). The monitoring of physical, chemical, and biological parameters and their effects on the soil capacity, plant growth, and environmental functions is necessary to evaluate the sustainability of a production system (Doran and Safley 1997). The sustainability of the production system depends on the soil characteristics and different land uses practices and management.

Soil fauna, flora, and microbes initiate the flux of nutrients across the trophic levels and got the top trophic level and thus, maintain the sustainability of natural and agricultural ecosystems (Chen et al. 2003). The soil communities help in maintaining the nutrient level and re-circulation of different nutrients in the soil. However, soil loses its nutrient due to leaching and erosion. Moreover, nutrients loss also occurs after the harvesting of crops. Therefore, it is of utmost importance to maintain the fertility of soil before sowing the next crop. In the traditional system, the soil was left for some time for the regaining of nutrients. This is called Jhoom cultivation. As the growing population is exerting pressure on soil, the Jhoom cultivation approach cannot be adopted (Brussard et al. 2004). Another option of increasing soil fertility is the application of soil amendments. Dynamic soil quality is related to the physical, chemical, and biological properties and their management. The amount of soil organic content, structure, and water holding capacity is greatly affected by soil management practices. The management practices rely on the quality of soil which is assessed by a soil indicator. The choice of soil indicator depends on the function of soil.

10.3.2 Soil Quality

The capacity of soil to function refers to soil quality. According to the Soil Science Society of America (1984), soil quality is more specifically defined as the capacity of a specific kind of soil to function within natural or managed ecosystem boundaries, to sustain plant or animal productivity, maintain and enhance water and air quality and support human health and habitation (Kerlen et al. 1997). According to the Soil Quality Institute (USDA 2006) "Inherent soil quality is a soil's natural ability to function." For example, clay type soil retains more water than the sandy soil, whereas sandy soil drains faster than a clay one. The rocky surface cannot provide sufficient space for the growth of deep roots as compared to soil. The quality of soil must be evaluated to decide its suitability for specific applications. The suitability of soil depends on the soil quality which depends on the functions of soil. The soil functions further affect the microbial, physical, and chemical properties of soil.

10.3.3 Soil Quality Indicator

The soil flora, fauna, and microbes are the main components of soil quality indicators. These are greatly influenced by soil use and management techniques. Further, environmental instability due to man-made or natural reasons poses a major influence on soil quality. A little change in the soil quality, i.e., physical, chemical, and biological rapidly changes the microbial population of soil (Figure 10.1). These changes are related to the ecological processes which are taking place in the soil and change with the variation in the soil composition and physical properties. These processes and microbial population decide the quality of the soil.

There is no general quality assessment parameter for all types of soil. The quality assessment completely depended on the soil usage and management and environmental conditions. The quality of soil is assessed by nutrient availability, water availability and holding capacity, water infiltration, and soil carbon storage. Soil characteristics are greatly affected by management practices like tillage and cropping system. Therefore, soil health is managed by maintaining soil water infiltration ability, soil carbon storage, and supply of nutrients to the plants. As soil quality is also affected by soil organic matter and microbial diversity, maintenance of soil organic matter and microbial diversity will add icing to the cake. The combined effect of all these factors poses a positive influence on the physical,

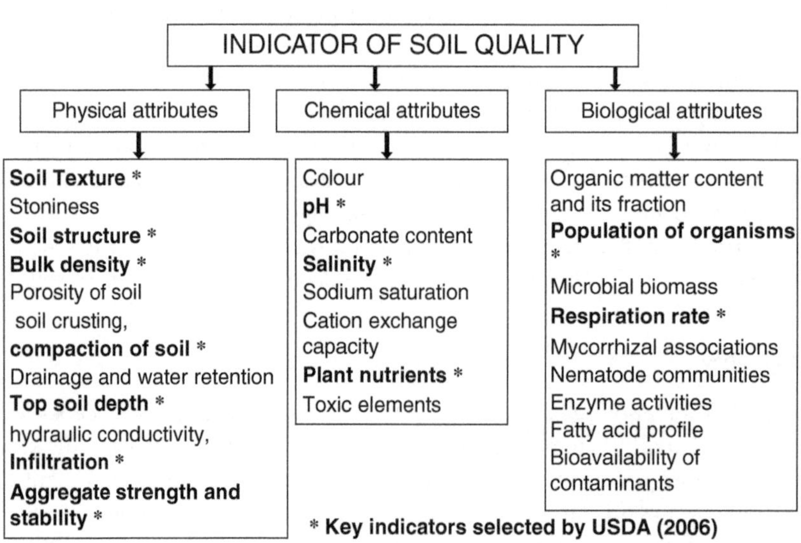

Figure 10.1 Physical, chemical, and biological indicators of soil quality. Key indicators selected by USDA (2006) are marked with an asterisk mark. *Source:* Based on USDA. (2006). Soil Quality Institute. Natural resources conservation service. Available at http://soils.usda.gov/sqi/.

chemical, and biological properties of soil. Soil quality can be documented after the detailed analysis of soil health indicators (Al-Kaisi et al. 2014).

According to Parr et al. (1992), soil quality must include environmental quality, animal and human health, food safety, and quality. The characterization of soil quality, physical, and chemical properties are easy to measure and quantify. Nowadays, soil degradation is the most destructive force to deteriorate soil quality which is giving rise to two major problems – the increase in the cost of food and shortage of food supply. Therefore, immediate action is needed on a global scale to conserve soil quality.

10.3.4 Soil Quality Index

Soil quality is measured by the soil quality index which is developed and used to predict the impact of farming system and management practices on soil productivity, food quality and safety, and human/animal health. Therefore, the soil quality index shows the mathematical model to quantify the effect of different farming practices on soil quality, human health, and environmental quality. Soil quality index is the function of soil properties, and can be assessed by using the following relationship:

$$\text{Soil Quality Index} = f\left(SP, P, E, H, ER, BD, FQ, MI\right)\left(\text{Parr et al.}\right)$$

Where, SP-soil properties, P-potential productivity, E-environmental factors, H-health, human/animal), ER-erodibility, BD-biological diversity, FQ-food quality/safety, and MI-management inputs.

Aforesaid indicators and their interaction and comparative weight are used to find out the quality of the soil. The quality of soil can be improved if the above-mentioned parameters are improved naturally or artificially. Soil amendment with specific nutrients is the way to enhance the quality of the soil. Soil amendment is the material that is introduced in the soil to overcome the problem of shortage of nutrients and to make it fertile. These amendments affect the physicochemical and biological properties of soil and make it fertile. Moreover, these amendments help in soil reclamation by improving its texture. The texture of the soil is also important to grow healthy crops. In case of improper aeration, the roots may suffocate and plants cannot survive. Therefore, the texture of the soil is also an important property to consider for healthy plant growth. The plant roots spread in the soil to acquire nutrients. In the tightly packed soil, air and water pockets are not present and hence, root cannot spread prolifically. Only healthy soil supports deep and strong root systems that are associated with healthy plant growth. The soil amendments help in reducing the compaction of soil and produce water and air pockets and increase aeration, maintenance of water, and nutrient-holding capacity. It also helps in the removal of an excessive amount of water.

Laishram et al. (2012) have reviewed the soil quality and soil health. It has been mentioned that the quality and health of soil determine agricultural sustainability and environmental quality, which jointly determine plant, animal, and human health (Haberern 1992; Doran 2002).The concept of soil quality or health is essentially an elaboration of the concept of soil productivity or fertility to deal with the multiple and complex problems faced by the world in the current scenario. This perspective of optimizing multiple functions makes soil health an integral dimension of agroecosystem health and sustainable development. Laishram et al. (2012) have compiled a profile of soil functions including ecological functions of the soil and their indicators. They have included the organic matter, topsoil depth, aggregation, texture, bulk density, infiltration, pH, soil respiration, etc., as key soil indicators for soil quality assessment. Soil quality indices are decision tools that effectively combine a variety of information for multi-objective decision making (Karlen and Stott 1994). Laishram et al. (2012) have explained carbon management Index (CMI), soil microbiological index (MDI), general indicator of soil quality (GISQ), QBS (Qualita Biologica del Suola) Index, farm-level environmental indicators, etc. Many soil quality and fertility indices have been proposed, however, the state of soil degradation that affects its functionality has not been addressed much.

The aforesaid discussion has focused on the fact that soil amendments are required for the growth of healthy plants and roots. Besides, the structure of the soil, water-holding capacity, nutrient cycling, and availability of nutrients affects the plants' growth. The soil properties can be improved by different types of amendments. Healthy plant and root growth are also associated with the prevention of soil erosion.

The soil amendments can be either organic or inorganic depending upon their source. The main differences between organic and inorganic amendments are given in Table 10.1.

In the rhizosphere of the plant, enzymes are present which control carbon, nitrogen, and phosphorous concentration in the soil by operating their natural cycles. This is the region where microbes are also present to maintain the balance of carbon and nitrogen and phosphorous. The presence of a microbe also supports the mechanism of nutrient transfer from soil to plants by increasing the bioavailability of nutrients. Urease, invertase, phosphatase, cellulose dehydrogenase, and amylase are some of the enzymes secreted by soil microorganisms such as bacteria, fungi, protozoa, and algae to increase the bioavailability of the nutrients to the plants (Antonious 2016).

10.4 Factors to Be Considered While Selecting a Soil Amendment

Four factors need to be considered before selecting a soil amendment:

- lasting period of the amendment in soil
- texture of the soil to be amended
- salinity levels of the soil in context
- the pH of the amendment

Before choosing the amendment for the soil, it is pertinent to note that the inorganic amendment has a fixed amount of N, P, K, while the organic amendment does not have. Therefore, the chemical composition of organic amendments must be assessed by using different laboratory tests before the application. Laboratory tests can determine the pH, salt content, and organic matter of the organic amendments. The quality and quantity of the amendments need to be assessed before application. These measured parameters helped in achieving this task for the small as well as big landscapes. The amendment to be chosen depends mainly on the tasks to be achieved. For a quick improvement of soil

Table 10.1 Differences between organic and inorganic amendments.

S. No.	Properties	Organic amendments	Inorganic amendments
1	characteristics	These are natural products derived from plants, plant products or the by-products	These are synthetic fertilizers that are manufactured artificially and contain mineral or synthetic chemicals
2	Fixed composition	Their chemical composition depends on the type of plant material and their ratio in the mixture and therefore, their chemical composition varies	They possess fixed chemical composition as they have a fixed amount of specific nutrient such as inorganic mineral, chemical, or synthetic fertilizer contains single nutrients including macro and micronutrient
3	Source of material and composition	These may contain animal-based materials such as worm castings, animal manure	These include vermiculite, perlite, tire chunks, pea gravel, sand, etc., and industrial discharges such as lime (from the lime factory), phosphate salt (from fertilizer industry), hydroxyapatite (from phosphorite), fly ash (from the thermal power plant), basic slag (from steel industry), Ca-montmorillonite (mineral), bentonite gypsum, etc.
4	Availability	They are readily available in the field sites/ households/farmhouses, and therefore, prove to be much cheaper and sustainable sources	These are readily available in the market. Specific nutrient containing amendments can be purchased as per requirement after assessing soil quality
5	Economic and sustainable option	Economic and sustainable	Expensive and nonsustainable option because energy is needed in their production
6	Time required for nutrient uptake	These can be uptaken by the plants only after the degradation of organic content	Inorganic amendments provide rapid nutrition to the plants as they need not to be decomposed. Hence, nutrient acquisition by plants can be forecasted very well
7	Soil quality improvement in long-term application	Soil quality improve in long-term application	These deteriorate the soil quality and cause environmental damage in long-term applications
8	Presence of pathogens	Organic amendments specifically the compost or manure or the household waste or farm animal waste need to be monitored properly for the possibilities of any smell or growth of pathogens	Most of these are relatively sterile in terms of plant pathogens
9	Requirement of composting	Composting is required	It is not required
10	C/N ratio	C/N ratio is high and therefore, composting is required	C/N ratio is not related to this as it possesses a specific formula
11	Storage	It can be stored in ground till its use	It can be stored in dry and dark place to protect it from moisture and weather changes so that it should be protected from chemical reaction

Source: Based on Saunders (2018) © University of New Hampshire.

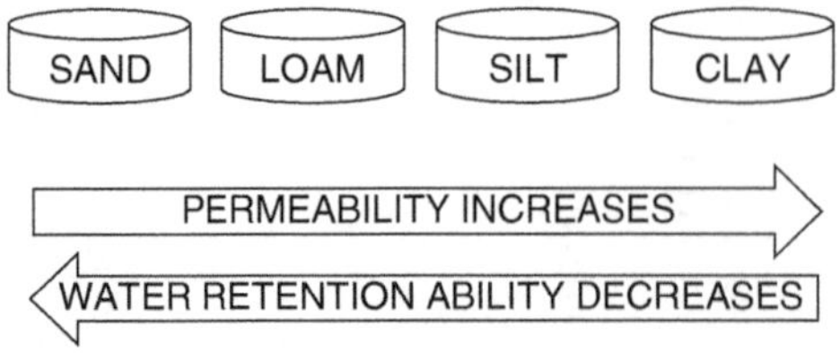

Figure 10.2 Permeability and water retention ability of different types of soil.

physical properties, a rapidly decomposable inorganic amendment should be chosen. Similarly, for the long-lasting improvement of soil, a slowly decomposable organic amendment should be used, whereas, for getting a long-lasting, quick improvement, a suitable combination of organic and inorganic amendments should be chosen. The permeability and water retention of various soil types is shown in Figure 10.2. Depending on crops, permeability, and water retention, the soils in question can be adjusted with the help of using certain soil amendments (Davis and Wilson 2005).

10.5 Advantages of Soil Amendments

The utilization of soil amendments is capable of protecting soil health as well as the environment by allowing remediation, revival, and reuse of polluted and exploited sites by reducing contaminant bioavailability. This can be done at a reasonable cost as compared to the other techniques. This technology may be an economically viable option for the reclamation of soil. Thus, the soil amendments have many benefits as mentioned below:

i) Restoration of soil health and structure and hence supports the growth of plants
ii) Recreation of the ecological functions of soils where amendments are used
iii) Reduction in the bioavailability of toxic pollutants
iv) Reduction in the leachability as well as the mobility of contaminants
v) Reduce soil erosion while improving the soil drainage property
vi) Cost-effective approach
vii) It can control the drainage from acid mines.

Besides, the above-mentioned benefits of restoring contaminated lands to natural habitats via soil amendments, the restoration process results in the creation of green space such as wildlife sanctuaries and an improvement in the aesthetic beauty of the land involved. Moreover, this may also cause a cultural stimulation for communities due to the improvement of economic value, cleaning air and water, reducing erosion caused by wind and water, and increasing evaporation or transpiration of water.

10.6 Land Modeling

Land modeling is an essential component for determining intrinsic soil quality. The model's construction is a suitable tool for predicting the changes in soil quality as a consequence of the soil amendment. The land evaluation modeling approaches are empirical-based modeling and process-based modeling. For this, expert systems, viz. computer programs that simulate the problem-solving skills of human experts in a given field, are used. They provide solutions to a problem expressing inferential knowledge through the use of decision trees. Land evaluation modeling can be categorized into two different classes based on the approaches used: land suitability or productivity and land vulnerability or degradation approaches. The clouds and the earth's radiant energy system (CERES) and gamma-ray observatory (GRO) models are the most widely known and used dynamic simulation models applied to agricultural production. They are also included in the "decision support system for agrotechnology transfer (DSSAT)" (ICASA 2006).

10.7 Major Applications of Soil Amendments

The salt-affected soils can be categorized into three types: saline, sodic, and saline-sodic. The soil which contains a high amount of soluble salts is called saline soils. The presence of excessive salts in such soils can not only impair plant productivity but also diminish the soil quality which cannot be improved by any chemical amendments or fertilizers. The only effective method to remove a high salt concentration is by leaching it using low saltwater. This could be possible only when there is a sufficient quantity of low saltwater available at source and if the soil drainage is good.

The sodic soil has low salt concentrations, nevertheless, a high amount of exchangeable sodium (Na^+). Its pH is higher than 8.5. It has poor soil structure, porosity due to salt deposition, and lack of air and water pockets and hence is not able to support the plant growth (Brady and Weil 1999). Sodium ions help in spreading the humus content in the soil and improves the characteristics of the soil. It helps in spreading the water content in the soil which is difficult in case of sodic soil due to poor permeability and slow water movement. The sodic soil shows a shrink-swell effect. The sodic soil swells due to the absorption of water while it shrinks due to the drying of soil particles. Water loss and drying of soil forms crack on its surface. Moreover, the deposition of salts makes it dried and cracked due to which the soil becomes very hard. The presence of humus and organic matter gives the black appearance to it because they move to the surface of the soil in an excessive water condition. This soil is known as black alkali soil. The soil amendment with organic or inorganic nutrients may improve its condition. These amendments can be incorporated by a physical method like deep plowing to improve the condition and texture of the soil. Besides this, the chemical method of incorporation of the amendment can be applied by the ion exchange method. Calcium ion-containing amendments can exchange sodium ions from the soil and thereby, sodium ions removed from the soil. Removal of sodium ions is necessary for sodic soil reclamation (Brady and Weil 1999; McCauley 2005).

All the aforesaid soil types can be reclaimed by using inorganic amendments. However, these amendments must be applied in an appropriate quantity.

10.7.1 Phyto-Stabilization in Polluted or Contaminated Soils

The polluted soil can be restored by the use of inorganic or organic amendments. Radziemska et al. (2017) have reported the application of both organic and inorganic amendments to boost the phytostabilization of trace elements in the polluted soils. The stabilization was achieved by introducing amendments, like lime, apatite, zeolite, materials with high organic matter content, manganese, and iron oxides. These materials can remove trace elements by absorption, complex formation, and co-precipitation (Bolan and Duraisamy 2003). Incorporation of low cost, abundantly available materials in the soil for immobilization of pollutants has several advantages. The advantages may be in terms of cost, easy methodology, and minimum effect on the environment. In addition to this, amendments may also improve soil fertility and microbial activity. It was reported that the use of amendments improved the quality of the soils polluted with trace elements. This was due to an increase in pH and organic matter content and the enhanced biochemical properties. Thus, the contamination effect of trace elements in the soils can be nullified or minimized through soil amendments, thereby attaining normal growth levels in plants.

10.7.2 Restoration of Soil

Different types of soil amendments have been reported to be used for remediation, revitalization, and re-use of the degraded lands. The commonly used amendments are discussed below.

10.7.2.1 Soil Acidity/pH Soil Amendments

The presence of pollutants in the soil damaged its quality. Heavy metals not only deteriorate the soil quality but also showed toxicity to soil microbial flora. Moreover, the bioavailability of heavy metals to the plants is also toxic. The pH of the polluted soil also falls in the acidic range. This pH, soil quality, and microbial population in the soil can be corrected by the addition of inorganic amendments. Acidic pH–bearing soil can be amended with alkaline substances. Such alkaline amendments may include the use of lime, wood ash, coal combustion products (CCPs), sugar beet lime, cement kiln dust, red mud, lime stabilized biosolids, etc. These amendments exert a good impact on the growth of plants. It also improves the runoff and leachate water quality. Common amendments used to correct lower soil pH (acidity) is elemental sulfur, ferrous sulfate, and aluminum sulfate.

10.7.2.2 Mineral Soil Amendments and Conditioners

A wide range of mineral by-products can be used as a significant soil amendment due to their conditioning properties. However, these can be used after proper characterization. They may include foundry sand (a by-product of metal casting industry), steel slag, dredged materials (found in abundance near commercial waterways and estuaries), gypsum (by-product during the manufacture of P fertilizers, titanium pigment production, etc.), water treatment residuals like alum, CCPs, etc.

The appropriate application rates of such amendments depend upon the specific concerns to be addressed. For example, higher application rates are required for soil reclamation compared to the amount required for increasing soil fertility.

The amount of moisture in soil amendments (commonly reported as % solids) is the predominant characteristic that dictates application procedure and timing. There may be a single application or multiple lifts. Individual soil amendments can also be combined with other residuals to produce characteristics optimal for the revitalization of a particular type of soil site. The process is called blending and needs specific blending equipment. There are two basic approaches to blending. On one hand, in situ mixing involves mixing of soil amendments into the receiving surface directly, whereas on the other hand priori blending involves the preparation of a soil mix made from amendments before their actual emplacement into the receiving surface.

10.7.2.3 Different Types of Inorganic Amendments

Lime amendments are used to raise the pH of acidic soil. Different types of limestone are used for the treatment of soil like ground limestone, dolomitic limestone, and burned limestone. The amendments for adjusting pH should be applied before the addition of fertilizers and cropping new crops. Phosphate is used in the form of phosphogypsum which is a by-product generated from phosphate rock (PR) in the form of phosphoric acid. This phosphoric acid accumulates near the PR-based production units.

Phosphogypsum is produced due to the generation of phosphoric acid from the phosphate-containing rock. The phosphogypsum accumulates in the adjacent area of production units. It contains phosphate, sulfate, and calcium which causes deterioration of soil quality. This soil becomes unsuitable for the growth of plants and therefore, phosphogypsum needs to remove from the soil. Several attempts were made to reduce the accumulation of phosphogypsum and its application in the soil to improve its quality.

Tobita and Nakabura (2018) have reported the implication of the direct application and residual effects of PR in the lowland rice system of Ghana. PR is reported to be a promising resource that can be used as an attractive alternative to water-soluble phosphorus fertilizers such as triple super phosphates (TSP). Direct application of PR has a positive impact on rice cultivation if used as a delayed-release fertilizer. PR solubility is found to be affected by soil and climatic conditions. Besides, the results of a two-year investigation showed a definite PR residual effect. The presence of residual effect continued for two years; thus, a single PR application can be used for three croppings. Although TSP application also showed residual effect at the rate of 270 kg P_2O_5 per hectare, in the case 60 kg P_2O_5 per hectare as the recommended level, the low residual effect was noted. These results indicated that PR direct application was effective, its direct or residual effect possibly varied owing to soil and/or meteorological conditions.

Grounded PR is used to ameliorate soil P deficiency and maintain crop productivity. Significant effectiveness of PR on rice production has been also observed in many studies (Fukuda et al. 2012; Nakamura et al. 2013) conducted under some specific soil conditions. The application of PR, along with organic matter such as cattle manure (CM) was reported to be effective (Fukuda et al. 2012).

Acidic soil was amended with low P of Ishigaki island containing 0, 21.8, and 87.3 mg P/kg soil when PR (Burkina Faso PR) or MCP (calcium dihydrogen phosphate) under flooded and upland conditions were used to grow rice (*Oryza sativa* L. cv. IR74) for 68 days with 4 replicates. It was concluded that the co-application of BPR and organic amendments (i.e. CM and RS) could be used to improve rice growth and soil fertility in lowland rice systems, including areas having acidic soils, low P content, and varying soil water contents from upland to flooded conditions. The effectiveness of these materials on rice growth has been influenced by the types and chemical properties of P sources, rates of application, methods of application, and management of total P fertilizer (Fukuda et al. 2018a, b)

The use of phosphogypsum is used as a soil amendment after composting with olive oil wastes and spent coffee grounds. The effect of two concentrations of phosphoric acid, i.e. 10% and 30% as compared with the unamended soil. This composted material was used by Kammoun et al. (2017) to evaluate their growth potential in potato plants. The potato yield increased by 55.17% due to phosphogypsum addition. The phosphogypsum can be used to develop high-quality edible products, soil quality improvement, and phosphogypsum valorization (Kammoun et al. (2017)). In another experiment, the effect of phosphogypsum was mixed with the wastes, like olive mill wastewater, olive pomace, and then composted. The compost with 10 and 30% phosphogypsum was added to the potato plant and growth was observed. The phosphogypsum at 10 or 30% concentration presented higher phosphorous and calcium content as compared to unamended control. Further, the presence of plant growth-promoting rhizobia did not affect by the addition of phosphogypsum in the compost (Samet et al. 2019).

Gypsum is one of the earliest forms of the agricultural amendment. Gypsum is a moderately soluble source of essential plant nutrients, calcium, and sulfur. The pure gypsum is hydrated calcium sulfate ($CaSO_4.2H_2O$) in which 79% calcium sulfate ($CaSO_4$) and 21% water (H_2O). Gypsum contains 23.3% calcium and 18.6% sulfur. Gypsum is moderately soluble in water which helps in spreading the calcium in the soil more readily than calcium of lime solution. It can increase plant growth. Gypsum amendments can also enhance the physical and chemical properties of soil, and thus, reduce the loss of nutrients caused by soil erosion and surface water run-off from agricultural land. Generally, it is used for sodic soil reclamation and can be included as a component in synthetic soils used in the nursery, greenhouse, and landscape applications. A special category of gypsum namely "Flue Gas Desulfurization (FGD) gypsum" which is similar to the mined gypsum can also be used to enhance crop production. A large amount of FGD gypsum is produced by the removal of sulfur dioxide (SO_2) from flue gas streams which is produced due to the burning of coal. Thus, agricultural applications of FGD gypsum can help in augmenting or replacing commercially mined gypsum, thus avoiding both energy-intensive and water-intensive mining activities associated with gypsum extraction. In a study, conducted by Chen and Dick (2011), the effect of the FGD gypsum amendment on the heavy metal distribution in sodic soil was studied. The soil amended with FGD gypsum is pH, electrical conductivity, and exchangeable sodium percentage in the amended soil were significantly reduced and falls in the range of water quality standard of China. Besides, the concentration of arsenic and mercury in the soil was positively correlated with the amount of FGD gypsum addition.

Dick et al. (2006) reported the application of gypsum in the agricultural land. Gypsum is the most economical amendment for the sodic soils. It is a long-lasting but slow-acting amendment for the sodic soils.

Gypsum is a very important inorganic soil amendment which helps in improving soil fertility in the following manner (Gypsum bulletin):

i) Gypsum provides nutrients to the plants.
ii) It can improve the physical and chemical properties of soil and make it suitable for the growth of plants.

iii) Gypsum is used to improve the quality of sodic soil. Gypsum exchanges the sodium ions with the calcium ions and helps in the reclamation of soil particles. This leads to the flocculation of soil particles which is a necessary condition for the formation and stabilization of soil structure.
iv) Gypsum is the most commonly used amendment for sodic soil reclamation because gypsum provides calcium that can exchange sodium and magnesium ions, thus leading to flocculation of soil particles. Flocculation is required for the formation and stabilization of soil structure. This increases water penetration and permeability of the soil and thus, reducing soil erosion and improving water quality.
v) It reduces the rate of surface drying and further surface crust formation due to reducing the dispersion of clay.
vi) Although agricultural lime is suggested for rectification of soil acidity and low pH the beneficial effects of calcite lime are mostly limited to the zone of incorporation. On the other hand, surface applications of gypsum may affect soil physical and chemical properties at depth due to gypsum's greater solubility than lime.
vii) Gypsum applications to calcium-deficient soil in humid regions have shown beneficial effects of calcium movement into the soil and thereby, improving root growth and lowering water stress.
viii) Gypsum decreases the storage rots of Cantaloupe and tomato (Sumner and Lerrimore 2006).

In a study, conducted by Lueders and Friedrich (2002), the effects of ferrihydrite and gypsum amendment on population dynamics of methanogens in Italian rice field soil slurries were studied using archaeal small subunit (SSU) ribosomal DNA (rDNA)- and rRNA-based terminal restriction fragment length polymorphism analysis and by quantitative SSU rRNA hybridization probing. The results showed an increase or decrease in the methanogenic population. This concluded that any change in the supplementation resulted in different responses of the methanogens (Lueders and Friedrich (2002)).

Soil amendment with gypsum along with mono-ammonium phosphate and elemental sulfur could remove the cadmium and improve the quality of the soil. Besides, it showed a positive effect on the growth of wheat (*Triticum aestivum* L.). The soil amendment with mono-ammonium phosphate and gypsum showed a significant increase in the growth and yield of wheat along with cadmium immobilization of soil while contrasting results were obtained in the case of sulfur supplementation. The soil amendment with sulfur increased in bioavailable cadmium in the soil and showed cadmium immobilization in a dose-dependent manner (Qayyum et al. 2017). Therefore, gypsum is proved as a good inorganic supplement for the growth of plants and cadmium immobilization and removal.

In contrast to the above findings, McBride et al. (2013) reported that the soil amendment with gypsum did not reduce lead and arsenic availability.

Wood ash has a long history of use in many countries like New Zealand and India. The people of New Zealand used to use wood ash for making soaps, to clean glass, silver or pewter and also as a soil amendment for the rectification of pH (Saunders 2018).

i) Wood ash contains significant amounts of calcium (25%) and potassium (5%) with smaller amounts of phosphorus (2%) and magnesium (1%) which are essential plant nutrients. It also contains micronutrients like zinc and copper, molybdenum, sulfur, boron.

ii) Wood ash is a natural substitute of lime to help maintain soil pH.
iii) Wood ash is more soluble and reactive than ground limestone and brings about a change in soil pH more quickly than lime. Lime can take six months to one year to make a similar effect.
iv) The amount of wood ash required should be determined through soil testing. Recommendations are based on the current soil pH, target pH, and the soil's buffering capacity (a measure of how readily pH changes in the soil depending on its clay and organic matter content). Some plants like blueberries, rhododendrons, and scab-susceptible potatoes varieties thrive in acid soils and should not be supplemented with wood ash. Too much wood ash can raise the soil pH higher than optimal, thereby posing a negative impact on plants.

There are several limitations in using wood ash for improving the soil quality. Wood ash is a liming material therefore caustic, protective equipment should be used at the time of its application. This includes eye protection, long sleeves, gloves, and a mask if very fine material.

Wood ash should never be mixed with nitrogen fertilizers such as ammonium sulfate, urea, or ammonium nitrate. They produce ammonia when placed in contact with high pH materials such as wood ash. The wood ashes from garbage, pressure-treated or painted wood should never be used as it can contain heavy metals and materials which are toxic to plants.

In a study, the wood ash amendment was compared with the calcium carbonate amendment. Wood ash not only provides nutrients to the soil but also acts as a liming agent. Wood ash increases the pH of the soil and provides calcium, magnesium, inorganic carbon, sulfate, and dissolved organic carbon and therefore, considered as a liming agent. Wood ash exchanges sulfate and hydroxyl ion and therefore increases the soil pH. It increases the bioavailability of nutrients to the plant. The precautions must be taken in the amendment of wood ash to the soil as a high amount of supplementation increases the concentration of sulfate and nitrate in the soil. It should be amended inappropriate amount for the best results in terms of soil quality and plant growth (Nkana et al. 2002). The presence of polycyclic aromatic hydrocarbon concentration in wood ash may adversely affect the soil quality. It is also possible that contaminants may remobilize to the soil by their application (Rey-Salgueiro et al. 2004). Therefore, their effect on the soil and plant must be studied before use as an amendment.

Wood ash amendment increases the calcium content of the leaves of forest trees. However, the growth of trees was not positively influenced by the amendment of wood ash due to a lack of nitrogen content. The amendment is beneficial in the case of potassium, calcium, and magnesium deficient plants. Ash may be rich in toxic heavy metals nevertheless, the bioavailability of this is very low if the forest ecosystem is considered. Its availability is limited to the forest litter and topsoil of the forest. The care must be taken in the amendment of wood ash. Wood ash should be applied only on mature plants rather than seedlings. It must be stabilized before application. It should only be applied to acidic soil (Augusto et al. 2008). Hannam et al. (2016) have reported the regulations and guidelines for its use as a soil amendment in Canadian forests. Ash derived from contaminant-free woody residues acts as a soil liming agent and is also a valuable source of many nutrients required by plants (e.g. calcium, magnesium, potassium, and phosphorus). Application of wood ash to forest soils has been used to replace the nutrients removed during timber and biomass harvesting; to counteract the acidifying effects of atmospheric deposition on soil and water, and to improve tree growth on sites where nitrogen does not limit productivity. Soil applications of wood ash could potentially be used as a management tool for emulating some of the effects of wildfire. Increased use of wood ash as a forest soil amendment would also decrease the social, financial, and environmental costs of landfilling.

Slag is a by-product of the process of steel production. A slag is a group of compounds removed during the smelting process. It can be a good source of silicon, calcium, and magnesium but it may also include other impurities such as metal oxides. Some slags are further processed to remove impurities. Silicate-based slags are variable in composition and fineness. Slags should be checked for the presence of any impurities of hazardous materials like heavy metals before being used as soil amendments (Torlon 2014).

These are the most common liming materials used as soil amendments (Havlin et al. 1999). It is used to correct soil acidity. Calcite (calcium carbonate) is more frequently used than dolomite (calcium magnesium carbonate). However, dolomite is used only when the soil is deficient in magnesium.

Silicon is the second most abundant substance in the earth's crust. It forms 28% of the earth's crust (Datnoff and Rodrgues 2005). Despite its abundance in nature, it is not considered an essential element for plants. However, its beneficiary effects have been reported in many crops (Elawad and Green 1979). Monocots are reported to be a higher accumulator of silicon than dicots. Crops such as rice, wheat, sugarcane, cucumber, and barley are responsive to silicon (Marschner 1995).

Silicone fertilizers are available in solid and liquid forms. Potassium silicate and sodium silicate are a liquid form of silicon. These are used as a foliar supplement in greenhouse and field experiments (Belenger et al. 1995). Liquids offer a rapid response as compared to the solid form. Calcium carbonate chemically reacts with silicone to form calcium silicate. Silicon is present in rock mineral, fertilizers, and rock phosphate. However, this is a non-bioavailable form of the mineral. Silicic acid is the plant absorbed from an in-soil solution. Additionally, rice hull dust and wheat straw are good sources of silicon after harvest. Silicon treatment of soil used to grow pumpkin plants has shown improved leaf retention, disease resistance, and increased yield (Heckman 2013; Torlon 2014).

Long-term cultivation of crops leads to the exhaustion of bioavailable silicon in the soil which may result in reduced yield. The silicon is also depleted due to soil erosion, leaching process, and absorption of silicon by plants. Naturally, tropical and sub-tropical soil lacks silicon, and therefore, the silicon amendment is beneficial for these soil types. The soil amendment with silicon may increase or decrease the yield of crops (Meena et al. 2014). Therefore, proper analysis of soil nutrients is recommended to get the maximum benefit in terms of crop yield. The following substances can be used as a silicon amendment:

Montana Grow: It is a natural silicon fertilizer listed by PMRI for organic use. It improves yield, suppresses foliage disease, and increase salt and drought resistance.

Glacier Rock Flour: It is commonly known as glacier rock dust, is a natural mineral source of silicon that provides mineral and trace elements that improve soil quality and enhance plant growth.

Wollastonite is named after the chemist and mineralogist William Hyde Wollaston (1766–1828). It is a calcium silicate mineral that may also contain small amounts of other elements such as magnesium, iron, manganese substituting for calcium. Pure Wollastonite is white and can be mined from naturally occurring deposits around the world. It is used in the manufacture of ceramics and numerous other products, which is less known as a soil amendment. When applied to the soil, wollastonite can neutralize the soil acidity and supply plant-available calcium and silicates. Wollastonite obtained as a naturally occurring mined product and without any synthetic additives, it may find use in organic farming as an alternative to carbonated liming materials such as a source of silicate (Torlon 2014). In a pot experiment, the effect of calcium silicate on the subcellular distribution of cadmium was assessed in the grains of amaranths. The soil amendment of calcium silicate reduces cadmium toxicity and its accumulation in amaranths by increasing the amount of pectate and pectin bound cadmium and decreasing the amount of water-soluble cadmium (Lu et al. 2017). The calcium silicate amendment is a good option to reduce the cadmium toxicity by increasing cadmium adsorption by soil and decreased cadmium uptake from the soil solution. It was proved to be better than the other quicklime and potassium dihydrogen phosphate amendments in reducing the cadmium toxicity (Lu et al. 2014).

The silicates can be applied in the foliar spray form. In 1990, the silicates were used the first time for the foliar spray. Later, silicic acid was used for foliar spray. Recently, silica nanoparticles are also applied as a foliar spray. These foliar sprays not only increase the yield and plant growth but also show their effectiveness as pesticides and reduce biotic stress (Laane 2018). Besides this, foliar spray of sodium selenite in combination with the calcium, magnesium, phosphorous amendment to the soil can reduce cadmium concentration lower than Chinese safety standards. This combined strategy can also reduce arsenic toxicity.

Different inorganic amendments pose a variety of effects on the growth of plants. Waddell et al. (2002) reported the different types of effects on the improvement in the crop on adding different amendments. While using the composted green waste as a sustainable source of nutrients on gray sandy soils in high rainfall areas, it has been reported that the use of inorganic amendments will be necessary not only to reduce phosphorus leaching but also to increase the buffering capacity of the soil. It increases the availability of nutrients to the plants, thereby support deep root growth, plant health, and stability. Moreover, it unraveled the problem of soil hydrophobicity and further, the problem of high water usage. The wastes produced by mining companies have substantial amounts of clay and basic cations like iron, manganese, calcium, and magnesium. This waste is considered as an important resource, rather than a waste, to augment the sandy soils. This amendment helps in improving soil quality. Such applications will not only minimize waste but also minimize the cost of maintaining sandy soil because it requires consistent nutrient and water applications to grow lawns, crops, vegetation, etc. The sources of inorganic amendments used were: red clay, yellow clay, chittering clay, gravel screenings, zeolite, Bentonite A, Bentonite B, etc.

10.8 Combination Strategy for Soil Quality Improvement

According to this concept, a combination of organic and inorganic substances is used as an amendment to improve the quality of the soil. Madejón et al. (2018) reviewed the effect of different amendments on the soil and plants along with their effectiveness concerning time and proposed that biomonitoring of trace elements and toxicants is required for assessing proper remediation. Ogunwole et al. (2010) have reported the effect of organic and inorganic soil amendments on soil physical and chemical properties in a West Arica savanna agroecosystem. Osundare et al. (2015) have reported the positive effects of organic and inorganic salt amendments on the growth performance of *Musa paradisiacal* L.

A combination strategy is not always successful in improving plant growth. In an experiment, a combination of biochar and gypsum was used to improve plant growth. The result of this study revealed the positive effect of biochar amendment. Biochar amendment improved the growth of *Abelmoschus esculentus* in terms of water use efficiency, plant height, and leaf area. However, the amendment of biochar in combination with gypsum reported being an unsuccessful strategy (Batool et al. 2015).

10.9 Conclusion and Future Prospects

The present chapter narrates the significance of maintaining proper soil health for sustainable agricultural practices. The soil quality can be improved through easily and naturally available sources of inorganic amendments and improved agricultural practices. This can result in improvement in crop yields besides maintaining proper soil health. The Bhoochetna movement of Karnataka provides clear evidence of the role of such improved practices in sustainable agricultural management.

References

Al-Kaisi, M., Douelle, A., and Kwaw-Mensah, D. (2014). Soil microaggregate and macroaggregate decay over time and soil carbon change as influenced by different tillage systems. *J. Soil Water Conserv.* 69 (6): 574–580. https://doi.org/10.2489/jswc.69.6.574.

Antonious, G.F. (2016). Soil amendments for agricultural production. In: Organic fertilizers from basic concepts to applied outcomes. doi:10.5772/63047

Augusto, L., Bakker, M.R., and Meredieu, C. (2008). Plant soil wood ash applications to temperate forest ecosystems - potential benefits and drawbacks. *Plant and Soil* 306 (1–2): 181–198. doi: 10.1007/s11104-008-9570-z.

Batool, A., Taj, S., Rashid, A. et al. (2015). Potential of soil amendments (biochar and gypsum) in increasing water use efficiency of *Abelmoschus esculentus* L. *Moench. Front. Plant Sci.* 6: 733. https://doi.org/10.3389/fpls.2015.00733.

Biswas, S., Ali, M.N., Goswami, R., and Chakraborthy, S. (2014). Soil health sustainability and organic farming: a review. *J. Food. Agric. Environ.* 12 (3&4): 237–243.

Bolan, N.S. and Duraisamy, V.P. (2003). Role of inorganic and organic soil amendments on immobilization and phytoavailability of heavy metals: a review involving specific case studies. *Aust. J. Soil. Res.* 41: 533–555.

Brady, N.C. and Weil, R.R. (1999). *The Nature and Properties of Soils*, 12e, vol. 1– 9, 453–556. London: Prentice Hall Publishers.

Brussard, L., Kuyper, T.W., Didden, W.A.M. et al. (2004). Biological soil quality: from biomass to biodiversity: importance and resilience to management stress and disturbance. In:

Managing Soil Quality: Challenges in Modern Agriculture (eds. P. Schjonning, S. Elmholt and B.T. Christensen), 139–161. Wallingford, UK: CAB International.

Chen, L. and Dick, W. (2011). Gypsum as an agricultural amendment. General use guidelines. Bulletin 945. Gypsum Bulletin. The OHIO State University: 1–35.

Chen, G., Zhu, H., and Zhang, Y. (2003). Soil activities and carbon and nitrogen fixation. *Res. Microbiol.* 154: 393–398.

Datnoff, L.E. and Rodrigues, F.A. (2005). The role of silicon in suppressing rice diseases. *APSnet Feature.* https://doi.org/10.1094/APSnetFeature-2005-0205.

Davis, J.G. and Wilson, C.R. (2005). Choosing a soil amendment. Fact sheet no. 7.235.

Dhanalakshmi, D., Rao, K.N., Prasad, S.L., and Chander, G. (2017). Impact assessment of bhoochetna-a soil test based nutrient management scaling out initiative in Karnataka. *Int. J. Pure Appl. Biosci.* 5 (2): 706–710.

Dick, W.A., Kost, D., and Nakano, N. (2006). A review of agricultural and other land application uses of flue gas desulfurization products. Report 101385, Electric Power Research Institute, Palo Alto, Calif. 97 p. doi: 10.1038/srep40583

Doran, J.W. (2002). Soil health and global sustainability: translating science into practice. *Agric. Ecosyst. Environ.* 88 (2): 119–127.

Doran, J.W. and Safley, M. (1997). Defining and assessing soil health and sustainable productivity. In: *Biological Indicators of Soil Health* (eds. C.E. Pankhurst, B.M. Doube and V.V.S.R. Gupta). Wallingford, UK: CAB International.

Elawad, S.H. and Green, V.E. Jr. (1979). Silicon and the rice plant environment- a review of recent research. *RISO* 28: 235–253.

Fukuda, M., Nagumo, F., Nakamuro, S., and Tobita, S. (2012). Burkinofaso phosphate rock is an affective phosphorus fertilizer for rice grown on acidic and low phosphorus Ghanian soil: incubation study. *Res. Trop. Agric.* 5 (extra issue 1): 73–74.

Fukuda, M., Dzomeku, I.K., Avornys, V.K., Nakamura, S., Nagumo, F., and Tobita, S. (2018a). Combination and timing of application of phosphate rock and organic amendments in the low land rice field of Ghana. JIRCAS working report no. 86: 39–44.

Fukuda, M., Nakamura, S., Issaka, R.N., Buri, M.M., Dzomoku, J.K., Avornyo, V.K., and Tobita, S. (2018b). Potential sources, applications and contributions of organic matter to soil fertility restoration for low land rice production in Ghana. JIRCAS working report no. 86: 7–12.

Haberern, J. (1992). Viewpoint: a soil health index. *J. Soil Water Conserv.* 47: 6.

Hannam, K.D., Deschamps, F., Kwaiton, M., Venier, L., and Hazlett, P.W. (2016). Regulations and guidelines of wood ash as a soil amendment in Canadian forests. Information Report GLC-X-17. Pub. Natural Resources in Canada, Canadian Forest Service, Great Lakes Forestry Centre.

Havlin, J.L., Beston, D., Tisdale, S.L., and Nelson, W.L. (1999). *Soil Fertility and Fertilizers: An Introduction to Nutrient Management*, 6e, 46, 59196–46, 59198. Princeton, NJ: Prentice Hall, Inc.

Heckman, J. (2013). Silicon: a beneficial substance. *Better crops* 97 (4): 14–16.

ICASA (2006). International Consortium for Agricultural Systems Application. Available at www.icasa.net.

Kammoun, M., Ghorbel, I., Charfeddine, S. et al. (2017). The positive effect of phosphogypsum-supplemented composts on potato plant growth in the field and tuber yield. *J. Environ. Manage.* 200: 475–483. https://doi.org/10.1016/j.jenvman.2017.06.016.

Karlen, D.L., Mausbach, M.J., Doran, J.W. et al. (1997). Soil quality: a concept, definition, and framework for evaluation. *Soil Sci. Soc. Am. J.* 61: 4–10.

Karlen, D.L. and Stott, D.E. (1994). A framework for evaluating physical and chemical indicators of soil quality. In: *Defining Soil Quality for a Sustainable Environment* (eds. J.W. Doran, D.C. Coleman, D.F. Bezdicek and B.A. Stewart). Madison, Wisconsin, USA: SSSA, Inc.

Laane, H.M. (2018). The effects of foliar sprays with different silicon compounds. *Plants* 7: 45. https://doi.org/10.3390/plants7020045.

Laishram, J., Saxena, K.G., Maikhuri, R.K., and Rao, K.S. (2012). Soil quality and soil health: a review. *Int. J. Ecol. Environ. Sci.* 38 (1): 19–37.

Lu, H.P., Zhuang, P., Li, Z.A. et al. (2014). Contrasting effects of silicates on cadmium uptake by three dicotyledonous crops grown in contaminated soil. *Environ. Sci. Pollut. Res. Int.* 21 (16): 9921–9930. https://doi.org/10.1007/s11356-014-2947-z.

Lu, H., Li, Z., Wu, J. et al. (2017). Influences of calcium silicate on chemical forms and subcellular distribution of cadmium in *Amaranthus hypochondriacus* L. *Sci. Rep.* 7: 40583. https://doi.org/10.1038/srep40583.

Lueders, T. and Friedrich, M.W. (2002). Effects of amendment with ferrihydrite and gypsum on the structure and activity of methanogenic populations in rice field soil. *Appl. Environ. Microbiol.* 68 (5): 2484–2494. https://doi.org/10.1128/AEM.68.5.2484-2494.2002.

Madejón, P., Domínguez, M.T., Madejón, E., Cabrera, F., Marañón, T., and Murillo, J.M. (2018). Soil-plant relationships and contamination by trace elements: a review of twenty years of experimentation and monitoring after the Aznalcóllar (SW Spain) mine accident.

Marschner, H. (1995). *Mineral Nutrition of Higher Plants*. San Diego: Academic Press.

McBride, M.B., Simon, T., Tam, G., and Wharton, S. (2013). Lead and arsenic uptake by leafy vegetables grown on contaminated soils: effects of mineral and organic amendments. *Water Air Soil Pollut.* 224 (1): 1378. https://doi.org/10.1007/s11270-012-1378-z.

McCauley, A. (2005). Salinity and sodicity management. Module 2 of soil and water management: a self-study course from the MSU Extension Service Continuing Education series.

Meena, V.D., Dotaniya, M.L., Coumar, V. et al. (2014). A case for silicon fertilization to improve crop yields in tropical soils. *Proc. Natl. Acad. Sci., India Sect. B Biol. Sci.* 84: 505–518. https://doi.org/10.1007/s40011-013-0270-y.

Nakamura, S., Issaka, R., Dzomeku, I. et al. (2013). Effect of Burkino Faso phosphate rock direct application on Ghanaian rice cultivation. *Afr. J. Agric. Res.* 8 (17): 1779–1799.

Nkana, J.C., Demeyer, A., and Verloo, M.G. (2002). Effect of wood ash application on soil solution chemistry of tropical acid soils: incubation study. *Bioresour. Technol.* 85 (3): 323–325. https://doi.org/10.1016/s0960-8524(02)00140-2.

Ogunwole, J.O., Iwuafor, E.N.O., Eche, N.M., and Diels, J. (2010). Effect of organic and inorganic soil amendments on soil physical and chemical properties in a West Africa savanna agroecosystem. *Trop. Subtrop. Agroecosyst.* 12: 247–255.

Osundare, T.O., Fajinmi, A.A., and Okonji, C.J. (2015). Effects of organic and inorganic soil amendments on growth performance of plantain (*Musa paradisiaca* L.). *Afr. J. Agric. Res.* 10 (3): 154–160.

Parr, J.F., Papendick, R.I., Hornick, S.B., and Meyer, R.E. (1992). Soil quality: attributes and relationship to alternative and sustainable agriculture. *Am. J. Altern. Agric.* 7 (1&2): 5–11.

Qayyum, M.F., Rehman, M.Z.U., Ali, S. et al. (2017). Residual effects of monoammonium phosphate, gypsum and elemental sulfur on cadmium phytoavailability and translocation from soil to wheat in an effluent irrigated field. *Chemosphere* 174: 515–523. https://doi.org/10.1016/j.chemosphere.2017.02.006.

Radziemska, M., Vaverková, M.D., and Baryła, A. (2017). Phytostabilization management strategy for stabilizing trace elements in contaminated soils. *Int. J. Environ. Res. Public Health* 14 (9): 1–15.

Rey-Salgueiro, L., García-Falcón, M.S., Soto-González, B., and Simal-Gándara, J. (2004). Procedure to measure the level of polycyclic aromatic hydrocarbons in wood ashes used as fertilizer in agroforestry soils and their transfer from ashes to water. *J. Agric. Food Chem.* 52 (12): 3900–3904.

Samet, M., Karray, F., Mhiri, N. et al. (2019). Effect of phosphogypsum addition in the composting process on the physico-chemical proprieties and the microbial diversity of the resulting compost tea. *Environ. Sci. Pollut. Res. Int.* 26 (21): 21404–21415. https://doi.org/10.1007/s11356-019-05327-3.

Saunders, O. (2018). Guide to using wood ash as an agricultural soil amendment. UNH Cooperate Extension (www.extension.unh.edu).

Soil Science Society of America. (1984). Glossary of Soil Science Terms. Soil Science Society of America, Madison, Wisconsin. 37 p.

Tobita, S. and Nakabura, S. (2018). Soil fertility improvement with indigenous resources in low land rice ecologies in Ghana. Working report no. 86. Japan International Research Centre for Agricultural Sciences (JIRCAS), Tasukaba, Ibaraka, Japan.

Torlon, J.L. (2014). Building soil fertility using soil amendments. M.Sc. thesis submitted under the guidance of Dr. J.R. Heckman to Plant Biology, Graduate School-New Brunswick Rutgers, State University of New Jersey.

Waddell, B., Mann, S., and Allen, D. (2002). 'Improvement of soil quality by waste amendment', *Waste and Recycle 2002 Conference,* Perth, 2–4 Oct. 2002.

USDA. (2006). Soil Quality Institute. Natural resources conservation service. Available at http://soils.usda.gov/sqi.

11

Improved Plant Resistance by Phytomicrobiome Community Towards Biotic and Abiotic Stresses

Neha Trivedi

Indian Agricultural Research Institute, New Delhi, India

11.1 Introduction

Plants are continuously exposed to different environmental stresses including both abiotic and biotic elements. Major abiotic stresses that affect plants and crops in the field include heat, high light intensity, drought, salinity, chilling, cold, freezing, nutrient, ozone (O_3), and anaerobic stresses (Suzuki et al. 2014). In addition to abiotic stresses, under natural conditions, plants also face the threat of infection by pathogens (including bacteria, fungi, viruses, and nematodes) and attack by herbivore pests (Atkinson and Urwin 2012), the major biotic stresses. Moreover, many abiotic stress conditions were shown to weaken the defense mechanisms of plants and enhanced their susceptibility to biotic stress, like pathogen infection as the habitat range of pests and pathogens affected by climate changes (temperature, rain, etc.). Plants respond to these stresses through the activation of the defense system (Amtmann et al. 2008;Mittler and Blumwald 2010).

11.2 Microbes and Plants

Microbes are the diverse environmental natural entities, having the potential to develop an intricate physiological and metabolic mechanism to manage and fight with potentially toxic reactive oxygen species (ROS) that are generated by environmental stresses. Subsequently, the interaction of microbial population with plants is essential for the ecosystem, and microbes are the natural partners that accommodate plants to combat with the opposed environment. To date, several intricate mechanisms inside the plant cell have been studied which develop by microbiome developed to deal with stresses. Further, biochemical, physiological, and molecular studies support the understanding of the intricate and integrated cellular processes of plant–microbe associations. During the persistent stress by increasing environmental variations, it is becoming more essential to characterize and decipher plant–microbe interaction concerning defense against

Phytomicrobiome Interactions and Sustainable Agriculture, First Edition. Edited by Amit Verma, Jitendra Kumar Saini, Abd El-Latif Hesham, and Harikesh Bahadur Singh.

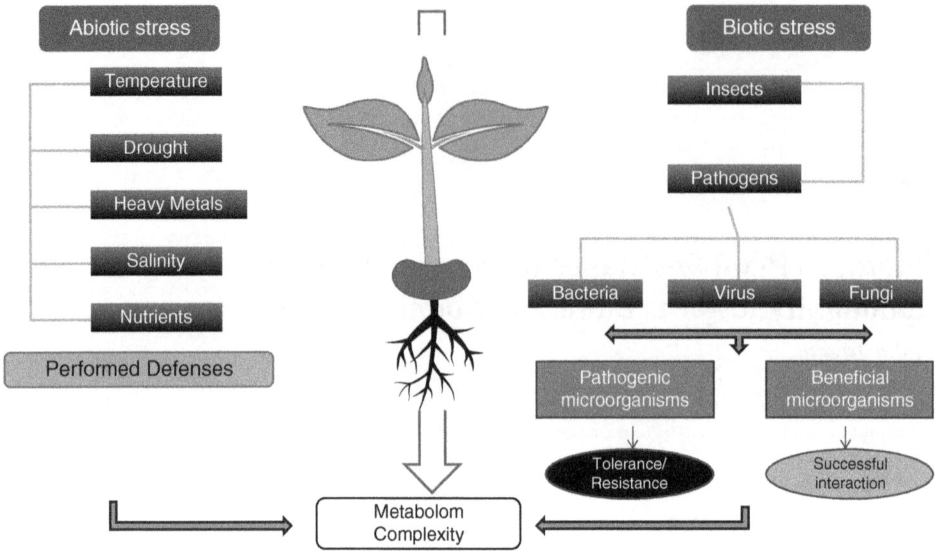

Figure 11.1 Plant response to different forms of biotic and abiotic stresses and stress-mediated alteration in the metabolome.

environmental challenges (Ansari 2018). A plant has to deal with several biotic and abiotic stresses during its life cycle which further imparts metabolome complexity and produces different secondary metabolites to combat stresses (Figure 11.1).

11.2.1 Abiotic Stress Responses and Microbe-Mediated Mitigation in Plants

Abiotic stresses are the primary limiting factors for agricultural productivity. Crop plants need to cope with adverse external pressure created by environmental and edaphic conditions with their intrinsic biological mechanisms, failing when their growth, development, and productivity suffer. Under the continuous pressure of increasing climatic alterations, it now becomes more imperative to define and interpret plant–microbe relationships in terms of protection against abiotic stresses. At the same time, it also becomes essential to generate deeper insights into the stress-mitigating mechanisms in crop plants for their translation into higher productivity. Microorganisms, the most natural inhabitants of diverse environments, exhibit enormous metabolic capabilities to mitigate abiotic stresses. Since microbial interactions with plants are an integral part of the living ecosystem, they are believed to be the natural partners that modulate local and systemic mechanisms in plants to offer defense under adverse external conditions. Plant–microbe interactions comprise complex mechanisms within the plant cellular system (Meena et al. 2017).

11.2.2 Microbial-Induced Response to Stresses

Microbes are the most fundamental living system on earth. Being an important living component of the soil system, naturally, they become an integral part of the crop production system as soon as a seed comes into the soil to start its life cycle. Microorganisms are important inhabitants of seeds also and proliferate as the seeds grow in the soil to form symbiotic associations at the surface or endophytic interactions inside the roots, stems, or leaves. A plant microbiome provides fundamental support to the plants in acquiring nutrients, tolerating stresses, and resisting against diseases (Turner et al. 2013). Work on plant–microbe interactions at biochemical, physiological, and molecular levels established that microbial associations largely direct plant responses toward stresses (Farrar et al. 2014).

11.3 Response of Abiotic Response on Plant

Plants are having an array of protective mechanisms acquired during evolution to combat adverse environmental situations (Yolcu et al. 2016). The adverse climatic condition creates abiotic stresses as the principal limiting factors in declining agricultural productivity (Padgham 2009; Grayson 2013).

A plant follows different mechanisms to deal with different stresses, some are included in Sections 11.3.1–11.3.2.3.

11.3.1 Induced Systemic Tolerance (IST)

Microbe-mediated induction of abiotic stress responses is known as induced systemic tolerance (IST). The role of microorganisms to lessen abiotic stresses in plants has been the area of great concern in the past few decades. Due to their potential intrinsic metabolic and genetic capabilities, microbes contribute to alleviate abiotic stresses in the plants (Meena et al. 2017). The role of several rhizospheric microbes which belongs to the genera *Pseudomonas* (Grichko and Glick 2001; Ali et al. 2009; Sorty et al. 2016), *Azospirillum* (Creus et al. 2004; Omar et al. 2009), *Azotobacter* (Sahoo et al. 2014), *Rhizobium* (Alami et al. 2000; Remans et al. 2008), *Bacillus* (Ashraf et al. 2004; Marulanda et al. 2007; Tiwari et al. 2011; Vardharajula et al. 2011), *Enterobacter* (Grichko and Glick 2001; Nadeem et al. 2014), *Pantoea* (Amellal et al. 1998; Egamberdiyeva and Höflich 2003), *Bradyrhizobium* (Figueiredo et al. 1999; Swaine et al. 2007; Tittabutr et al. 2013), *Burkholderia* (Barka et al. 2006; Oliveira et al. 2009), *Trichoderma* (Ahmad et al. 2015) and cyanobacteria (Singh et al. 2011) in plant growth promotion and mitigation of multiple kinds of abiotic stresses has been documented. Rhizobacteria-induced drought endurance and resilience (RIDER) that includes changes in the levels of phytohormones, defense-related proteins and enzymes, antioxidants, and epoxypolysaccharides have been observed for microbe-mediated plant responses. Such strategies make plants tougher toward abiotic stresses (Kaushal and Wani 2016).

11.3.2 Metabolic Changes in Plants Induced by Microbes During Stress

Plants can sense the changing environmental conditions smartly and can induce defense according to the change. Plants have to cope frequently with the rapidly changing adverse environmental conditions because of their intrinsic metabolic capabilities (Simontacchi et al. 2015). Changing the outside environment could disturb the natural metabolic homeostasis of the plant (Foyer and Noctor 2005), and make the plant needful for some advanced genetic and metabolic mechanisms within its cellular system (Apel and Hirt 2004; Gill and Tuteja 2010). Microbes associated with plants play an important role in the defense against biotic and abiotic stress. Microbes associated with the plant's roots are enormously diverse. The complex microbial communities associated with plant species are cited as the second genome of the plant which is considered to be highly important for the plant's health and development. Recent studies on plant–microbe interaction have revealed that plants shape their rhizospheric microbiome by secreting root exudates (Berendsen et al. 2018). The microbial consortium in the plant system activates the antioxidant enzyme activities and the phenylpropanoid pathway leading to the accumulation of total phenolics, proline, and pathogenesis-related (PR) proteins after the pathogen challenge (Jain et al. 2012).

11.3.2.1 Metabolic Cross-Talk in Plants After Stress Induction

Plants use different approaches in response to stresses which involve an interactive metabolic cross-talk within various biosynthetic pathways. Root architecture is sensitive in sensing soil-borne stress signals and responding accordingly in the soil. The plant microbiome particularly the root microbiota is engaged in the protection against biotic stresses, by going about as a defensive shield against soil-borne pathogens. The components are incorporating different direct communications with plant pathogens and also back-handed associations through the plant by incitement of the immune system of plants. The rhizosphere is occupied by different microorganisms, which also participate in the induction of defense against soil-borne stress (Ansari 2018). *Trichoderma harzianum,* found to play an important role in combating stresses in plants, by upregulation of stress-related genes, namely, dehydrin, malondialdehyde, and aquaporin genes including physiological parameters. Several microbes induce plant responses which altered the level of several polysaccharides, defense proteins, antioxidant enzymes, and phytohormones, for example, RIDER (Pandey et al. 2016).

11.3.2.2 Activation of Antioxidant Mechanism

One of the most universal and earliest resistance reactions against invading pathogens by plants is oxidative burst and is considered to be required for plant defense. Plants regulate the levels of ROS in a way that the levels of ROS such as $H_2O_2$are sensitive to the pathogen but not to the host. H_2O_2 further provides strength to the cell wall through helping the process of lignifications, contributing in cross-linking of cell wall polymers and proline or hydroxyproline-rich proteins during the pathogen-challenged conditions and thereby suppress or limit the growth and development of the invading pathogens directly. Proper selection of microbes for developing a consortium may lead to enhanced antioxidant activities in plants that can not only reduce pathogen damages but also protect plants from potential oxidative damages due to pathogen attack (Cessna et al. 2000).

11.3.2.3 Activation of Systemically Induced Resistance

Systemic acquired resistance (SAR) is an appropriate and suitable form of resistance which increases the level of resistance of host plants for a few months after pathogen challenge. Following the onset of pathogen infection, defense responses start at the place of infection, and systemic response is activated in the entire plant to defend the uninjured tissues and restrict further infection (Pieterse et al. 2009). SAR triggers resistance in the entire foliage and enhances the production of the defense signal molecule salicylic acid (SA) for further signaling (Kachroo and Robin 2013). While the SA pathway is stimulated against biotrophic pathogens, the jasmonic acid (JA) pathway is stimulated against necrotrophic pathogens and herbivores (Thaler et al. 2012). The regulatory protein NPR1 (non-expresser of pathogenesis-related gene 1 protein) is crucial for SA signal transduction and functions as transcriptional co-activator of PR gene expression. It has been demonstrated previously that chitinases and β-1, 3 glucanases (both PR proteins) of different plants restrict the fungal pathogens synergistically (Van Loon 1997). Various BCAs are reported to induce synthesis of these PR proteins in plants particularly under pathogen stress (Sarma et al. 2015).

11.4 Role of Phytohormones in Increasing Abiotic and Biotic Stress Tolerance

Plant roots are heavily colonized with microorganisms (compared to soil and other habitats) because of the rich nutrient component of root exudates (Schlaeppi and Bulgarelli 2015). The rhizosphere is a relatively nutrient-rich environment containing amino acids, sugars, fatty acids, and other organic compounds, which attract microbes (Vorholt 2012) that utilize various nutrients released by the root. In turn, the microbes synthesize biologically active compounds, including phytohormones (auxins, gibberellins cytokinins, and abscisic acid), enzymes, antifungal compounds, and compatible solutes. These microbial metabolites play a vital role in plant growth, nutrition, and development (Ruiz-Lozano et al. 2012). They can stimulate plant growth development, provide resistance to various abiotic and biotic stress factors, improve nutrient acquisition and protect plants from various soil-borne pathogens (Grover et al. 2013; Cho et al. 2015). The microbes mitigate stress responses by regulating the nutritional and hormonal balance in plants and inducing systemic tolerance to stress. One of the mechanisms of improvement of plant growth and stress tolerance by microbes is their phytohormone synthesizing ability in the rhizosphere or root tissue (Etesami et al. 2015). Microbial phytohormones affect the metabolism of endogenous growth regulators in plant tissue and play a major role in changing root morphology upon exposure to drought, salinity, extreme temperature, and heavy metal toxicity (Spaepen et al. 2008; Khan et al. 2011). Root-associated microbes, including free-living, symbiotic, or endophytic microbes, can produce various type of phytohormones and belong to different genera and species (Sgroy et al. 2009).

Following are some examples of phytohormones producing microbes: *Acinetobacter, Enterobacter, Sinorhizobium, Pseudomonas, Bacillus, Rhizobium, and Pantoea,* from halotolerant weed (*Psoralea corylifolia L.*) (Sorty et al. 2016), and *Arthrobacter, Bacillus, Paenibacillus, Enterobacter, Pseudomonas, Cellulosimicrobium, Rhizobium, Brevibacillus, Mycobacterium, Ochrobactrum,* and *Pseudoxanthomonas* associated with soybean root (Egamberdieva et al. 2016).

11.5 Gene Transfer in Plants

Different plants may have different tolerance capacities to biotic or abiotic stresses, growth at different latitudes or altitudes, or nutrient-poor soils, where, in many cases, their microbiome allows them to survive in such conditions (Santoyo et al. 2017). Therefore, the general interest is to transfer these capabilities to plants that lack them, including for the biocontrol of potential pathogens (Wicaksono et al. 2018; Yuan et al. 2016). By transferring one or more microbial species, mainly bacteria, a greater amount of genetic material or gene numbers can be transferred, compared to single gene transfer (i.e. genes coding for *Bacillus thuringiensis* cry toxins) (Romeis et al. 2006), improving several functions at the same time.

11.6 Conclusion and Future Prospects

Microbes play a major role in every living system along with their importance as single entities. The microbiome is supposed to be the "second genome" of different organisms by many groups (plants, animals, and insects). Plant microbiota interactions are highly diverse and several factors shape community assembly and functioning. The plant microbiome includes not only the bacterial communities but also viruses, fungi, archaea, and nematodes which have important effects on host biology. In some cases, the microorganisms may develop forced symbioses with their host. There are several studies that have been done that highlighted the role of the endophytic microbiome in response to abiotic stress. Microorganisms have a tremendous potential to improve stress resilience and health, as well as plant growth. Plant microbiota affects plants by changing phytohormones concentration, secreting different antioxidants, and modification in plant metabolome. Several approaches may be applied in the field to improved crop success, such as designing smart microbial consortia, the selection of agricultural management practices favoring microbiota with beneficial functions, or a new generation of plant breeding approaches. Metabolic cross-talk plays an important role in establishing a mutual relationship between plant roots and plant growth promoting rhizobacteria (PGPR). As plants have been much exposed for their changes by induced biotic stress and plant–microbe interactions, but details regarding associated microorganisms are very less. Therefore, future research should be focused more on the characterization of internal plant microbiota and their impact on gene expression in plants which ultimately changes the functioning of the plant at both a genomics and proteomics level.

References

Ahmad, P., Hashem, A., Abd-Allah, E.F. et al. (2015). Role of *Trichoderma harzianum* in mitigating NaCl stress in Indian mustard (*Brassica juncea* L) through antioxidative defense system. *Front. Plant Sci.* 6: 868.

Alami, Y., Achouak, W., Marol, C. et al. (2000). Rhizosphere soil aggregation and plant growth promotion of sunflowers by an exopolysaccharide-producing Rhizobium sp. Strain isolated from sunflower roots. *Appl. Environ. Microbiol.* 66 (8): 3393–3398.

Ali, S.Z., Sandhya, V., Grover, M. et al. (2009). *Pseudomonas sp.* strain AKM-P6 enhances tolerance of sorghum seedlings to elevated temperatures. *Biol. Fertil. Soils* 46 (1): 45–55.

Amellal, N., Burtin, G., Bartoli, F. et al. (1998). Colonization of wheat roots by an exopolysaccharide-producing *Pantoeaagglomerans* strain and its effect on rhizosphere soil aggregation. *Appl. Environ. Microbiol.* 64 (10): 3740–3747.

Amtmann, A., Troufflard, S., and Armengaud, P. (2008). The effect of potassium nutrition on pest and disease resistance in plants. *Physiol. Plant.* 133 (4): 682–691.

Ansari, M.I. (2018). Plant microbiome and its functional mechanism in response to environmental stress. *Int. J. Green Pharm.* 12 (01).

Apel, K. and Hirt, H. (2004). Reactive oxygen species: metabolism, oxidative stress, and signal transduction. *Annu. Rev. Plant Biol.* 55: 373–399.

Ashraf, M., Hasnain, S., Berge, O. et al. (2004). Inoculating wheat seedlings with exopolysaccharide-producing bacteria restricts sodium uptake and stimulates plant growth under salt stress. *BiolFert Soils.* 40 (3): 157–162.

Atkinson, N.J. and Urwin, P.E. (2012). The interaction of plant biotic and abiotic stresses: from genes to the field. *J. Exp. Bot.* 63 (10): 3523–3543.

Barka, E.A., Nowak, J., and Clément, C. (2006). Enhancement of chilling resistance of inoculated grapevine plantlets with a plant growth-promoting rhizobacterium, *Burkholderia phytofirmans* strain PsJN. *Appl. Environ. Microbiol.* 72 (11): 7246–7252.

Berendsen, R.L., Vismans, G., Yu, K. et al. (2018). Disease-induced assemblage of a plant-beneficial bacterial consortium. *ISME J.* 12 (6): 1496.

Cessna, S.G., Sears, V.E., Dickman, M.B. et al. (2000). Oxalic acid, a pathogenicity factor for *Sclerotinia sclerotiorum*, suppresses the oxidative burst of the host plant. *Plant Cell* 12 (11): 2191–2199.

Cho, S.T., Chang, H.H., Egamberdieva, D. et al. (2015). Genome analysis of *Pseudomonas fluorescens* PCL1751: a rhizobacterium that controls root diseases and alleviates salt stress for its plant host. *PLoS One* 10 (10).

Creus, C.M., Sueldo, R.J., and Barassi, C.A. (2004). Water relations and yield in *Azospirillum*-inoculated wheat exposed to drought in the field. *Can. J. Bot.* 82 (2): 273–281.

Egamberdieva, D., Wirth, S., Behrendt, U. et al. (2016). Biochar treatment resulted in a combined effect on soybean growth promotion and a shift in plant growth promoting rhizobacteria. *Front. Microbiol.* 7: 209.

Egamberdiyeva, D. and Höflich, G. (2003). Influence of growth-promoting bacteria on the growth of wheat in different soils and temperatures. *Soil Biol. Biochem.* 35 (7): 973–978.

Etesami, H., Alikhani, H.A., and Hosseini, H.M. (2015). Indole-3-acetic acid (IAA) production trait, a useful screening to select endophytic and rhizosphere competent bacteria for rice growth promoting agents. *MethodsX* 2: 72–78.

Farrar, K., Bryant, D., and Cope-Selby, N. (2014). Understanding and engineering beneficial plant-microbe interactions: plant growth promotion in energy crops. *Plant Biotechnol. J.* 12 (9): 1193–1206.

Figueiredo, M.V.B., Vilar, J.J., and Burity, H.A. (1999). Alleviation of water stress effects in cowpea by *Bradyrhizobium* spp. inoculation. *Plant and Soil* 207 (1): 67–75.

Foyer, C.H. and Noctor, G. (2005). Oxidant and antioxidant signaling in plants: a re-evaluation of the concept of oxidative stress in a physiological context. *Plant Cell Environ.* 28 (8): 1056–1071.

Gill, S.S. and Tuteja, N. (2010). Reactive oxygen species and antioxidant machinery in abiotic stress tolerance in crop plants. *Plant Physiol. Biochem.* 48 (12): 909–930.

Grayson, M. (2013). Agriculture and drought. *Nature* 501 (7468): S1–S1.

Grichko, V.P. and Glick, B.R. (2001). Amelioration of flooding stress by ACC deaminase-containingplant growth-promoting bacteria. *Plant Physiol. Biochem.* 39 (1): 11–17.

Grover, A., Mittal, D., Negi, M., and Lavania, D. (2013). Generating high temperature tolerant transgenic plants: achievements and challenges. *Plant Sci.* 205: 38–47.

Jain, A., Singh, S., Kumar Sarma, B. et al. (2012). Microbial consortium–mediated reprogramming of defense network in pea to enhance tolerance against *Sclerotinia sclerotiorum*. *J. Appl. Microbiol.* 112 (3): 537–550.

Kachroo, A. and Robin, G.P. (2013). Systemic signaling during plant defense. *Curr. Opin. Plant Biol.* 16 (4): 527–533.

Kaushal, M. and Wani, S.P. (2016). Plant-growth-promoting rhizobacteria: drought stress alleviators to ameliorate crop production in drylands. *Ann. Microbiol.* 66 (1): 35–42.

Khan, A.L., Hamayun, M., Kim, Y.H. et al. (2011). Gibberellins producing endophytic *Aspergillus fumigatus sp.* LH02 influenced endogenous phytohormonal levels, isoflavonoids production and plant growth in salinity stress. *Process Biochem.* 46 (2): 440–447.

Marulanda, A., Porcel, R., Barea, J.M. et al. (2007). Drought tolerance and antioxidant activities in lavender plants colonized by native drought-tolerant or drought-sensitive *Glomus* species. *Microb. Ecol.* 54 (3): 543.

Meena, K.K., Sorty, A.M., Bitla, U.M. et al. (2017). Abiotic stress responses and microbe-mediated mitigation in plants: the omics strategies. *Front. Plant Sci.* 8: 172.

Mittler, R. and Blumwald, E. (2010). Genetic engineering for modern agriculture: challenges and perspectives. *Annu. Rev. Plant Biol.* 61: 443–462.

Nadeem, S.M., Ahmad, M., Zahir, Z.A. et al. (2014). The role of mycorrhizae and plant growth promoting rhizobacteria (PGPR) in improving crop productivity under stressful environments. *Biotechnol. Adv.* 32 (2): 429–448.

Oliveira, C.A., Alves, V.M.C., Marriel, I.E. et al. (2009). Phosphate solubilizing microorganisms isolated from rhizosphere of maize cultivated in an oxisol of the Brazilian Cerrado Biome. *Soil Biol. Biochem.* 41 (9): 1782–1787.

Omar, M.N.A., Osman, M.E.H., Kasim, W.A. et al. (2009). Improvement of salt tolerance mechanisms of barley cultivated under salt stress using *Azospirillum brasilense*. In: *Salinity and Water Stress* (eds. M. Ashraf, M. Ozturk and H.R. Athar), 133–147. Dordrecht: Springer.

Padgham, J. (2009). Agricultural development under a changing climate: opportunities and challenges for adaptation. World Bank.

Pandey, V., Ansari, M.W., Tula, S. et al. (2016). Dose-dependent response of Trichoderma harzianum in improving drought tolerance in rice genotypes. *Planta* 243 (5): 1251–1264.

Pieterse, C.M., Leon-Reyes, A., Van der Ent, S. et al. (2009). Networking by small-molecule hormones in plant immunity. *Nat. Chem. Biol.* 5 (5): 308.

Remans, R., Ramaekers, L., Schelkens, S. et al. (2008). Effect of Rhizobium–Azospirillumcoinoculation on nitrogen fixation and yield of two contrasting Phaseolus vulgaris L. genotypes cultivated across different environments in Cuba. *Plant Soil.* 312 (1–2): 25–37.

Romeis, J., Meissle, M., and Bigler, F. (2006). Transgenic crops expressing *Bacillus thuringiensis* toxins and biological control. *Nat. Biotechnol.* 24 (1): 63.

Ruiz-Lozano, J.M., Porcel, R., Azcón, C. et al. (2012). Regulation by arbuscular mycorrhizae of the integrated physiological response to salinity in plants: new challenges in physiological and molecular studies. *J. Exp. Bot.* 63 (11): 4033–4044.

Sahoo, R.K., Ansari, M.W., Dangar, T.K. et al. (2014). Phenotypic and molecular characterization of efficient nitrogen-fixing *Azotobacter* strains from rice fields for crop improvement. *Protoplasma* 251 (3): 511–523.

Santoyo, G., Pacheco, C.H., Salmerón, J.H. et al. (2017). The role of abiotic factors modulating the plant-microbe-soil interactions: toward sustainable agriculture. A review. *Span. J. Agric. Res.* 15 (1): 13.

Sarma, B.K., Yadav, S.K., Singh, S. et al. (2015). Microbial consortium-mediated plant defense against phytopathogens: readdressing for enhancing efficacy. *Soil Biol. Biochem.* 87: 25–33.

Schlaeppi, K. and Bulgarelli, D. (2015). The plant microbiome at work. *Mol. Plant-Microbe Interact.* 28 (3): 212–217.

Sgroy, V., Cassán, F., Masciarelli, O. et al. (2009). Isolation and characterization of endophytic plant growth-promoting (PGPB) or stress homeostasis-regulating (PSHB) bacteria associated to the halophyte *Prosopis strombulifera*. *Appl. Microbiol. Biotechnol.* 85 (2): 371–381.

Simontacchi, M., Galatro, A., Ramos-Artuso, F. et al. (2015). Plant survival in a changing environment: the role of nitric oxide in plant responses to abiotic stress. *Front. Plant Sci.* 6: 977.

Singh, D.P., Prabha, R., Yandigeri, M.S. et al. (2011). Cyanobacteria-mediated phenylpropanoids and phytohormones in rice (*Oryza sativa*) enhance plant growth and stress tolerance. *Antonie van Leeuwenhoek* 100 (4): 557–568.

Sorty, A.M., Meena, K.K., Choudhary, K. et al. (2016). Effect of plant growth promoting bacteria associated with halophytic weed (*Psoralea corylifolia* L.) on germination and seedling growth of wheat under saline conditions. *Appl. Biochem. Biotechnol.* 180 (5): 872–882.

Spaepen, S., Dobbelaere, S., Croonenborghs, A. et al. (2008). Effects of *Azospirillum brasilense* indole-3-acetic acid production on inoculated wheat plants. *Plant and Soil* 312 (1–2): 15–23.

Suzuki, N., Rivero, R.M., Shulaev, V. et al. (2014). Abiotic and biotic stress combinations. *New Phytol.* 203 (1): 32–43.

Swaine, E.K., Swaine, M.D., and Killham, K. (2007). Effects of drought on isolates of *Bradyrhizobium elkanii* cultured from *Albizia adianthifolia* seedlings of different provenances. *Agr. Syst.* 69 (2): 135–145.

Thaler, J.S., Humphrey, P.T., and Whiteman, N.K. (2012). Evolution of jasmonate and salicylate signal crosstalk. *Trends Plant Sci.* 17 (5): 260–270.

Tittabutr, P., Piromyou, P., Longtonglang, A. et al. (2013). Alleviation of the effect of environmental stresses using co-inoculation of mungbean by *Bradyrhizobium* and rhizobacteria containing stress-induced ACC deaminase enzyme. *Soil Sci. Plant Nutr.* 59 (4): 559–571.

Tiwari, S., Singh, P., Tiwari, R. et al. (2011). Salt-tolerant rhizobacteria-mediated induced tolerance in wheat (*Triticum aestivum*) and chemical diversity in rhizosphere enhance plant growth. *Biol. Fert. Soils* 47 (8): 907.

Turner, T.R., Ramakrishnan, K., Walshaw, J. et al. (2013). Comparative metatranscriptomics reveals kingdom level changes in the rhizosphere microbiome of plants. *ISME J.* 7 (12): 2248.

Van Loon, L.C. (1997). Induced resistance in plants and the role of pathogenesis-related proteins. *Eur. J. Plant Pathol.* 103 (9): 753–765.

Vardharajula, S., Zulfikar Ali, S., Grover, M. et al. (2011). Drought-tolerant plant growth promoting Bacillus spp.: effect on growth, osmolytes, and antioxidant status of maize under drought stress. *J. Plant Interact.* 6 (1): 1–14.

Vorholt, J.A. (2012). Microbial life in the phyllosphere. *Nat. Rev. Microbiol.* 10 (12): 828.

Wicaksono, W.A., Jones, E.E., Casonato, S. et al. (2018). Biological control of Pseudomonas syringaepv. actinidiae (Psa), the causal agent of bacterial canker of kiwifruit, using endophytic bacteria recovered from a medicinal plant. *J. Biol. Control* 116: 103–112.

Yolcu, S., Ozdemir, F., Güler, A. et al. (2016). Histone acetylation influences the transcriptional activation of POX in *Beta vulgaris* L. and *Beta maritima* L. under salt stress. *Plant Physiol. Biochem.* 100: 37–46.

Yuan, Z., Druzhinina, I.S., Labbé, J. et al. (2016). Specialized microbiome of a halophyte and its role in helping non-host plants to withstand salinity. *Sci. Rep.* 6: 32467.

12

Bioprospecting

At the Interface of Plant and Microbial Communities

Madan L. Verma[1], *Varsha Rani*[2], *Reena Kumari*[3], *Deepka Sharma*[3], *Sanjeev Kumar*[4], *and Rekha Kushwaha*[5]

[1] *Department of Biotechnology, Indian Institute of Information Technology, Una, Himachal Pradesh, India*
[2] *Department of Biotechnology, Shoolini University, Solan, Himachal Pradesh, India*
[3] *Department of Biotechnology, College of Horticulture and Forestry, Dr. YSP University of Horticulture and Forestry, Hamirpur, Himachal Pradesh, India*
[4] *Department of Basic Science, College of Horticulture and Forestry, Dr. YSP University of Horticulture and Forestry, Hamirpur, Himachal Pradesh, India*
[5] *University of Missouri, Columbia, MO, USA*

12.1 Introduction

An increase in population and urbanization has created stress on the already present resources to fulfill their demand for food, animal feed, and other energy sources. Presently bioprospecting is the alternative solution which is not only used to search the animal and plant species, or used to extract the commercially valuable compounds but also develops new sources of microorganisms, macro-organisms, chemical compounds, and genes, as well as other useful products (Strobel and Daisy 2003; Tikhonovich and Provorov 2011). These resources or compounds can be important in one way or the other way to the human race. It is well-known that the plant and its microbiome have a strong symbiotic relationship (Bulgarelli et al. 2012; Hirsch and Mauchline 2012; Lundberg et al. 2012). The microbiome helps the host plant to tolerate the stress conditions, provide defense against the pathogens and support the development, growth, and supply of nutrients (Berg 2009; Bragina et al. 2014; Grube et al. 2015). There are numerous examples of successful bioprospecting; one of the best examples is of opium poppy for which morphine, a powerful pain reliever is extracted. Similarly, white willow tree, spoiled sweet clover, Madagascar periwinkle from which aspirin (a pain reliever and lifesaver), warfarin (an anticoagulant [blood thinner]) and vinblastine, an anticancerous drug, simultaneously are being prospected. If well managed, bioprospecting can be advantageous since, in addition to generating new products, it can generate income and provide an incentive for the conservation of biodiversity. However, if not managed properly, it can create environmental problems due to over-exploitation or habitat destruction, and socioeconomic issues related to unfair sharing of benefits. Bioprospecting consists of mainly four phases: sample collection; culturing of organisms, specific compounds isolation

Phytomicrobiome Interactions and Sustainable Agriculture, First Edition. Edited by Amit Verma, Jitendra Kumar Saini, Abd El-Latif Hesham, and Harikesh Bahadur Singh.

as well as characterization; screening for specified uses; development of products, and their commercialization, including trials, patenting, marketing, and sales.

Plants have been recognized from the past of having a close relationship with microorganisms like bacteria, archaea, protists, and fungi which facilitates the growth of plant and productivity (Berendsen et al. 2012; Buee et al. 2009; Lindow and Brandl 2003; Thrall et al. 2007; Vorholt 2012). These interactions are not in a particular direction but the microbial associates are provided with novel metabolic capabilities by their host plant. For instance, bacterial communities showed important biocontrol potential in the soil which is infected. In the infected soil, surrounding microbiota protects the plants from the pathogenic microbes. Santhanam et al. (2015) reported that bacteria that are associated with roots rescued *Nicotiana attenuata* from the wilt disease. Plant and bacterial relationships have been studied intensively which benefits directly the health of plant hosts. Rascovan et al. (2016) found in habituation of certain bacteria promotes the plant growth in soybean and wheat plant roots. Some bacteria modulate the hormones levels in plants and induce systemic resistance in the plants (Berg and Koskella 2018).

This chapter aims to discuss the plant-associated microbial communities, beneficial effects from the microorganisms which are associated with the plants and role of microbial processing (signals) for facilitation the plant growth.

12.2 Plant-Associated Microbial Communities

Plants are being evolved with a plethora of microorganisms which are playing a very important role in the growth and health of the plant. Because of the interesting capabilities of the functionality of microbiota residing with plants, it is needed to research new microbes so that the current challenge of crop production has been achieved. Production of crops is facing challenges like demographic development, production in a sustainable way, climate changes, etc. Microorganisms have the capability to be used as biopesticides or as biofertilizers, so the demand to use these microorganisms is increasing instead of using chemicals in agricultural practices (Mendes et al. 2013; Mitter et al. 2016). Recruitment of the microorganisms via plants takes place from the surrounding reservoir of microbes such as phyllosphere (leaf surface concerning the external environment), soil/rhizosphere, spermosphere (exterior of germinated seed), anthosphere (flowers external environment) and carposphere (external fruit environment) (Hardoim et al. 2015). Horizontal transfer of root microbiota takes place most of the time, containing highly diverse microbes that are dominated by Verrucomicrobia, Acidobacteria, Proteobacteria, Actinobacteria, Planctomycetes, and Bacteroidetes. However, vertical transfer of the bacteria may also take place via seeds. Seed, proliferating in the roots of the developing plant, represents an important source of the microorganisms (Liu et al. 2012; Hardoim et al. 2012). The root system of plants provides a unique ecological niche for the microorganisms which colonize the roots, rhizosphere, and above-ground parts. The rhizosphere is one, out of the most composite ecosystems, and is considered as a hot spot for the activity of microbes (Hiltner 1904). Plant productivity has been regulated by the soil microbes, especially in the ecosystems which are poor in the nutrients. Plant productivity has also been strongly regulated by the free-living microbes through mineralization. Microorganisms present in the soil play an important role in different ecosystems and influence many ecological procedures, such as the acquisition of nutrients (Smith and Read 1997; Sprent 2001), cycling of carbon (Hogberg et al. 2001), soil formation (Rillig and Mummey 2006) and nitrogen cycling (Kowalchuk and Stephen 2001; Tiedje 1988).

Root-associated microorganisms are producing direct effects on the plants via forming mutualistic or pathogenic relationships and root microbiota is also affecting plants indirectly, via the action of free-living microbes which are altering the nutrient supply rate and resource partitioning. Microorganisms residing in the soil live symbiotically with the plants and enhance the productivity in plants via supplying nutrients in limited concentrations to the plants. One of the best-studied plant symbiotic associations is with the nitrogen-fixing bacteria, which is converting atmospheric nitrogen into ammonia (Sprent 2001). Nitrogen, along with potassium and phosphorous, serves as the main element for plant productivity so nitrogen-fixing bacteria act as very important regulators for plant productivity which can transform the atmospheric nitrogen into ammonia so that plants can utilize nitrogen (Chapin 1980). Nitrogen-fixing bacteria contribute toward plant productivity in the grasslands, tropical savannah, and legumes-dominated tropical forests. Twenty percent of the nitrogen required by the plants annually has been contributed by the nitrogen-fixing bacteria residing in the legumes (Van der Heijden et al. 2006).

Microorganisms associated with plants fulfill important functions for promoting health and growth in plants. Microorganisms are promoting plant growth via hormonal stimulation and nutrient acquisition. Microbiota of plants has diverse mechanisms for suppressing the plant pathogens which is indirectly related to plant growth. *Rhizobium* and *Azospirillum* are among the most studied bacterial genera for the promotion of plant growth. Bacteria like *Bacillus, Pseudomonas, Serratia, Stenotrophomonas, Streptomyces;* and fungi, like *Trichoderma, Coniothyrium,* and *Ampelomyces,* are the model organisms, influencing the plant health. Microbial inoculants can be developed based on these beneficial plant–microbe interactions to be used in agricultural biotechnology. Microorganisms can be utilized as plant strengtheners, biofertilizers, biopesticides, and phytostimulators (Figure 12.1).

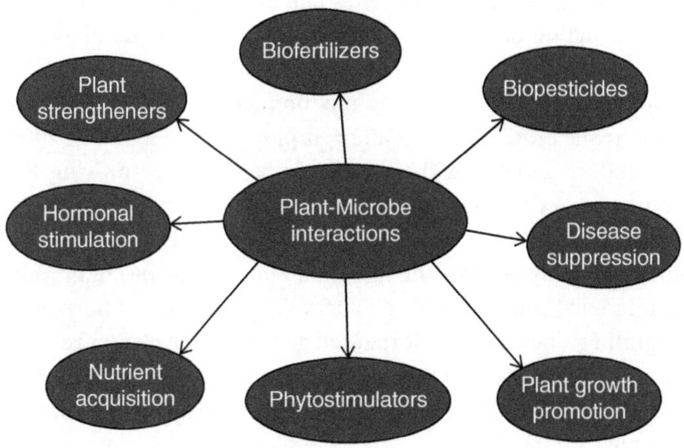

Figure 12.1 Plant microbiota: promoting plant growth and health, suppressing plant pathogens, act as plant strengtheners, phytostimulators, biopesticides, and biofertilizers.

Successful plant–microbe interactions can be achieved by the competency of microorganisms to set up colonies in the plant habitats (Lugtenberg et al. 2002; Kamilova et al. 2005). Colonization follows some steps like adherence, recognition, invasion in the form of endophytes or pathogens, growth, and colonization, followed by various strategies for establishing the interactions. Roots of the plants start cross-talks with the microorganisms present in the soil by the production of signals which can be recognized by the microorganisms. After receiving the signal from the roots, microbes also produce a signal in return, to start the process to colonize (Bais et al. 2006). Motile organisms have given preference to participate in these kinds of cross-talks (Lugtenberg et al. 2002). Duration, intensity and the result of plant–microbial interactions is highly affected by the microbial confirmation (Danhorn and Fuqua 2004).

Endophytic bacteria colonize the interior of the plants without producing any negative or harmful effects on the host plant (Schulz and Boyle 2006). The primary site for entering the endophytes inside the host is via roots except for the endophytes residing the seeds (Pan et al. 1997; Germaine et al. 2004). Endophytes colonize the plant via the vascular system or the apoplast or can exist inside the plant in the tissues such as the xylem or the cortex of the root (Mahaffee et al. 1997; Quadt-Hallmann et al. 1997). Endophytic bacteria interact with their respective host plants as compared to the phyllosphere and rhizosphere bacteria. Plants provide nutrients and space to live in, for the endophytes. while in return, plant growth, as well as health, is improved (Mastretta et al. 2006). Endophytes save their host by preventing the growth of plant pathogens via competing for nutrients and space, inducing plant defense mechanisms, hydrolytic enzymes production, antibiosis, and by inhibiting the enzymes or toxins produced by the pathogens. Besides, endophytes are also being employed in the process of phytoremediation very efficiently. Plants can further benefit from the endophytes, during the process of phytoremediation of organic contaminants via degradative pathways and capabilities of metabolism which lead to the degradation of contaminants very efficiently and reduce the phytotoxicity as well as evapotranspiration of the volatile compounds. Endophytes possess a resistance system to metals which can reduce the phytotoxicity of metals and translocate metal to the parts of plants that are present above the ground. As depicted in Figure 12.2, endophytes can also degrade the contaminants and improve the metal extraction process, thus improving the phytoremediation of mixed pollutants.

Engineered endophytes can be used to improve the phytoremediation process and to complement the host plant metabolic properties. The yellow lupine plant (Barac et al. 2004) and poplar (Taghavi et al. 2005), inoculated with endophytic bacteria, are proving this concept of engineered endophytes via degrading toluene. Toluene phytotoxicity as well as toluene evapotranspiration has been decreased significantly (Taghavi et al. 2005).

Inoculated strains will not necessarily become the integrated portion of the community of endophytes. Endogenous microbial communities can adapt via the process of horizontal gene transfer instead of integrating a new strain. Horizontal gene transfer plays a key role in adapting the communities of microorganisms toward the factors responsible for environmental stress, including endophytic (Taghavi et al. 2005) and rhizospheric communities (Devers et al. 2005; Ronchel et al. 2000).

Environmentally sustainable and sound production of crops is the major challenge of the twenty-first century. As the human population is increasing, industrial processes demand

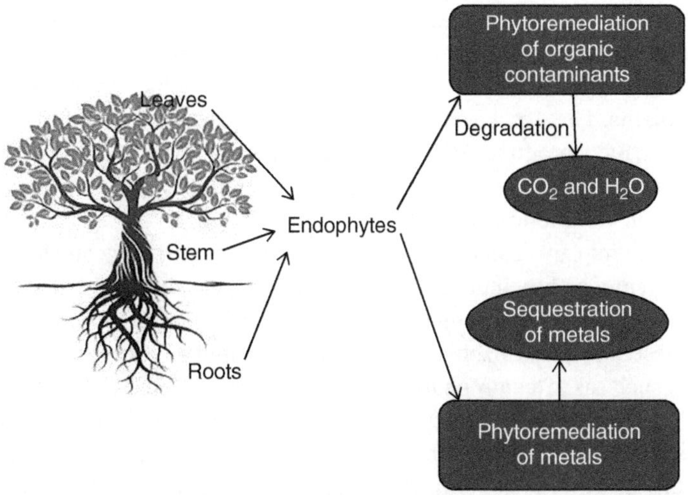

Figure 12.2 Role of endophytes in the process of phytoremediation: endophytes present in the plant parts, like stem, leaves, and roots, help in the process of phytoremediation of organic contaminants and metals. *Source:* Based on Weyens et al. (2009).

enhanced production of the crops. Improper use of chemicals, pesticides, and fertilizers are posing a threat to the environment as well as to our health (Gunnell et al. 2007; Leach and Mumford 2008).

Furthermore, endemic, emerging, and re-emerging pathogens of plants are continuously challenging the ability to safeguard the health and growth of plants (Miller et al. 2009). So, it is needed to develop ecologically sound and compatible strategies to be applied in the field of agriculture. Beneficial plant–microbe interactions fulfill important ecological functions of soil as well as plants (Smith et al. 1999). Plant-associated microorganisms enhance nutrient availability, stress tolerance ability, and the disease resistance power of the host plants, thus promoting the biodiversity (Lugtenberg et al. 2002; Morrissey et al. 2004). Among all the microenvironments associated with plants, rhizospheres are the regions where microbes colonize in abundance (Berg et al. 2005). Generally, the proportion of the microbial isolates which expresses plant growth–promoting traits was found to be two-thirds of the cultivated population which is very high (Fürnkranz et al. 2009). Promotion of growth in plants can be attained directly via beneficial interaction between the host plant and microbes and indirectly by the antagonistic activity of microbes against the plant pathogens.

Microbial inoculants have several advantages over the chemically synthesized fertilizers and pesticides, such as microbial inoculants, are much safer, showed targeted activity, lesser damage to the environment, show effectiveness in smaller quantities, decomposes very quickly as compared to the chemically synthesized pesticides, multiply themselves but under the control of plant and by the indigenous population of microorganisms, reduce resistance development and can be utilized in the pest management systems (Berg 2009). The balance of hormones in the plant can also be affected by the microorganisms associated with that plant. An important example of hormonal balance is ethylene. Ethylene has been reported to promote the growth of several plant species including *Arabidopsis thaliana* at the lower level but normally, ethylene is considered as an inhibitor of the growth of the plant (Pierik et al. 2006). 1-Aminocyclopropane-1-carboyclic acid (ACC) is a compound, acting like an ethylene precursor in the case of plants. The uptake as well as the level of ACC is reduced in the roots because of bacteria which produce ACC deaminase degrade ACC (Glick 2005). Ethylene can also act as a stress hormone so the bacteria which produce ACC deaminase protects the plants by abiotic and biotic stresses (Saleem et al. 2007).

Direct promotion of growth in plants can be achieved via improved nutrient acquisition. Macro as well as micronutrients can be supplied by microorganisms associated with plants. An important example of plant growth-promoting rhizobacteria (PGPR) is a symbiotic association between the host plant and rhizobia. Nitrogen fixation ability is also present in the free-living bacteria such as *Stenotrophomonas, Burkholderia,* and *Azospirillum* (Dobbelare et al. 2003). Oxidation of bacteria can also provide another nutrient, namely sulfate to the plant (Banerjee and Yesmin 2002). Bacteria liberate phosphorous from organic compounds like phytates promote the growth of plants (Unno et al. 2005). Inorganic nutrients which are poorly soluble can be made available to the plant via the bacterial siderophores. Werra et al. (2009) reported the ability of *Pseudomonas fluorescens* CHA0 to solubilize mineral phosphate and acidify its environment. Multifunctionality in the terrestrial ecosystems is positively correlated with the microbial biodiversity and strongly influences the important ecological processes which are highly important for our society's future (Delgado-Baquerizo et al. 2016). Forests act as essential biomes for the global biogeochemical cycling and the microorganisms residing below the ground are playing a key role in providing relevant ecosystem services (Llado et al. 2018).

The warm climate and other changes in the rainfall are the phenomena that require the understanding of surface water storage and land surface routes. Availability as well as the water movement inside an ecosystem act as a control on the geophysical and biological activity and also influence the feedback in the climate. Hydrophobicity of the soil and water repellency is the key phenomenon affecting the infiltration of water into the land surface. It has been reported by Seaton et al. (2019) that water repellency of the soil is driven by the changes in microbial communities of the soil and plant, responding to the stresses of the environment. Stress tolerance of the soil community and plant microbial community changes are related to the repellency changes as compared to the soil physiochemical properties (Seaton et al. 2019). Thus, soil microbial communities are playing a key role in the maintenance of the water repellency in the soil.

12.3 Beneficial Effects of Plant-Associated Microbial Communities

All the plants on earth characterized morphologically into two systems that are shoot and root. The microorganisms interact with plants through either the rhizosphere or phyllosphere. The rhizosphere consists of roots of the plants and the soil surrounding the roots. In the rhizosphere, many important interactions occur between the soil, plant, and microbes. The rhizospheric microorganisms interact with the plants and thus help improve plant health and growth (Hartmann et al. 2009). The competition for colonization with diverse plant habitats is important for the successful plant–microbe interaction (Kamilova et al. 2005). Highlights of different beneficial plant–microbial interactions are discussed in the following sections.

12.3.1 Rhizoremediation

This is a type of rhizosphere phytoremediation that can be used to remove organic pollutants or metals from the soil. The removal of metals from the environment is more tedious than the removal of organic pollutants as that could be degraded inside the plant system or in the soil (Glick 2010; Thijs 2015; Weyens et al. 2009). Mycorrhizal fungi, free-living as well as endophytic fungi, been reported to enhance plant growth and development and their degradative capabilities to remove the organic pollutants (Coninx et al. 2017; Deng and Cao 2017; Lenoir et al. 2016; Thijs 2015). *Pseudomonas* sp. is known to release the different enzymes which are responsible or having a role in the degradation of complex organic compounds. Through molecular studies, it was reported that these species contain a large size of plasmids (~ 50–200 kb) having genes coded for the enzymes involved in biodegradative pathways. In many studies, the contamination of soil by petroleum products has been removed by the combined action of plants with biodegradative bacteria, a variety of halogenated compounds, polycyclic and other aromatic hydrocarbons.

In many studies, it was reported that in the metal-contaminated site, rhizobacteria is present on the rhizosphere, which enhanced the growth of plants by producing growth hormones, secondary metabolites, and other compounds like biosurfactants, osmolytes, organic acids, metal chelating siderophores, nitric oxide, enzymes, and antibiotics. The synthesis of ACC deaminase via these beneficial compounds decreases the level of ethylene production, provides resistance against biotic and abiotic stress, and increases the uptake of N, P and K. (Rajkumar et al. 2012; Ma et al. 2015; Ullah et al. 2015). Many microbes have been reported to present inside the plants without affecting the host. These bacteria help the plant to resist the stress caused by different biotic and abiotic factors (Ma et al. 2011, 2015). These endophytes make the plant tolerant to high metal concentration and help in the enhancement of their growth through biocontrol mechanism and induced resistance against phytopathogens (Ma et al. 2011, 2015).

12.3.2 Plant Growth-Promoting Rhizobacteria (PGPR)

Different PGPR strains have been utilized successfully for the process of crop inoculations. These comprise members of the bacterial genera *Bacillus*, *Rhizobium*, *Azospirillum*, *Pseudomonas*, *Serratia*, *Streptomyces* and *Stenotrophomonas*. Some fungi belonging to the genera *Coniothyrium*, *Ampelomyces*, and *Trichoderma* are found to be beneficial for the host plant (Harman et al. 2004). PGPR uses complex mechanisms for the promotion of plant growth, protection, and development. Phytostimulation (plant growth promotion, usually by producing the phytohormones), biocontrol (controlling diseases, mainly by producing antifungal metabolites and antibiotics, lytic enzymes, and by inducing plant defense responses) and bio fertilization (increasing the nutrients available to the plant) are important among them. The most commonly studied PGPR are *Bacillus* and *Pseudomonas* genera which dominated the bacterial groups in the rhizosphere.

Production of phytohormones like cytokinin, auxin, and gibberellins is the main function of the plant growth–promoting bacteria (Garcia de Salamone et al. 2005). Plant growth–promoting bacteria which possess enzyme, ACC deaminase has been reported by Ma et al. 2003, Madhaiyan et al. 2006, and Mayak et al. 2004. Ethylene hormone in plants has been found to play an important role in root initiation and elongation, nodulation, abscission, senescence, and ripening as well as in stress tolerance.

12.3.3 Biotic and Abiotic Stress Resistance

The interactions among microbial communities and plants are very complex. The microbial association with plants is not only useful for improving plant growth under normal conditions but also protects plants from the adverse environment by promoting plant growth under stress conditions. Several bacterial strains have been reported that cause significant effects on the development and growth of the plant under stressed conditions including drought, salinity, heavy metal, temperature, and pathogen (Belimov et al. 2005; Trivedi and Sa 2008; Singh et al. 2010; Nadeem et al. 2016; Numan et al. 2018). Mycorrhizae fungi facilitate the plant growth under various kinds of stresses by mechanisms like enhancing antioxidant system and osmolytes production in addition to the supply of nutrients to the host plant (Nadeem et al. 2013; Habibzadeh 2015).

Mycorrhizae is the fungal association with the roots of plants used in horticulture and forestry (Bonfante and Genre 2010; Delavaux et al. 2019). It helps plants to increase nutrients and water uptake and resistance against various pathogens (Ziedan et al. 2011; Van der Heijden et al. 2017). Some fungal hyphae invade the cortical cells of roots and form specialized feeding hyphae called arbuscules, however, sometimes foods storing hyphal swelling are formed as vesicles. Endomycorrhizal fungi were reported to absorb nutrients from the soil and deliver these to the plant via branched arbuscule (Parniske 2008). A *Trichoderma* fungus was reported to produce antibiotics and act as a biocontrol agent to provide resistance against various pathogens (Harman et al. 2004; Srinivas et al. 2017). *Trichoderma* sp. also affects the plant genes involved in ethylene biosynthesis. Mycorrhizal fungi were reported to minimize the release of nitrous oxide (N_2O) from the soil and reduces the risk of depletion of theozone layer by N_2O thereby reduces global warming (Bender et al. 2014).

Crop destructive disease management had been controlled by beneficial plant microbial interactions (Romeralo et al. 2015). Plant–AMF associations have been reported to improve the tolerance from the pathogenic infections (Bach et al. 2016). The AMF has also reported having biocontrol potential against different kinds of nematodes (Vos et al. 2012). Amendment of bioorganic fertilizer by using *Trichoderma harzianum* and *Glomus mosseae* alone was reported to reduce the significant level of *Ralstonia solanacearum* in the soil present in the rhizosphere. Although the combined use of (*T. harzianum* with *G. mosseae*) was found to show a high level of induced systemic resistance and has the potential to increase the host plant biomass (Yuan et al. 2016).

Potassium and phosphate solubilizing bacteria were reported to increase uptake of phosphorus and potassium by the plant. Besides these nutrients, the availability of calcium was reported to increase in the plants inoculated with PGPRs. This PGPR was also reported to provide resistance against various pathogens by limiting the nutrients available to the plants or via breaking down the cell wall of pathogens. In soil, there is limited availability of various trace elements like iron, copper, zinc, manganese, etc. Among all these trace elements the availability of iron in its bioavailable form is essential to plant growth. Siderophores are the compounds having lower molecular weight, produced by the microorganisms when the iron is in a limited amount. Siderophores act like a ferric ion transport vehicle with high membrane permeability into the cells of microbes (Butler and Theisen 2010). Hydrocarboxylic acid, catechols, and hydroxamic acid are the common iron-binding substances for the siderophores (Ahmed and Holmström 2014). Siderophores are benefiting the plants by supplying iron and also suppresses the soil-borne diseases (Tank et al. 2012). *Pseudomonas putida* has been found to overexpress siderophores more efficiently against *Fusarium* wilt in the case of a tomato plant. Siderophores produced by *Bacillus subtilis* exhibited antagonistic effects on the wilt and dry root rot causal fungus in chickpea (Patil et al. 2014).

These plant growth–promoting bacteria also provides the plant's resistance against various diseases by producing various enzymes like chitinases, cellulases, glucanases, proteases, and lipases which were reported to degrade the cell wall of pathogens (Beneduzi et al. 2012). In plant–bacteria interactions, it was also reported to release antimicrobial and antioxidative substances which strengthen the plant resistance against microbial infection (Van loon 2007).

Omics studies can be used to understand the plant microbial interaction and its impact on crop improvement studies. Timmusk and Wagner (1999) reported that inoculation of *Paenibacillus polymyxa* B2 with *A. thaliana* plants resulted in resistance against the drought conditions as compared to the control plant. In drought stress conditions the expression of genes encoding for the drought stress, viz. dehydrin-like protein, early nodulin ENOD18, S-adenosylmethionine synthetase, pathogenesis-related protein, vacuolar H+-ATPase, and adenosine kinase, was reported in pepper plants inoculated with *Bacillus licheniformis* by Lim and Kim (2013). In addition to the combined inoculation of plants with rhizobacteria and AMF showed a high level of resistance against plants. The enhancement in the growth and stress tolerance against drought was observed in common bean when it was inoculated with a mixture of *Rhizobium tropici*-CIAT 899, *P. polymyxa*-DSM36, and *P. polymyxa-Loutit* strain as compared to single inoculation with Rhizobium species (Figueiredo et al. 2008).

Wang et al. (2012) treated cucumber plants with a consortium of *B. subtilis* SM21, *Bacillus cereus* AR156 and *Serratia* sp. and reported to increase the activity of enzyme superoxide dismutase, ribulose-1,5-bisphosphate carboxyl/oxygenase and ascorbate peroxidase in cucumber plant leaves. In 2014, the consortium of microbes of exopolysaccharides producing bacterial isolates has a greater tolerance for drought in maize than the single bacterial strain (Naseem and Bano 2014). The increase in plant antioxidant activity was reported under drought stress conditions when plants are inoculated with AMF which results in the reduction of oxidative damages in stress conditions (Hashem et al. 2015; Kumar et al. 2015). Further combined application of PGPR and AMF improved the water and nutrient uptake capacity of rice plants grown in the arid zone (Ruız-Sanchez et al. 2011). In saline environments, the application of PGPR and AMF was reported to be beneficial for the plants in context to their uptake capacity of nutrients (Nadeem et al. 2014).

12.3.4 Signalomics

The studies on beneficial plant–microbe interaction are becoming important to develop and commercialize microbial inoculation to improve plant health. The interaction between plants and microorganisms is achieved by chemical signals in the rhizosphere. The benefits of mutual interaction between plant and microbes are understood by characterizing the chemicals and metabolites either present in the root exudates or secreted by the microorganisms. The signal omics is the studies deal with the identification of different signal and chemical communication occurs between plant and microorganism in the rhizospheres. Such communication is playing an important role in the following ways, like the formation of biofilm from microbial communities of soil (Rosier et al. 2016; Sasse et al. 2017) soil-borne pathogen inhibition (Bertin et al. 2003; Li et al. 2013). In the rhizosphere, a wide range of chemicals are present due to the presence of different microbial species. To characterize the profile of all the chemicals, present in the complex chemical environment of the rhizosphere, the metabolomics approach has been used. Metabolomics is an approach by which hundreds of compounds could be quantified at a time (Lloyd et al. 2015), and widely used to analyze different responses that occur due to complex plant microbial interaction in rhizospheres.

Pseudomonas and *Bacillus* sp. are important and beneficial for bioremediation, disease control, and plant growth promotion. There are many reports available on the role of PGPR about the induction of resistance in plants against various fungal and bacterial diseases (Beneduzi et al. 2012; Pineda et al. 2010; Saraf et al. 2014). To establish mutual interaction with plants, rhizobacteria produces molecules that enhance gene expression in plants. However, the plants release phytohormones. The PGPR could also secrete these phytohormones (Fahad et al. 2015). Besides this, organic compounds volatile in nature are also produced by PGPR and have the potential to improve soil nutrients, plant growth development, and induced resistance (Wei-Wei et al. 2008; Kai et al. 2009). Volatile organic compounds production via PGPR like 2-endecanone, 2-heptanol, and pentadecane are reported to be beneficial for plant growth. For example, when *A. thaliana* was co-cultivated with PGPR strains (*Bacillus amyloliquefaciens* IN937a and *B. subtilis* GB03) in petri dishes was found to result in enhancing the growth of *A. thaliana* (Ryu et al. 2003).

Mechanism of quorum sensing involves the interaction of microbial communities as well as sensing their environment. *N*-acyl homoserine lactones (AHL) are the QS signals produced by gram-negative bacteria (Hassan et al. 2016). This was reported to establish a symbiotic relationship between plants and microbes (Schikora 2016).

12.4 Role of Microbial Processing (Signals) in Facilitating Plant Growth

Soil microbes are significant in biogeochemical cycles and have been utilized for harvest generation for quite a long time. Free-living soil bacteria valuable to plant development are fit for promoting plant growth by colonizing the plant root. Phosphate solubilizing *Bacillus* sp. and soil rhizospheric nitrogen-fixing bacteria enhanced the yield in plants like rice (Sudha et al. 1999), sugar beet (Çakmakçi et al. 2006), maize (Pal 1998), canola, and conifer species (Bent et al. 2002). PGPRs are known to work in three distinct ways (i) by combining specific mixes for the plants, (ii) by enhancing mineral uptake from the soil, and (iii) protecting plants from infections (Ortíz-Castro et al. 2008). PGPR are likewise named PHPR or then again knob advancing rhizobacteria. Advantageous nitrogen-fixing microbes incorporate the cyanobacteria, *Bradyrhizobium*, *Allorhizobium*, *Rhizobium*, *Azorhizobium*, *Mesorhizobium,* and *Sinorhizobium*. Cooperative nitrogen fixers or free-living nitrogen-fixing microbes are the model microbes having a place with the animal types *Enterobacter*, *Azospirillum*, *Pseudomonas,* and *Klebsiella* had append roots and effectively colonize the surface of roots.

Plant–bacteria affiliations have been ready for a long time. Notwithstanding, a total comprehension of the instruments used by plant development–advancing microbes had remained to some degree subtle, regularly making it hard to exploit these complexes. A modern understanding of plant–microbe interaction reveals the "selective" potential of plants to have useful microbiome which affects the plant development in a certain positive manner (Hardoim et al. 2008; Marasco et al. 2012; Rashid et al. 2012). The number of inhabitants in this whole world will be more than 10 billion stamps by 2050. This increasing population is responsible for putting extra pressure on the land territory for nourishment, fiber, fuel, and crude materials. Novel and new answers for the improvements in plant development are needed to facilitate the weight forced on our condition and different assets. A gander at potential answers for these issues by analyzing a portion of the exploration directed concerning the natural uses of free-living PGPR is addressed. PGPR are the microscopic organisms gathering which encourages the plant growth and yield through different plant development–advancing substances. The major applications of bacteria for better and improved plant growth include horticulture, agriculture, forestry, and renewal of the environment. The utilization of PGPR as biofertilizers, phytostimulators, and biocontrol specialists would be an alluring choice to diminish the utilization of substance manures which lead to natural contamination, well-being perils, and the demolishing of soil efficiency. Plant development improved by PGPR is measured by seedling rise, energy, biomass, multiplication of framework of roots, and yield among different species of plants.

Joshi et al. (2012) analyzed the mechanism, application, advantages, and disadvantages of the plant growth–promoting rhizobacteria (PGPR). Rhizobacteria are the microbes found in the rhizosphere which has constructive outcomes on plants. These microorganisms which can either live freely or in advantageous interaction with roots of plants invigorate the development and strength of the plants and named as PGPR. The role of these microbes underlies the fact that they enable crop improvement by enriching the soil with essential nutrients and inhibiting plant pathogenic activity. The entire microbial network in rhizosphere specialty should be adjusted to intercede in the plant development advancement by PGPR. PGPR advances the plant development by either encouraging acquirement of phosphorus, nitrogen, and basic minerals, and by controlling levels of plant hormones, in a roundabout way by diminishing the impacts of inhibition of different pathogens on the plant development and improvement (Glick 2012).

The direct or indirect role of soil bacteria in facilitating the growth of plants and the mechanism thereof was explained by Glick (2012). He explained the effect of the growth of the plant. These PGPs influence plant growth by acquisition of resources (nitrogen, phosphorus, and iron) from the environment or by the regulation or provision of growth hormones to the plants such as cytokinins, auxins, and ethylene. Indirectly, they support the growth of the plant by inhibiting the damaging effect of many pathogenic microbes, fungi, and nematodes by producing antibiotics and enzymes which degrades the cell wall, reducing levels of ethylene in plants, by systemic resistance which is induced, by lowering the availability of iron to the pathogens (Glick 2015). Rhizosphere bacteria (surrounding the roots of plants) and endophytes (present in the interior of the plant) are known for having the capability to act as plant growth–promoting bacteria (Lacava and Azevedo 2013).

Plant growth promoting bacteria (PGPB) benefited the plants directly by supplementing them with the required nutrients from the environment like nitrogen, iron, and phosphorous (Glick 2012). Various free-living microscopic organisms, like *Azospirillum* sp., can fix the atmospheric nitrogen and make them available to the plants (Bashan and Levanony 1990). Nitrogenase is the key enzyme required by these microbes for nitrogen fixation. Nitrogenase reduces nitrogen (N_2) to ammonia (NH_3). Nitrogenase is encoded by the Nif genes or homolog. N if genes are required for nitrogen fixation incorporate auxiliary qualities, qualities associated with iron protein initiation, biosynthesis of a cofactor for iron-molybdenum, electron gift, administrative qualities needed for the amalgamation and chemical capacity. The *nif* genes are generally present in clusters in the case of diazotrophic microorganism shaving a size of 20–24 kb and 7 is the number of operons which encodes for 20 unique proteins. Low concentration of fixed nitrogen as well the oxygen concentrations induced *nif* genes expression, thus higher concentration of oxygen has both inhibitory effects on nitrogenase activity as well the *nif* gene expression articulation; be that as it may, it is required for *Rhizobium* sp. bacteroid breath. Regardless of the way that measuring phosphorus is commonly very high, the greater part of the phosphorus in the soil is insoluble and hence not accessible by the plants.

Nitrogen is one of the essential structural as well as functional components of plants. Symbiotic association of the nitrogen-fixing microbes with that of legumes converts the atmospheric nitrogen into ammonia (Shiferaw et al. 2004). Rhizobial species that belong to different genera (*Mesorhizobium*, *Rhizobium*, *Azorhizobium*, *Allorhizobium*, *Bradyrhizobium*, and *Sinorhizobium*) are known to form a symbiotic association with the roots of leguminous plants in the response of flavonoid signals produced by the legume host by expressing the *nod* genes in the rhizobia. Expression of nod genes leads to the production of lipo-chitooligosaccharide (LCO) which are responsible for the activation of mitotic cell division inside the roots leading further to the formation of nodules (Matiru and Dakora 2004). There are a series of interactions that take place between the leguminous plants and rhizobia for the formation of nodules. Notwithstanding N_2-obsession in vegetables, *Rhizobia,* such as types of *Bradyrhizobium* and *rhizobium,* produce particles (cytokinins, auxins, lumichrome, LCO, abscisic acids, riboflavin, and nutrients) that advance plant development (Hayat et al. 2008a, 2008b).

Scope of plant development advancing rhizobacteria take part in cooperation with C4 (e.g. rice, maize, wheat, cotton, and sugarcane) and C3 plants, and essentially increment the vegetative development and grain yield in these plants. *Azotobacter* sp. are free-living (*Azotobacter vinelandii, Azotobacter chroococcum*) heterotrophic diazotrophs which rely upon a sufficient stockpile of decreased C mixes, for example, sugars for vitality (Kennedy and Tchan 1992). *Azotobacter* movement in the culture of rice can be expanded by applying the straw, apparently because of the breakdown of cellulose into glucose and cellobiose by microbes.

Phosphorus is a significant basic macronutrient present at levels of 400-1200 $mg\,kg^{-1}$ in the soil and is directly related to plant development and improvement. Phosphorus is present in two forms in soil, as natural and as an inorganic form of phosphates. To change over phosphates which are insoluble (natural and inorganic) mixes in a structure open for the plant is a significant characteristic for plant growth–promoting bacteria for enhancing the yield in plants. HPO_4^{2-} and $H_2PO_4^{-}$ are the two forms of phosphorus generally used by plants. Soil pH and soil type are mainly responsible for the precipitation and phosphate obsession in the soil. Reports have been archived about the P discharge from the microbes from the natural phosphate sources. Many bacterial strains (*Pseudomonas*, *Burkholderia*, *Bacillus*, *Achromobacter*, *Micrococcus*, *Rhizobium*, *Erwinia*, *Agrobacterium*, *Aerobacter*, and *Flavobacterium*) are known to solubilize insoluble inorganic phosphate (tricalcium phosphate, dicalcium phosphate, hydroxylapatite, and shake phosphate) (Abd-Alla 1994; Gügi et al. 1991; Richardson et al. 2001; Ryu et al. 2005; Singh and Satyanarayana 2010, 2011; Skrary and Cameron 1998). Iron is the fourth most abundant element on the earth. But it is not acclimatized by either microscopic organisms or plants because ferric particle or Fe^{3+}, the prevalent structure in nature, is just dissolvable to measure the iron accessible for digestion by the living life forms and are very low.

Bacterial siderophores, direct benefits for the development in plants, have been studied extensively, such as (i) use of radiolabeled ferric-siderophores demonstrated the type of iron utilized by the plants (Crowley et al. 1988; Duijff et al. 1994; Robin et al. 2008); (ii) vaccination of mung bean plants with *Pseudomonas strain* GRP3 with limited iron, demonstrated decreased chlorotic manifestations and an improved level of chlorophyll (Sharma et al. 2003) (iii) complex of Fe-pyoverdine orchestrated by *Pseudomonas fluorescence* C7, taken by the *A. thaliana* plants, prompting iron expansion inside the tissues of plants and improving the development of the plants (Vansuyt et al. 2007). Phytohormones play an important role in plant growth as well as plant development during normal as well as stressful conditions (Davies 2004). Certain environmental stress conditions, biotic as well as abiotic, affect plant development (Glick et al. 2007). At the point when plants experience development constraining ecological conditions, they often endeavor to change the endogenous phytohormones levels to diminish the ill effects of the natural stressors (Salamone et al. 2005). This technique is now and again effective, and rhizospheric microbes may likewise deliver or adjust phytohormones (in vitro) to change the phytohormone levels via PGPB and in this manner influence the plant's hormonal distribution as well further plant growth.

Many soil microscopic organisms, PGPB specifically, can deliver either gibberellins or cytokinins or both (Salamone et al. 2001). Along these lines, cytokinins are recognized in the sans cell mode of certain strains like *Rhizobium* spp., *Azotobacter* sp., *Pantoea agglomerans*, *Rhodospirillum rubrum*, *P. fluorescens*, *B. subtilis*, and *P. polymyxa*. Many *Rhizobium* strains that have been studied by Badenoch-Jones et al. (1984) were found to deliver IAA. This increased level of auxins inside the host plant is important for knob development. Thus, this further explained the presence of fewer knobs on the roots of soybean plant as compared to the wild strain of the bacterium *Bradyrhizobium elkanii* that had a diminished degree of IAA (Fukuhara et al. 1994). According to Glick (2012), the biocontrol bacterial ability to promote the plant growth indirectly has been studied extensively in terms of the science interest (mechanism used by the biocontrol bacteria) as well as in terms of the commercial interest, i.e. utilizing them instead of harmful chemical pesticides.

The amalgamation of the different anti-infection agents is a PGPB feature which is often related to the bacterium's ability to stop plant pathogens spreading (Compant et al. 2005; Mazurier et al. 2009). Many of these anti-infection agents are being studied in-depth along with their mode of action as well as their specificity and a portion of them has been commercialized. The problem of a high degree of reliance on antimicrobial development of microorganisms as biocontrol specialists is using these strains excessively; resistance against the antibiotics can be developed by certain phytopathogens. Some biocontrol microbes are known to control the pathogenic agent by producing certain lytic enzymes like cellulases, chitinases, ββ-1, 3 glucanases, lipases, and proteases. These lytic enzymes disrupt the cell structure of pathogenic agents and hence the pathogenic growths. *Fusarium oxysporum, Botrytis cinerea, Sclerotium rolfsii, Rhizoctonia solani, Phytophthora* spp., and *Pythium ultimum* are some of the examples of the bacteria using these methods for the biocontrol (Kim et al. 2008).

Some bacterial strains are known to use the biocontrol potential of the siderophores to control some phytopathogens (Dowling et al. 1996). Biocontrol PGPB develops siderophores that are much more iron-friendly than infectious pathogens so that, in the absence of iron, parasitic pathogens cannot replicate in the rhizosphere of the host plant's underlying structures (Sullivan and Gara 1992). Therefore, in this model, biocontrol PGPB significantly competes with contagious pathogens for iron uptake without affecting the plant's development (Sullivan and Gara 1992). Thus, siderophore-producing bacteria impart induced systemic resistance to the plants (Sindhu et al. 1997; Sayyed and Patel 2011).

12.5 Conclusion and Future Prospects

Complex and diverse microbial communities live in an intimate relationship with plants. There is an acute need to explore different microbial groups for fulfilling our future demands. From the literature of centuries ago, it is evident that there is a close relationship between plants and microorganisms. Microorganisms play important role in improving the growth, development, and the removal of stress of plants. So, a better understanding is needed to know the different potentials of the microbiome and in which way it can prove beneficial for fulfilling human needs.

Acknowledgments

Dr Madan Verma would like to thank the Director, Indian Institute of Information Technology Una, Himachal Pradesh, India for the encouragement and support to pursue the present work.

References

Abd-Alla, M.H. (1994). Use of organic phosphorus by *Rhizobium leguminosarum* biovar. viceae phosphatases. *Biol. Fertil. Soils.* 18: 216–218.

Ahmed, E. and Holmström, S.J.M. (2014). Siderophores in environmental research: roles and applications. *Microb. Biotechnol.* 7: 196–208.

Bach, E., Seger, G.D.S., Fernandes, G.C. et al. (2016). Evaluation of biological control and rhizosphere competence of plant growth promoting bacteria. *Appl. Soil Ecol.* 99: 141–149.

Badenoch-Jones, J., Summons, R.E., Rolfe, B.G. et al. (1984). Phytohormones, *Rhizobium* mutants, and nodulation in legumes. IV. Auxin metabolites in pea root nodules. *J. Plant Growth Regul.* 3 (1–3): 23–39.

Bais, H.P., Weir, T.L., Perry, L.G. et al. (2006). The role of root exudates in rhizosphere interactions with plants and other organisms. *Annu. Rev. Plant Biol.* 57: 234–266.

Banerjee, M. and Yesmin, L. (2002). Sulfur-oxidizing plant growth promoting Rhizobacteria for enhanced canola performance. US Patent.

Barac, T., Taghavi, S., Borremans, B. et al. (2004). Engineered endophytic bacteria improve phytoremediation of water soluble, volatile, organic pollutants. *Nat. Biotechnol.* 22: 583–588.

Bashan, Y. and Levanony, H. (1990). Current status of *Azospirillum* inoculation technology: *Azospirillum* as a challenge for agriculture. *Can. J. Microbiol.* 36 (9): 591–608.

Belimov, A.A.N., Hontzeas, V.I., Safronova, S.V. et al. (2005). Cadmium-tolerant plant growth-promoting bacteria associated with the roots of Indian mustard (*Brassica juncea* L. Czern.). *Soil Biol. Biochem.* 7: 241–250.

Bender, S.F., Plantenga, F., Neftel, A. et al. (2014). Symbiotic relationships between soil fungi and plants reduce N_2O emissions from soil. *J. Microb. Eco.* 8: 1336–1345.

Beneduzi, A., Ambrosini, A., and Passaglia, L.M.P. (2012). Plant growth-promoting rhizobacteria (PGPR): their potential as antagonists and biocontrol agents. *Genet. Mol. Biol.* 4: 1044–1051.

Bent, E., Breuil, C., Enebak, S. et al. (2002). Surface colonization of lodgepole pine (*Pinus contorta* var *lati folia* [Dougl. Engelm.])roots by *Pseudomonas fluorescens* and *Paenibacilluspolymyxa* under gnotobiotic conditions. *Plant Soil* 241: 187–196.

Berendsen, R.L., Pieterse, C.M.J., and Bakker, P.A.H.M. (2012). The rhizosphere microbiome and plant health. *Trends Plant Sci.* 17: 478–486.

Berg, G. (2009). Plant–microbe interactions promoting plant growth and health: perspectives for controlled use of microorganisms in agriculture. *Appl. Microbiol. Biotechnol.* 84: 11–18.

Berg, M. and Koskella, B. (2018). Nutrient- and dose-dependent microbiome-mediated protection against a plant pathogen. *Curr. Biol.* 28: 2487–2492.

Bertin, C., Yang, X., and Weston, L.A. (2003). The role of root exudates and allelochemicals in the rhizosphere. *Plant Soil* 256: 67–83.

Bonfante, P. and Genre, A. (2010). Mechanisms underlying beneficial plant-fungus interactions in mycorrhizal symbiosis. *Nature Commun.* 1: 48.

Bragina, A., Oberauner-Wappis, L., Zachow, C. et al. (2014). The *Sphagnum* microbiome supports bog ecosystem functioning under extreme conditions. *Mol. Ecol.* 23: 4498–4510.

Buee, M., De Boer, W., Martin, F. et al. (2009). The rhizosphere zoo: an overview of plant-associated communities of microorganisms, including phages, bacteria, archaea, and fungi, and of some of their structuring factors. *Plant Soil* 321: 189–212.

Bulgarelli, D., Rott, M., Schlaeppi, K. et al. (2012). Revealing structure and assembly cues for *Arabidopsis* root-inhabiting bacterial microbiota. *Nature* 488: 91–95.

Butler, A. and Theisen, R.M. (2010). Iron (III)-siderophore coordination chemistry: reactivity of marine siderophores. *Coord. Chem. Rev.* 254: 288–296.

Çakmakçi, R., Dönmez, F., Aydın, A. et al. (2006). Growth promotion of plants by plant growth-promoting rhizobacteria under greenhouse and two different field soil conditions. *Soil Biol. Biochem.* 38: 1482–1487.

Chapin, F.S. (1980). The mineral nutrition of wild plants. *Annu. Rev. Ecol. Syst.* 11: 233–260.

Compant, S., Duffy, B., Nowak, J. et al. (2005). Use of plant growth-promoting bacteria for biocontrol of plant diseases: principles, mechanisms of action, and future prospects. *Appl. Environ. Microbiol.* 71 (9): 4951–4959.

Coninx, L., Martinova, V., and Rineau, F. (2017). Mycorrhiza-assisted pytoreme-diation. In: *Adv. Bot. Res.*, vol. 83 (eds. A. Cuypers and J. Vangronsveld), 127–188. Amsterdam (The Netherlands): Elsevier.

Crowley, D.E., Reid, C.P.P., and Szaniszlo, P.J. (1988). Utilization of microbial siderophores in iron acquisition by oat. *Plant Physiol.* 87: 685–688.

Danhorn, T. and Fuqua, C. (2004). Biofilm formation by plant-associated bacteria. *Annu. Rev. Microbiol.* 61: 401–422.

Davies, P.J. (2004). *Plant Hormones: Biosynthesis, Signal Transduction, Action!* Dordrecht, The Netherlands: Kluwer Academic.

De Werra, P., Péchy-Tarr, M., Keel, C. et al. (2009). Role of gluconic acid production in the regulation of biocontrol traits of *Pseudomonas fluorescens* CHA0. *Appl. Environ. Microbiol.* 75: 4162–4174.

Delavaux, C.S., Weigelt, P., Dawson, W. et al. (2019). Mycorrhizal fungi influence global plant biogeography. *Nat. Ecol. Evol.* 3: 424.

Delgado-Baquerizo, M., Maestre, F.T.,.R.,.P. et al. (2016). Microbial diversity drives multifunctionality in terrestrial ecosystems. *Nat. Commun.* 7: 10541.

Deng, Z. and Cao, L. (2017). Chemosphere fungal endophytes and their interactions with plants in phytoremediation: a review. *Chemosphere* 168: 1100–1106.

Devers, M., Henry, S., Hartmann, A. et al. (2005). Horizontal gene transfer of atrazine-degrading genes (atz) from *Agrobacterium tumefaciens* St96-4 pADP1:Tn5 to bacteria of maize-cultivated soil. *Pest Manage. Sci.* 61: 870–880.

Dobbelare, S., Vanderleydern, J., and Okon, Y. (2003). Plant-growth promoting effects of diazotrophs in the rhizosphere. *Crit. Rev. Plant Sci.* 22: 107–149.

Dowling, D.N., Sexton, R., Fenton, A. et al. (1996). Iron regulation in plant-associated Pseudomonas fluorescens M114: implications for biological control. In: *Molecular Biology of Pseudomonads* (eds. T. Nakazawa, K. Furukawa, D. Haas and S. Silver), 502–511. Washington, DC, USA: American Society for Microbiology Press.

Duijff, B.J., Bakker, P.A.H.M., and Schippers, B. (1994). Ferric pseudobactin 358 as an iron source for carnation. *J. Plant Nutr.* 17 (12): 2069–2078.

Fahad, S., Hussain, S., Bano, A. et al. (2015). Potential role of phytohormones and plant growth-promoting rhizobacteria in abiotic stresses: consequences for changing environment. *Environ. Sci. Pollut. Res.* 22: 4907–4921.

Figueiredo, M.V., Burity, H.A., Martínez, C.R. et al. (2008). Alleviation of drought stress in the common bean (Phaseolus vulgaris L.) by coinoculation with *Paenibacillus polymyxa* and *Rhizobium tropici*. *Appl. Soil Ecol.* 40: 182–188.

Fukuhara, H., Minakawa, Y., Akao, S. et al. (1994). The involvement of indole-3-acetic acid produced by *Bradyrhizobium elkanii* in nodule formation. *Plant Cell Physiol.* 35 (8): 1261–1265.

Fürnkranz, M., Müller, H., and Berg, G. (2009). Characterization of plant growth promoting bacteria from crops in Bolivia. *J. Plant Dis. Protect.* 116: 149–155.

Germaine, K., Keogh, E., Borremans, B. et al. (2004). Colonization of poplar trees by gfp expressing endophytes. *FEMS Microbiol. Ecol.* 48: 109–118.

Glick, B.R. (2005). Modulation of plant ethylene levels by the bacterial enzyme ACC deaminase. *FEMS Microbiol. Lett.* 252: 1–7.

Glick, B.R. (2010). Using soil bacteria to facilitate phytoremediation. *Biotechnol. Adv.* 28: 367–374.

Glick, B.R. (2012). *Plant Growth-Promoting Bacteria: Mechanisms and Applications*. Waterloo, Canada: Hindawi Publishing Corporation, Scientifica.

Glick, B.R. (2015). *Beneficial Plant-Bacterial Interactions*. Heidelberg: Springer.

Glick, B.R., Cheng, Z., Czarny, J. et al. (2007). Promotion of plant growth by ACC deaminase-producing soil bacteria. *Eur. J. Plant Pathol.* 119 (3): 329–339.

Grube, M., Cernava, T., Soh, J. et al. (2015). Exploring functional contexts of symbiotic sustain within lichen-associated bacteria by comparative omics. *ISME J.* 9: 412–424.

Gügi, B., Orange, N., Hellio, F. et al. (1991). Effect of growth temperature on several exported enzyme activities in the psychrotropic bacterium *Pseudomonas fluorescens*. *J. Bacteriol.* 173: 3814–3820.

Gunnell, D., Eddleston, M., Phillips, M.R. et al. (2007). The global distribution of fatal pesticide self-poisoning: systematic review. *BMC Public Health* 21: 357.

Habibzadeh, Y. (2015). The effect of arbuscular mycorrizal fungi and phosphorus levels on dry matter production and root traits in cucumber (*Cucumis sativis* L.). *Afric. J. Environ. Sci. Technol.* 9: 65–70.

Hardoim, P.R., Van, O., Verbeek, L.S., and van Elsas, J.D. (2008). Properties of bacterial endophytes and their proposed role in plant growth. *Trends Microbiol.* 16 (10): 463–471.

Hardoim, P.R., Hardoim, C.C., van Overbeek, L.S. et al. (2012). Dynamics of seedborne rice endophytes on early plant growth stages. *PLoS One* 7: e30438.

Hardoim, P.R., van Overbeek, L.S., Berg, G. et al. (2015). The hidden world within plants: ecological and evolutionary considerations for defining functioning of microbial endophytes. *Microbiol. Mol. Biol. Rev.* 79: 293–320.

Harman, G.E., Howell, C.R., Viterbo, A. et al. (2004). *Trichoderma* species: opportunistic, avirulent plant symbionts. *Nat. Rev. Microb.* 2: 43–56.

Hartmann, A., Schmid, M., van Tuinen, D., and Berg, G. (2009). Plant-driven selection of microbes. *Plant Soil* 321: 235–275.

Hashem, A., Allah, E.F., Alqarawi, A.A. et al. (2015). Arbuscular mycorrhizal fungi enhance salinity tolerance of *Panicum turgidum* Forssk by altering photosynthetic and antioxidant pathways. *J. Plant Interact.* 10: 230–242.

Hassan, R., Shaaban, M.I., Abdel Bar, F.M. et al. (2016). Quorum sensing inhibiting activity of *Streptomyces coelicoflavus* isolated from soil. *Front. Microbiol.* 7: 659.

Hayat, R., Ali, S., Siddique, M.T. et al. (2008a). Biological nitrogen fixation of summer legumes and their residual effects on subsequent rainfed wheat yield. *Pak. J. Botany* 40 (2): 711–722.

Hayat, R., Ali, S., Ijaz, S.S. et al. (2008b). Estimation of N_2 – fixation of mung bean and mash bean through xylem uriede technique under rainfed conditions. *Pak. J. Botany* 40 (2): 723–734.

Hiltner, L. (1904). Uber nevereErfahrungen und Probleme auf dem Gebiet der Boden Bakteriologie und unterbesondererBeurchsichtigung der Grundungung und Broche. *Soil Biol. Biochem.* 98: 59–78.

Hirsch, P.R. and Mauchline, T.H. (2012). Who's who in the plant root microbiome? *Nat. Biotechnol.* 30: 961–962.

Hogberg, P., Nordgren, A., Buchmann, N. et al. (2001). Large-scale forest girdling shows that current photosynthesis drives soil respiration. *Nature* 411: 789–792.

Joshi, A.U., Kavan, N., Andharia, K.N. et al. (2012). *Plant Growth Promoting Rhizobacteria: Mechanism, Application, Advantages and Disadvantages* (eds. A.S. Tomar, B. Vijay and V.B. Mandaliya), 13–40. New Delhi: Daya Publishing House.

Kai, M., Haustein, M., Molina, F. et al. (2009). Bacterialvolatiles and their action potential. *Appl. Microbiol. Biotechnol.* 81: 1001–1012.

Kamilova, F., Validov, S., Azarova, T. et al. (2005). Enrichment for enhanced competitive plant root tip colonizers selects for a new class of biocontrol bacteria. *Environ. Microbiol.* 7: 1809–1817.

Kennedy, I.R. and Tchan, Y. (1992). Biological nitrogen fixation in no leguminous field crops: recent advances. *Plant Soil* 141: 93–118.

Kim, Y.C., Jung, H., Kim, K.Y. et al. (2008). An effective biocontrol bioformulation against *Phytophthora* blight of pepper using growth mixtures of combined chitinolytic bacteria under different field conditions. *Eur. J. Plant Pathol.* 120 (4): 373–382.

Kowalchuk, G.A. and Stephen, J.R. (2001). Ammonia-oxidizing bacteria: a model for molecular microbial ecology. *Annu. Rev. Microbiol.* 55: 485–529.

Kumar, A., Sharma, S., Mishra, S. et al. (2015). Arbuscular mycorrhizal inoculation improves growth and antioxidative response of *Jatropha curcas* (L.) under Na_2SO_4 salt stress. *Plant Biosyst.* 149: 260–269.

Lacava, P.T. and Azevedo, J.L. (2013). Endophytic bacteria: A biotechnological potential in agrobiology system. In: *Bacteria in Agrobiology: Crop Productivity* (eds. D.K. Maheshwari, M. Sarah and A. Aeron), 1–44. Berlin: Springer-Verlag.

Leach, A.W. and Mumford, J.D. (2008). Pesticide environmental accounting: a method for assessing the external costs of individual pesticide applications. *Environ. Pollut.* 151: 139–147.

Lenoir, I., Lounes, H., Sahraoui, A. et al. (2016). Arbuscular mycorrhizal fungal-assisted phytoremediation of soil contaminated with persistent organic pollutants: a review. *Eur. J. Soil Sci.* 67: 624–640.

Li, X., Zhang, T., Wang, X. et al. (2013). The composition of root exudates from two different resistant peanut cultivars and their effects on the growth of soil-borne pathogen. *Int. J. Biol. Sci.* 9: 164–173.

Lim, J.H. and Kim, S.D. (2013). Induction of drought stress resistance by multi-functional PGPR *Bacillus licheniformis* K11 in pepper. *Plant Pathol. J.* 29: 201–208.

Lindow, S.E. and Brandl, M.T. (2003). Microbiology of the phyllosphere. *Appl. Environ. Microbiol.* 69: 1875–1883.

Liu, Y., Zuo, S., Xu, L. et al. (2012). Study on diversity of endophytic bacterial communities in seeds of hybrid maize and their parental lines. *Arch. Microbiol.* 194: 1001–1012.

Llado, S., Lopez-Mondejar, R., and Baldrian, P. (2018). Drivers of microbial community structure in forest soils. *Appl. Microbiol. Biotechnol.* 102: 4331–4338.

Lloyd, N., Johnson, D.L., and Herderich, M.J. (2015). Metabolomics approaches for resolving and harnessing chemical diversity in grapes, yeast and wine. *Aust. J. Grape Wine Res.* 21: 723–740.

Lugtenberg, B.J.J., Chin-A-Woeng, T.F.C., and Bloemberg, G.V. (2002). Microbe plant interactions: principles and mechanisms. *Antonievan Leeuwenhoek* 81: 373–383.

Lundberg, D.S., Lebeis, S.L., Paredes, S.H. et al. (2012). Defining the core *Arabidopsis thaliana* root microbiome. *Nature* 488: 86–90.

Ma, W., Guinel, F.C., and Glick, B.R. (2003). *Rhizobium leguminosarum* biovar *viciae* 1-aminocyclopropane-1-carboxylate deaminase promotes nodulation of pea plants. *Appl. Environ. Microbiol.* 69: 4396–4402.

Ma, Y., Rajkumar, M., Luo, Y.M. et al. (2011). Inoculation of endophytic bacteria on host and nonhost plants effects on plant growth and Ni uptake. *J. Hazard. Matter* 196: 230–237.

Ma, Y., Oliveira, R.S., Nai, F. et al. (2015). The hyperaccumulator *Sedum plumbizincicola* harbors metal-resistant endophytic bacteria that improve its phytoextraction capacity in multi-metal contaminatedsoil. *J. Environ. Manag.* 156: 62–69.

Madhaiyan, M., Poonguzhali, S., Ryu, J. et al. (2006). Regulation of ethylene levels in canola (*Brassica campestris*) by 1-aminocyclopropane-1carboxylate deaminase-containing *Methylobacterium fujisawaense*. *Planta* 224: 268–278.

Mahaffee, W.F., Kloepper, J.W., Van Vuurde, J.W.L. et al. (1997). Endophytic colonization of *Phaseolus vulgaris* by *Pseudomonas fluorescens* strain 89B-27 and *Enterobacter asburiae* strain JM22. In: *Improving Plant Productivity in Rhizosphere Bacteria* (eds. M.H.R. Ryder, P.M. Stevens and G.D. Bowen), 180. Inst in Basic Youthilionos.

Marasco, R., Rolli, E., Ettoumi, B. et al. (2012). A drought resistance promoting microbiome is selected by root system under desert farming. *PLoS One* 7: e48479.

Mastretta, C., Barac, T., Vangronsveld, J. et al. (2006). Endophytic bacteria and their potential application to improve the phytoremediation of contaminated environments. *Biotech. Gen. Eng. Rev.* 23: 175–207.

Matiru, V.N. and Dakora, F.D. (2004). Potential use of rhizobial bacteria as promoters of plant growth for increased yield in landraces of African cereal crops. *Afr. J. Biotechnol.* 3 (1): 1–7.

Mayak, S., Tirosh, T., and Glick, B.R. (2004). Plant growth-promoting bacteria that confer resistance to water stress in tomatoes and peppers. *Plant Sci.* 166: 525–530.

Mazurier, S., Corberand, T., Lemanceau, P., and Raaijmakers, J.M. (2009). Phenazine antibiotics produced by fluorescent pseudomonads contribute to natural soil suppressiveness to *Fusarium* wilt. *ISME J.* 3 (8): 977–991.

Mendes, R., Garbeva, P., and Raaijmakers, J.M. (2013). The rhizosphere microbiome: significance of plant beneficial, plant pathogenic, and human pathogenic microorganisms. *FEMS Microbiol. Rev.* 37: 634–663.

Miller, S.A., Beed, F.D., and Harmon, C.L. (2009). Plant disease diagnostic capabilities and networks. *Annu. Rev. Phytopathol.* 47: 15–38.

Mitter, B., Pfaffenbichler, N., and Sessitsch, A. (2016). Plant–microbe partnerships in 2020. *Microb. Biotechnol.* 9: 635–640.

Morrissey, J.P., Dow, J.M., Mark, L. et al. (2004). Are microbes at the root of a solution to world food production? *EMBO Rep.* 5: 922–926.

Nadeem, S.M., Naveed, M., Zahir, Z.A. et al. (2013). Plant-microbe interactions for sustainable agriculture: fundamentals and recent advances. In: *Plant Microbe Symbiosis: Fundamentals and Advances* (ed. N. Arora), 53–103. India: Springer.

Nadeem, S.M., Ahmad, M., Zahir, Z.A. et al. (2014). The role of mycorrhizae and plant growth promoting rhizobacteria (PGPR) in improving crop productivity under stressful environments. *Biotechnol. Adv.* 32: 429–448.

Nadeem, S.M., Ahmad, M., Naveed, M. et al. (2016). Relationship between in vitro characterization and comparative efficacy of plant growth-promoting rhizobacteria for improving cucumber salt tolerance. *Arch. Microbiol.* 198: 379–387.

Naseem, H. and Bano, A. (2014). Role of plant growth-promoting rhizobacteria and their exopolysaccharide in drought tolerance of maize. *J. Plant Interact.* 9: 689–701.

Numan, M., Bashir, S., Khan, Y. et al. (2018). Plant growth promoting bacteria as an alternative strategy for salt tolerance in plants: a review. *Microb. Research* 209: 21–32.

Ortíz-Castro, R., Valencia-Cantero, E., and López-Bucio, J. (2008). Plant growth promotion by *Bacillus megaterium* involves cytokinin signaling. *Plant Signal. Behav.* 3 (4): 263–265.

Pal, S.S. (1998). Interactions of an acid tolerant strain of phosphate solubilizing bacteria with a few acid tolerant crops. *Plant Soil* 198: 169–177.

Pan, M.J., Rademan, S., Kuner, K. et al. (1997). Ultrastructural studies on the colonization of banana tissue and *Fusarium oxysporum* f. sp. cubense race 4 by the endophytic bacterium *Burkholderia cepacia*. *J. Phytopathol.* 145: 479–486.

Parniske, M. (2008). Arbuscular mycorrhiza: the mother of plant root endosymbioses. *Nat. Rev. Microbiol.* 6: 763–775.

Patil, S., Bheemaraddi, M.C., and Shivannavar, C.T. (2014). Biocontrol activity of siderophore producing *Bacillus subtilis* CTS-G24 against wilt and dry root rot causing fungi in chickpea. *J. Agric. Vet. Sci.* 7: 63–68.

Pierik, R., Tholen, D., Poorter, H. et al. (2006). The Janus factor of ethylene: growth inhibition and stimulation. *Trends Plant Sci.* 11: 176–183.

Pineda, A., Zheng, S.J., Van Loon, J.J.A. et al. (2010). Helping plants to deal with insects: the role of beneficial soil-borne microbes. *Trends Plant Sci.* 15: 507–514.

Quadt-Hallmann, A., Benhamou, N., and Kloepper, J.W. (1997). Bacterial endophytes in cotton: mechanisms entering the plant. *Can. J. Microbiol.* 43: 577–582.

Rajkumar, M., Sandhya, S., Prasad, M.N.V. et al. (2012). Perspectives of plant-associated microbes in heavy metal phytoremediation. *Biotechnol. Adv.* 30: 1562–1574.

Rascovan, N., Carbonetto, B., Perrig, D. et al. (2016). Integrated analysis of root microbiomes of soybean and wheat from agricultural fields. *Sci. Rep.* 6: 280–284.

Rashid, S., Charles, T.C., and Glick, B.R. (2012). Isolation and characterization of new plant growth-promoting bacterial endophytes. *Appl. Soil Ecol.* 61: 217–224.

Richardson, A.E., Hadobas, P.A., Hayes, J.E. et al. (2001). Utilization of phosphorus by pasture plants supplied with myo-inositol hexaphosphate is enhanced by the presence of soil micro-organisms. *Plant Soil* 229: 47–56.

Rillig, M.C. and Mummey, D.L. (2006). Mycorrhizas and soil structure. *New Phytol.* 171: 41–53.

Robin, G., Vansuyt, P., Hinsinger, J.M. et al. (2008). Iron dynamics in the rhizosphere. Consequences for plant health and nutrition. *Adv. Agron.* 99: 183–225.

Romeralo, C., Witzell, J., Romeralo, T.R. et al. (2015). Antagonistic activity of fungal endophyte filtrates against *Gremmeniella abietina* infections on Aleppo pine seedlings. *Eur. J. Plant Pathol.* 143: 691–704.

Ronchel, M.C., Ramos-Diaz, M.A., and Ramos, J.L. (2000). Retrotransfer of DNA in the rhizosphere. *Environ. Microbiol.* 2: 319–323.

Rosier, A., Bishnoi, U., and Lakshmanan, V. (2016). A perspective on inter-kingdom signaling in plant-beneficial microbe interactions. *Plant Mol. Biol.* 90: 537–548.

Ruiz Sánchez, M., Armada, E., Munoza, Y. et al. (2011). *Azospirillum* and arbuscular mycorrhizal colonization enhance rice growth and physiological traits under well-watered and drought conditions. *J. Plant Physiol.* 168: 1031–1037.

Ryu, C.M., Farag, M.A., Hu, C.H. et al. (2003). Bacterial volatiles promote growth in Arabidopsis. *Proc. Natl. Acad. Sci. U.S.A.* 100: 4927–4932.

Ryu, C.M., Hu, C.H., Locy, R.D. et al. (2005). Study of mechanisms for plant growth promotion elicited by rhizobacteria in *Arabidopsis thaliana*. *Plant Soil* 268: 285–292.

Salamone, I.E.G., Hynes, R.K., and Nelson, L.M. (2001). Cytokinin production by plant growth promoting rhizobacteria and selected mutants. *Can. J. Microbiol.* 47 (5): 404–411.

Salamone, I.E.G., Hynes, R.K., and Nelson, L.M. (2005). Role of cytokinins in plant growth promotion by rhizosphere bacteria. In: *PGPR: Biocontrol and Biofertilization* (ed. Z.A. Siddiqui), 173–195. Amsterdam, The Netherlands: Springer.

Saleem, M., Arshad, M., Hussain, S. et al. (2007). Perspective of plant growth promoting Rhizobacteria (PGPR) containing ACC deaminase in stress agriculture. *J. Ind. Microbiol. Biotech.* 34: 635–648.

Santhanam, R., Weinhold, A., Goldberg, J. et al. (2015). Native root-associated bacteria rescue a plant from a sudden-wilt disease that emerged during continuous cropping. *Proc. Natl. Acad. Sci. U. S. A.* 112: 13–20.

Sasse, J., Martinoia, E., and Northen, T. (2017). Feed your friends: do plant exudates shape the root microbiome? *Trends Plant Sci.* 23: 25–41.

Sayyed, R.Z. and Patel, P.R. (2011). Biocontrol potential of siderophore producing heavy metal resistant *Alcaligenes* sp. and *Pseudomonasaeruginosa* RZS3 Vis-à-Vis organophosphorus fungicide. *Indian J. Microbiol.* 51 (3): 266–272.

Schikora, A. (2016). Beneficial effects of bacteria-plant communication based on quorum sensing molecules of the N -acyl homoserine lactone group. *Plant Methods* 90: 605–612.

Schulz, B. and Boyle, C. (2006). What are endophytes? In: *Microbial Root Endophytes* (eds. B.J.E. Schulz, C.J.C. Boyle and T.N. Sieber), 1–13. Springer-Verlag Berlin Heidelberg.

Seaton, F.M., Jones, D.L., Creer, S. et al. (2019). Plant and soil communities are associated with the response of soil water repellency to environmental stress. *Sci. Total Environ.* 687: 929–938.

Sharma, A., Johri, B.N., Sharma, A.K. et al. (2003). Plant growth-promoting bacterium *Pseudomonas* sp. strain GRP3 influences iron acquisition in mung bean (*Vigna radiata* L. Wilzeck). *Soil Biol. Biochem.* 35 (7): 887–894.

Shiferaw, B., Bantilan, M.C.S., and Serraj, R. (2004). Harnessing the potential of BNF for poor farmers: technological policy and institutional constraints and research need. In: *Symbiotic Nitrogen Fixation: Prospects for Enhanced Application in Tropical Agriculture* (ed. R. Serraj), 3. New Delhi: Oxford and IBH publishing Co. Pvt. Ltd.

Sindhu, S.S., Suneja, S., and Dadarwal, K.R. (1997). Plant growth promoting rhizobacteria and their role in crop productivity. In: *Biotechnological Approaches in Soil Microorganisms for Sustainable Crop Production*, vol. 1997 (ed. K.R. Dadarwal), 149–193. Jodhpur: Scientific Publisher.

Singh, B. and Satyanarayana, T. (2010). Plant growth promotion by an extracellular HAP-phytase of a thermophilic mold *Sporotrichum thermophile*. *Appl. Biochem. Biotechnol.* 160 (5): 1267–1276.

Singh, B. and Satyanarayana, T. (2011). Microbial phytases in phosphorus acquisition and plant growth promotion. *Physiol. Mol. Biol. Plants.* 17 (2): 93–103.

Singh, G., Sachdev, B., and Sharma, N. (2010). Interaction of *Bacillus thuringiensis* vegetative insecticidal protein with ribosomal S2 protein triggers larvicidal activity in *Spodoptera frugiperda*. *Appl. Environ. Microbiol.* 60: 7202–7209.

Skrary, F.A. and Cameron, D.C. (1998). Purification and characterization of a *Bacillus licheniformis* phosphatase specific for D-alpha-glycerphosphate. *Arch. Biochem. Biophys.* 349: 27–35.

Smith, S.E. and Read, D.J. (1997). *Mycorrhizal Symbiosis*, 2e. London, UK: Academic Press.

Smith, K.P., Handelsman, J., and Godman, R.M. (1999). Genetic basis in plants for interactions with disease-suppressive bacteria. *Proc. Natl. Acad. Sci. U. S. A.* 9: 4786–4790.

Sprent, J.I. (2001). *Nodulation in Legumes*. Gardens, Kew, UK: Royal Bot.

Srinivas, A., Pushpavathi, B., Lakshmi, B.K.M. et al. (2017). Efficacy of biocontrol agents against seed mycoflora of sunflower at different storage periods. *Int. J. Pure Appl. Biosci.* 5: 818–824.

Strobel, G. and Daisy, B. (2003). Bioprospecting for microbial endophytes and their natural products. *Microbiol. Mol. Biol. Rev.* 67: 491–502.

Sudha, S.N., Jayakumar, R., and Sekar, V. (1999). Introduction and expression of the cry1Ac gene of bacillus thuringiensis in a cereal-associated bacterium, *Bacillus polymyxa*. *Curr. Microbiol.* 38: 163–167.

Sullivan, D.J.O. and Gara, F.O. (1992). Traits of fluorescent *Pseudomonas* spp. involved in suppression of plant root pathogens. *Microbiol. Rev.* 56 (4): 662–676.

Taghavi, S., Barac, T., Greenberg, B. et al. (2005). Horizontal gene transfer to endogenous endophytic bacteria from poplar improves phytoremediation of toluene. *Appl. Environ. Microbiol.* 71: 8500–8505.

Tank, N., Rajendran, N., Patel, B. et al. (2012). Evaluation and biochemical characterization of a distinctive pyoverdin from a *Pseudomonas* isolated from chickpea rhizosphere. *Braz. J. Microbiol.* 16: 39–48.

Thijs, S. (2015). Rhizoremediation. In: *Princ Plant-Microbe Interact Microbes Sustain Agric* (ed. B.J.J. Lugtenberg), 277–286. New York: Springer.

Thrall, P.H., Hochberg, M.E., Burdon, J.J. et al. (2007). Coevolution of symbiotic mutualists and parasites in a community context. *Trends Ecol. Evol.* 22: 120–126.

Tiedje, J.M. (1988). Ecology of denitrification and dissimilatory nitrate reduction to ammonium. In: *Biology of Anaerobic Microorganisms* (ed. A.J.B. Sehnder), 179–244. New York: Wiley.

Tikhonovich, I.A. and Provorov, N.A. (2011). Microbiology is the basis of sustainable agriculture: an opinion. *Ann. Appl. Biol.* 159: 155–168.

Timmusk, S. and Wagner, H. (1999). The plant-growth-promoting rhizobacterium *Paenibacillus polymyxa* induces changes in Arabidopsis thaliana gene expression: A possible connection between biotic and abiotic stress responses. *Mol. Plant Microbe. Interact.* 12: 951–959.

Trivedi, P. and Sa, T. (2008). *Pseudomonas corrugate* (NRRL B-30409) mutants increased phosphate solubilization, organic acid production and plant growth at lower temperatures. *Curr. Microbiol.* 56: 140–144.

Ullah, A., Heng, S., Munis, M.F.H. et al. (2015). Phytoremediation of heavy metals assisted by plant growth promoting (PGP) bacteria: a review. *Environ. Exp. Bot.* 117: 28–40.

Unno, Y., Okubo, K., Wasaki, J. et al. (2005). Plant growth promotion abilities and microscale bacterial dynamics in the rhizosphere of lupin analyzed by phytate utilization ability. *Environ. Microbiol.* 7: 396–404.

Van der Heijden, M.G.A., Bakker, R., Verwaal, J. et al. (2006). Symbiotic bacteria as a determinant of plant community structure and plant productivity in dune grassland. *FEMS Microb. Ecol.* 56: 178–187.

Van der Heijden, M., Dombrowski, G.A.N., and Schlaeppi, K. (2017). Continuum of root–fungal symbioses for plant nutrition. *Proc. Natl. Aacd. Sci. U. S. A.* 114: 11574–11576.

Van Loon, L.C. (2007). Plant responses to plant growth promoting rhizobacteria. *Eur. J. Plant Path.* 119: 243–254.

Vansuyt, G., Robin, A., Briat, J.F. et al. (2007). Iron acquisition from Fe-pyoverdine by *Arabidopsis thaliana*. *Mol. Plant Microbe Interact.* 20 (4): 441–447.

Vorholt, J.A. (2012). Microbial life in the phyllosphere. *Nat. Rev. Microbiol.* 10: 828–840.

Vos, C.M., Tesfahun, A.F., Panis, B. et al. (2012). Arbuscular mycorrhizal fungi induce systemic resistance in tomato against the sedentary nematode Meloidogyne incognita and the migratory nematode *Pratylenchus penetrans*. *Appl. Soil Ecol.* 61: 1–6.

Wang, C.J., Yang, W., and Wang, C. (2012). Induction of drought tolerance in cucumber plants by a consortium of three plant growth-promoting rhizobacterium strains. *PLoS One* 7: 52–56.

Wei-wei, L.I.U., Wei, M.U., Bing-Yu, Z.H.U. et al. (2008). Antagonistic activities of volatiles from four strains of *Bacillus* spp. and *Paenibacillus* spp. against soil-borne plant pathogens. *Agric. Sci. China* 7: 1104–1114.

Weyens, N., Van, D., Lelie, D. et al. (2009). Exploiting plant–microbe partnerships to improve biomass production and remediation. *Trends Biotechnol.* 27: 591–598.

Yuan, S.F., Li, M.Y., Fang, Z.Y. et al. (2016). Biological control of tobacco bacterial wilt using *Trichoderma harzianum* amended bio-organic fertilizer and the arbuscular mycorrhizal fungi *Glomus mosseae*. *Biol. Control.* 92: 164–171.

Ziedan, E.H., Elewa, I.S., Mostafa, M.H. et al. (2011). Application of mycorrhizae for controlling root diseases of sesame. *J. Plant Prot. Res.* 51: 355–361.

13

Advances in Omics and Bioinformatics Tools for Phyllosphere Studies

Hina Bansal

Amity Institute of Biotechnology, Amity University, Noida, Uttar Pradesh, India

13.1 Introduction

Advances in genome sequencing of various microbes and their analysis through the advanced techniques used in microbial research revealed that biotic functions are natively conserved between the species in all the pathways during evolution. The information became a doorway for the initial development of cutting-edge bioinformatics that's not simplest restricted to the organic database improvement and organization but on the withdrawal of valuable data as an awful lot as probable from such huge databases. The investigations of the plant-related microbiota based on DNA give understanding into the synthesis and physiological capability of plant-related microorganisms, whereas the studies based on RNA can outspread to illuminate the real metabolic activities and regulatory mechanisms of the microbial cells underneath specific environments. By the approach of monstrous genome sequencing ventures of microorganisms has become a vigorously "information-driven," quickly developing, exceptionally unpredictable, and regularly reliant organic research. Utilizing bioinformatics servers and tools, a more prominent comprehension of complex metabolic system changes inside the natural procedures can be gotten. The focal research topic of the molecular science of life constantly urged to see how living frameworks gather, procedure, and utilize organic data inside macromolecules. The proposed chapter talks about the coordination of databases and genomic assets for the examination of microorganisms that live in relationships with plants.

The massive amount of terrestrial biomass on Earth is made up of plants and microbes (Bar-On et al. 2018). It does not astonish, at the point, that plant phenotypes are inseparably molded by their connections with microorganisms. A few organisms cause ailment in plants, whereas others suggest insurance in contrast to these microbial pathogens and nonmicrobial foes (Innerebner et al. 2011; Mansfield et al. 2012). Although others assist plants with using their abiotic condition using the procedures, like supplement procurement. Endeavors to portray the molecular science, biology, and advancement of these collaborations have uncovered that they change among plant species and genotypes, and in

Phytomicrobiome Interactions and Sustainable Agriculture, First Edition. Edited by Amit Verma, Jitendra Kumar Saini, Abd El-Latif Hesham, and Harikesh Bahadur Singh.

this way have a hereditary premise (Corwin and Kliebenstein 2017; Haney et al. 2015; Smith and Goodman 1999). Through investigations on communications of plants with each or a couple of microbial variants, in turn, this has to a great extent appeared regularly in controlled and streamlined conditions in the research facility. Although these endeavors have been enormously productive, the appearance of ecological sequencing has prompted the revelation that leaves harbor a complex microbial biological system. From highland bushes to seagrass and from subarctic grassland to tropical woodland shelters (Fahimipour et al. 2017; Lambais et al. 2006; Ruiz-Perez et al. 2016; Uroz et al. 2016), the photosynthesizing tissues of plants are populated by a huge and different arrangement of microbial homesteaders, including microscopic organisms, parasites and phages (Lindow and Brandl 2003; Morella et al. 2018; Rastogi et al. 2013; Sapp et al. 2018). Plant scholars and reproducers are presently put up to the test of deciphering the abundance of information picked up from the investigation of individual plant–organism associations to a comprehension of variables molding the microbiome. This will help unravel the host microorganisms from the plant–microbial association, besides coordinating the job of ecological variety.

13.2 Recent Trends and Approaches

The ongoing approach in sequencing in the mix with a variety of "omics" systems enables specialists to recognize microbiome assembly and elements alongside have associations on an uncommon level. Present-day sequencing strategies give inside and outside data depending on the behavior and relative plenitude of plants and their accomplices with microbes (Epstein 2013; Hug et al. 2016). The computational investigation has considerably added to the comprehension of bacteriological jobs and their communication with plants (Cha et al. 2016; Koberl et al. 2013; Marasco et al. 2012; Spence et al. 2014). Ongoing research incorporates phyllosphere studies with computational displaying and engineered network tests joined with multi-omics (Scheuring and Yu 2012; Vorholt et al. 2017).

Advancements in biocomputing created parallel to the natural research empowered researchers to connect, incorporate and decipher the perplexing idea of any organic framework, e.g. their cell procedures, development and improvement, administrative systems for digestion and catabolism, protein structure, capacity and compliance, changes and articulations at the hereditary level, entire genome, proteome and metabolite constitution, post-transcriptional and post-translational changes and their effects, reactions of life forms to the earth, pathogens and abiotic stresses, connections with different living being phenotypic, genotypic and metabolic qualities.

Indeed, data abstraction from complex information is an extraordinary issues in organic research where computational frameworks, biostatistics, and data advances are searching for their expanding applications. The array and reconciliation of every one of these advances in taking care of the issues identified with the natural frameworks have been named "bioinformatics" in the mid-1980s. From that point forward it has not exclusively been alluded to the computational techniques for near examination of genomic information, however, it has been characterized as the investigation of the informatic forms in the biotic frameworks (Hagen 2000; Hogeweg 2011).

Bioinformatics research can be ordered under the following significant methodologies: investigation dependent on the information coming from laboratory experiments, the utilization of scientific displaying to infer new data, and a coordinated methodology that incorporates search strategies with mathematical and statistical demonstrating. The significant effect of bioinformatics looked into has been to mechanize the genome sequencing, computerized advancement of coordinated genomics and proteomics databases, robotized genome correlations with distinguishing the genome work, computerized deduction of metabolic pathways, quality articulation examination to determine administrative pathways, the improvement of factual strategies, bunching procedures and information mining systems to infer protein–protein and protein–DNA connections, and displaying of the 3D structure of proteins and 3D docking among proteins and biochemicals for a sane medication plan, distinction investigation among pathogenic and non-pathogenic strains to recognize up-and-comer qualities for immunizations and hostile to microbial specialists, and the entire genome correlation with comprehending the microbial development (Figure 13.1).

The current bioinformatics systems can encourage the disclosure of reasons for maladies, immunization and objective medication structure, and improved financially savvy specialists for bioremediation by pruning out the impasses. The improvement of a huge number of databases, web tools, and servers has been seen in the previous years to encourage bioinformatics-related assignments and this has encouraged wide applications for analysts.

The accessibility of genomics and proteomics information and upgraded bioinformatics and biochemical devices has increased the desire for humankind to have the option to control the hereditary qualities by controlling the current organisms. The points of interest

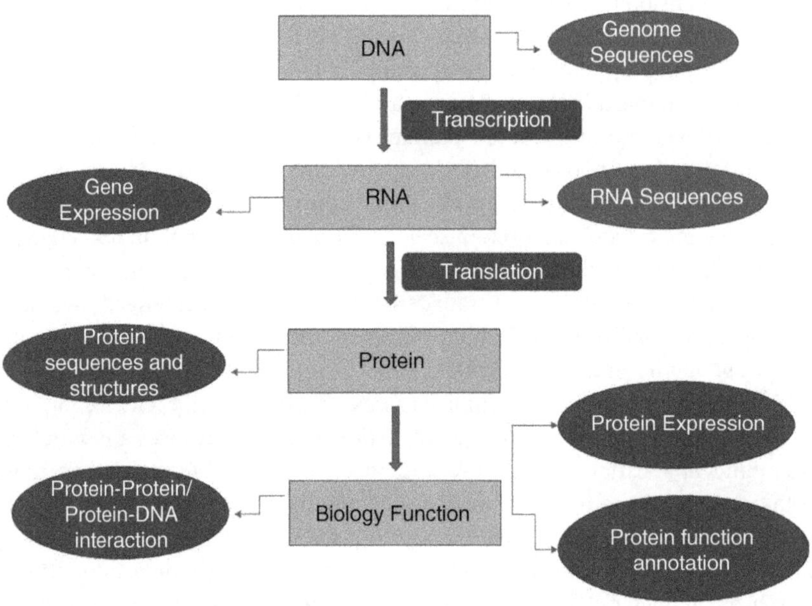

Figure 13.1 Schematic representation of central dogma with biological tasks.

are better finding of the ailments using protein biomarkers, assurance against maladies utilizing practical immunizations (Robinson et al. 2003; Zagursky et al. 2003) and objective medication structure, upgrading in farming quality and amount, and the advancement of methods that assist us with imagining and comprehend the definite microbial machine at the foundational level, the plan hostile to microbial specialists (Liu et al. 2004). Furthermore, the computational investigation has improved our understandings of the genome structure and the microorganism rebuilding process.

After the sequencing of *Haemophilus influenzae* in 1995 (Fleischmann et al. 1995), many microbial genomes have been sequenced and filed for open research in GenBank with the purposeful exertion of, NIH, EMBL, DDBJ, national labs, scholarly colleges, numerous medication advancement organizations, such as TIGR, and organizations engaged with the farming industry and bioremediation. The sequencing of the human genome (Venter et al. 2001) and other significant eukaryotic genomes has developed the desire for understanding host–pathogen association for the improvement of better antibodies and discerning medications to control the quality level and pathway level variations that are answerable for pathogenesis.

13.3 Computing for Biology

The improvement of computational strategies dependent on the sorted-out calculations, interpretational abilities and high stockpiling limits encouraged the examination of whole genomes, and along these lines, grant researchers contemplate progressively complex transformative patterns (Nakashima et al. 2005). The data elements in a solitary trial that incorporates the investigation of quality articulation includes an examination of qualities, assurances of protein assemblies encoded by the qualities and subtleties of how these items collaborate. The simplicity with which PCs handle huge amounts of different information at once and test the perplexing elements seen in nature makes them irreplaceable to help biotic research. Calculations, the processable arrangement of steps to accomplish wanted outcomes, are likely at the core of bioinformatics (Heng 2011; Nakashima et al. 2005). Calculations are utilized to look at genome successions at different levels, to discover comparative districts for genes, find their functions, examine their regulation, and evaluate how they and whole genomes have developed over time. Far beyond, the point is to give naturally significant expectations from clarified information and changes or controls of these datasets to discover most fittingly anticipated outcomes. Biocomputing techniques provide various practices shown in Figure 13.2.

13.4 Bioinformatics in Microbial Research

In the current era the augment, development, and collection of the macromolecule information and the information on the omics' innovations advanced and enhanced undertakings have risen to assist distinctive interconnected regions in biology frameworks (Juretic et al. 2005). Bioinformatics has encouraged scientists to think about microbial biodiversity on account of its immediate intercessions in atomic distinguishing proof, information

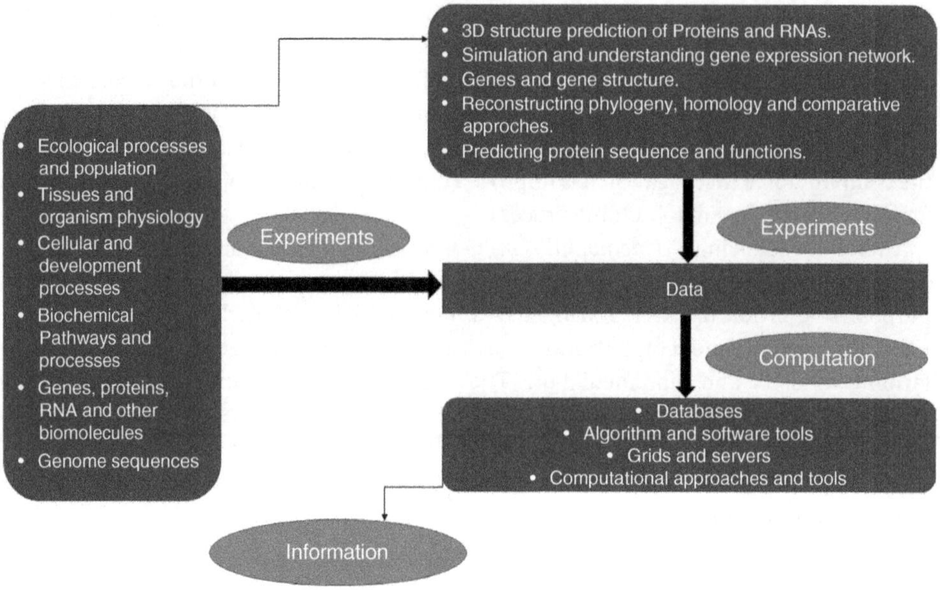

Figure 13.2 Structural organization of information extraction in bioinformatics.

stockpiling and recovery framework that was the stuff of bad dreams. The bioinformatics-driven methodologies empowered individuals to work proficiently on microbial assorted varieties, recognizable proof, portrayal, sub-atomic scientific classifications, and network examinations (Rogozin et al. 2002). Indexing and digitization of microbial decent varieties are one of the most significant undertakings for bioinformaticians. The portrayal of new species, genera, and even sub-atomic taxa rose significantly in the writing after the 1990s and these endeavors are to a great extent driven by progress in sequencing advances. Even though microbiologists accept that 99% of the microbial world still can't seem to be revealed, thus, bioinformatics must assume a significant job (Curtis et al. 2002). Different tasks that can be done through bioinformatics in microbial research are shown in Figure 13.3 (Juretic et al. 2005; Prabha et al. 2014).

The following approaches have been utilized in bioinformatics: use of computational searches and alignment techniques (Altschul et al. 1990, 1997; Pearson and Lipman 1988; Waterman 2018) for the comparison of new genomes to identify its structure and function, the identification of common features, patterns, and functions can be done by using mathematical modeling techniques, and the integration of above techniques using computational modeling. Computational resources in bioinformatics are requisite in life science and a key element to help test genomic contemplates to respond to unrequited biological queries (Searls 2000; Vassilev et al. 2005). Combined assortment, stockpiling, and simultaneous examination of coordinated natural and hereditary data can be applied to quality-based medication revelations, microbial ID and network investigation, bioremediation and biodegradation processes, crop improvement programs (McDaniel et al. 2005; Ouzounis 2002; Pearson and Lipman 1988) and ecological and agrarian advancement. Comparative microbial genomics research is coming to fruition at a quicker rate by using bioinformatics tools (Varshney et al. 2005).

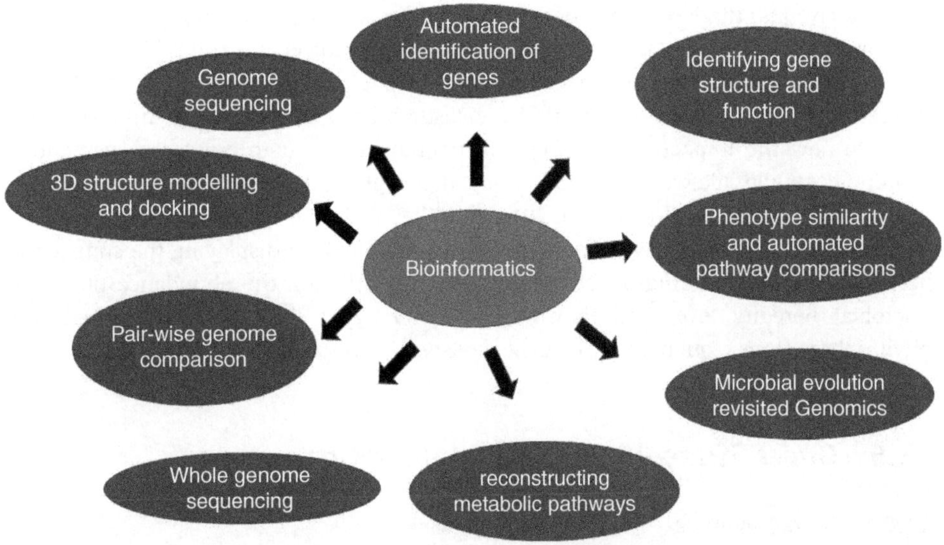

Figure 13.3 Various tasks that can be done through bioinformatics in microbial research. *Sources:* Juretic et al. (2005); Prabha et al. (2014).

After the rapid increase in genome sequencing, a few challenges have come up associated with the identification of protein structure and function (Gomez et al. 2008). By predicting the structures of encoded proteins, the function of associated genes can be identified (Ideker et al. 2001; Jones 2000). Further, the chromosomal positioning of the genes can be located by comparing the complete genome sequences of different microbes (Prabha et al. 2014). These techniques are progressively being used in the annotation of the genome (Fulekar and Sharma 2008). Currently, the annotation of gene functions using comparative genomics led to the identification of accurate functional annotations for various organisms' proteins (Callister et al. 2008; Petrosino et al. 2009).

13.5 Phyllosphere Microbiome Studies Based on Genome-Wide Association

Natural sequencing demonstrates that plants harbor complex networks of organisms that fluctuate crosswise over conditions. Be that as it may, numerous methodologies for mapping plant hereditary variety to microorganism-related attributes were created in the moderately basic setting of twofold host–organism collaborations under precise surroundings.

Ongoing developments in sequencing and insights make genome-wide affiliation considers (GWAS) a progressively encouraging methodology for recognizing the plant hereditary variety related to microorganisms in a network setting (Aflitos et al. 2014). GWAS allude to the use of affiliation mapping, a strategy that tests for the measurable relationship between hereditary variety at a locus and the phenotypes of organisms, to a great many loci all through the genome (Lu et al. 2016). Since GWAS just requires genotype and phenotype information crosswise, it can in principle be utilized for any property of the phyllosphere that can be evaluated and is affected by plant genotype (Brachi et al. 2011).

Promisingly, quantitative records of network synthesis and decent variety change among plant genotypes developed in like manner situations (Horton et al. 2014; Peiffer et al. 2013), making them reasonable phenotypes to examine through GWAS. A key quality of GWAS in this setting is the adaptability of its hidden measurable models, which can think about and evaluate how the impacts of genotype on phenotype fluctuate crosswise over natural conditions, while likewise controlling for the ecological commotion that bewilders relationship in the field (Korte et al. 2012; Sasaki et al. 2015). Current computational and factual advances permit GWAS to pick up influence through displaying the connections between qualities, offering an energizing road for utilizing the abundance of various microbial heredity level and network-level phenotypes that can be removed from phyllosphere DNA sequencing (Curtin et al. 2017; Hackinger and Zeggini 2017).

13.6 Omics Strategies and Their Integration

13.6.1 Metagenomics

The traditional metagenome approach includes the cloning of DNA into a vector with the assistance of host strain. The subsequent clone libraries are along these lines separated either for gene marker or metabolic functions (Riesenfeld et al. 2004). For screening proteins, the methodology driven by functions is as yet vital with possibilities in biotechnology (Ekkers et al. 2012), though in microbial biology, expanding throughput and reducing expenses for DNA sequencing have condensed the sequence driven methodology to a great extent out of date. These days to examine the gene inventories' microbial world, DNA sequencing is used. By joining the subsequent metagenomic information with biodiversity data, a new kind of all-inclusive ecosystem studies has become viable (Simon and Daniel 2011). So also, it can be incorporated with metabolome information (Turnbaugh and Gordon 2008). Such integrative "biological systems science" thinks about (Shi et al. 2011; Teeling et al. 2012) present plenty of difficulties regarding the trial plan and bioinformatic downstream preparing. These include contemplations about the natural surroundings, testing system, sequencing innovation, biodiversity estimation, work forecasts and examinations, information joining and ensuing translation and information affidavit.

13.6.2 Metatranscriptomics

In a microbiome, the microbial activity and function can be examined by looking at the number of transcript sequences from the organism. The most popular method for the analysis of transcriptome is RNA sequencing, which was for the first time carried out in a root microbiome *Arabidopsis thaliana* rhizosphere at its various growing phases (Chaparro et al. 2014). Currently, bioinformatic investigation of meta-transcriptomic sequencing data is done by high-throughput sequencing in the form of short read. Generally, the data produced from Illumina sequencing technology mostly functional for meta-transcriptome studies, mainly when deep coverage and numerous samples and deep coverage are essential, whereas long-read sequencing can sequence complete mRNAs that can differentiate between various isoforms and provide long stretches of sequence for similarity searches

(Pollard et al. 2018). Reference-based assemblies cover the genome of those species which could be freezed and have a high quality, while reference-free assemblies have restrictions because of homologous regions of the gene (Gongora-Castillo and Buell 2013). As far as the measurable investigation, transcripts and genes can be evaluated with explicit statistical strategies intended for such reason (Bray et al. 2016). For the investigation of genes, those are differentially expressed now as DEG among different samples under various circumstances, and R language can be used (Core Team 2017). For gene annotation, reconstruction of metabolic pathways, gene ontology studies, assembly of databases, and many software tools are available. Co-expression networks can be feasible for further analysis and can be used for large-scale experimental setups (Robinson et al. 2010; Schlapfer et al. 2017; Serin et al. 2016). Trinotate, a method used for automatic functional annotation, is offered for an inclusive transcriptome's annotation (Bryant et al. 2017). In-plant microbial-related investigations, meta-transcriptomic information was utilized for the comprehension of plant rhizosphere microbiome developed in similar soil and for evaluating bacterial gene expression growth (Lambais et al. 2017; Levy et al. 2018; Turner et al. 2013).

13.6.3 Metabolomics

Metabolomics thinks about a target, seeing little atom metabolites of the biological framework under explicit conditions. Normally, the metabolome comprises primary and secondary metabolites. Contrasted with other complex organic frameworks, a large variety of secondary metabolites is involved in microbes defense mechanisms (Wink 2003). The examination of metabolomic mixes brings fingerprints and metabolic profiles up to the identification of novel biomarkers, and this can be coordinated for the investigations and progressively all-encompassing comprehension of the plant microbiome.

Most regular advances utilized in metabolomics are NMR, GC-MS and LC-MS. Mass spectrometry (MS)-based methods identify metabolites with a lot more advanced affectability than NMR (Emwas 2015). However, MS tests require a detailed readiness, and the discovery is restricted uniquely to metabolites that can ionize into the perceivable mass range. The benefits of utilizing NMR stand apart for intensify that are hard to disintegrate (Markley et al. 2017). Arrangements of openly accessible tools dependent on their usefulness and innovation utilized are accessible in Spicer et al. (2017). The application of multi-omics data received, metabolomics concentrating on metabolic demonstrating in plants has been surveyed as of late (Rai et al. 2017). Various online assets are accessible also, giving programming devices, instructional exercises, conventions and rules on preparing, measurable examination, and perception of metabolomic information.

13.6.4 Proteomics

Proteomics is the investigation of the proteins in a microbial network from a natural example. As opposed to other omics procedures, proteomics gives direct proof to alterations, associations, and interactions (Hettich et al. 2013). As a rule, proteomics generally uses techniques beginning from MS-based proteomics. MS examination in an enormous scale base up analysis promptly brings about a large number of spectra that require mechanized mass otherworldly translation. Significant strides in the computational work process comprise of range pre-processing, identification of peptides, evaluation, a grouping of protein (Huson et al. 2007; Mesuere et al. 2018). In proteomics study identification of peptides have a crucial and important role to derive most of the ingredients of a microbe. Searching of the database and de novo sequencing are the most mainstream ways to assign a peptide sequence to a spectrum (Lam et al. 2007; Ma et al. 2003). Maximum protein sequence databases are assembled from different omics sources, however, at the center use gene prediction from essential genome congregations. Separately, the quality of the genome quality and the genome assembly together extraordinarily impact the substance in reference databases (O'Leary et al. 2016; Pundir et al. 2017; Zerbino et al. 2018). The features need to be adjusted in proteomics study, i.e. the complexity, completeness, and size of the database (Zerbino et al. 2018). In the study of the plant microbiome, proteomics was used to assess bacterial networks in the phyllosphere (Kierul et al. 2015; Lambais et al. 2017; Lin et al. 2013). Regardless of the effective use, proteomics is restricted because of the lesser appearance of proteins in plant microbes.

13.7 Conclusion and Future Prospects

Development of advanced techniques recently has extensively enhanced and accelerated the examination of plant-related microorganisms, profound comprehension of roots microbiome, and distinguishing proof of bacterial taxa and genes. An inauspicious way to deal with the comprehending of corresponding impacts of microbes with their associated plants is in dismantling them and setting up manufactured microbial networks for building again to consider their collaborations. Hence, the utilization of genomes that are sequenced and completely portrayed microbes would be taken into consideration foreseeing practical interrelations.

Because of the huge number of the variety of plants with their sequences and gathering difficulties, numerous plants have been investigated for their genome, where as many plant genome arrangements are yet spoken to as a draft. In this manner, tests are directed in model plants. Moreover, various Plant Genomes Projects can lessen this breaking point by sequencing agent species from each significant microbe. Regardless of the distinctions in the microbial network dependent on many factors, such as on plant species, soil, and environmental condition, it is essential to ponder if center microbiome capacities explicit to phyllosphere and rhizosphere exist, and, provided that this is true, are able to comprehend connection systems between center microorganisms and plants. These bits of knowledge will be tested by our comprehension of microbiome commitments to plant well-being and the advancement of utilizations in horticulture. With the decreased expense of sequencing, an immense measure of omics information from plant microbial network can be normalized. The improvement of such databases should be organized to empower the useful and natural elucidation of the up-and-coming huge-scale multi-omics plant microbiome information.

References

Aflitos, S., Schijlen, E., De Jong, H. et al. (2014). Exploring genetic variation in the tomato (*Solanum* section *Lycopersicon*) clade by whole-genome sequencing. *Plant J.* 80: 136–148.

Altschul, S.F., Gish, W., Miller, W. et al. (1990). Basic local alignment search tool. *J. Mol. Biol.* 215 (3): 403–410.

Altschul, S.F., Madden, T.L., Schäffer, A.A. et al. (1997). Gapped BLAST and PSI-BLAST: a new generation of protein database search programs. *Nucleic Acids Res.* 25 (17): 3389–3402.

Bar-On, Y.M., Phillips, R., and Milo, R. (2018). The biomass distribution on earth. *Proc. Natl. Acad. Sci. U. S. A.* 115: 6506–6511.

Brachi, B., Morris, G.P., and Borevitz, J.O. (2011). Genome-wide association studies in plants: the missing heritability is in the field. *Genome Biol.* 12: 232.

Bray, N.L., Pimentel, H., Melsted, P., and Pachter, L. (2016). Near-optimal probabilistic RNA-seqquantification. *Nat. Biotechnol.* 34: 525–527.

Bryant, D.M., Johnson, K., Di-Tommaso, T. et al. (2017). A tissue-mapped axolotl De Novo transcriptome enables identification of limb regeneration factors. *Cell Rep.* 18: 762–776.

Callister, S.J., McCue, L.A., Turse, J.E. et al. (2008). Comparative bacterial proteomics: analysis of the core genome concept. *PLoS One* 3 (2): e1542.

Cha, J.Y., Han, S., Hong, H.J. et al. (2016). Microbial and biochemical basisof a *Fusarium* wilt-suppressive soil. *ISME J.* 10: 119–129.

Chaparro, J.M., Badri, D.V., and Vivanco, J.M. (2014). Rhizosphere microbiome assemblage is affected by plant development. *ISME J.* 8: 790–803.

Core Team (2017). *R: A Language and Environment for Statistical Computing*. Vienna, Austria: Foundation for Statistical Computing.

Corwin, J.A. and Kliebenstein, D.J. (2017). Quantitative resistance: more than just perception of a pathogen. *Plant Cell* 29: 655–665.

Curtin, S.J., Tiffin, P., and Guhlin, J. (2017). Validating genome-wide association candidates controlling quantitative variation in nodulation. *Plant Physiol.* 173: 921–931.

Curtis, T.P., Sloan, W.T., and Scannell, J.W. (2002). Estimating prokaryotic diversity and its limits. *Proc. Natl. Acad. Sci. U. S. A.* 99: 10494–10499.

Ekkers, D.M., Cretoiu, M.S., Kielak, A.M., and van Elsas, J.D. (2012). The great screen anomaly – a new frontier in product discovery through functional metagenomics. *Appl. Microbiol. Biotechnol.* 93 (3): 1005–1020.

Emwas, A.H. (2015). The strengths and weaknesses of NMR spectroscopy and mass spectrometry with particular focus on metabolomics research. *Methods Mol. Biol.* 1277: 161–193.

Epstein, S.S. (2013). The phenomenon of microbial uncultivability. *Curr. Opin. Microbiol.* 16: 636–642.

Fahimipour, A.K., Kardish, M.R., Lang, J.M. et al. (2017). Global-scale structure of the Eelgrass microbiome. *Appl. Environ. Microbiol.* 83: e03391–e03316.

Fleischmann, R.D., Adams, M.D., White, O. et al. (1995). Whole-genome random sequencing and assembly of *Haemophilus influenzae* Rd. *Science* 269 (5223): 496–512.

Fulekar, M.H. and Sharma, J. (2008). Bioinformatics applied in bioremediation. *Innov. Rom. Food Biotechnol.* 3: 28.

Gomez, S.M., Choi, K., and Wu, Y. (2008). Prediction of protein-protein interaction networks. *Curr. Protoc. Bioinformatics* 22 (1): 8–2.

Gongora-Castillo, E. and Buell, C.R. (2013). Bioinformatics challenges in de novo transcriptome assembly using short read sequences in the absence of a reference genome sequence. *Nat. Prod. Rep.* 30: 490–500.

Hackinger, S. and Zeggini, E. (2017). Statistical methods to detect pleiotropy in human complex traits. *Open Biol.* 7 (170): 125.

Hagen, J.B. (2000). The origins of bioinformatics. *Nat. Rev. Genet.* 1: 231–236.

Haney, C.H., Samuel, B.S., Bush, J., and Ausubel, F.M. (2015). Associations with rhizosphere bacteria canconfer anadaptive advantage to plants. *Nat. Plants* 1 (15): 051.

Heng, L.S. (2011). The digital side of biology. *Asia Research News* (www.asiaresearchnews.com).

Hettich, R.L., Pan, C., Chourey, K., and Giannone, R.J. (2013). Metaproteomics: harnessing the power of high-performance mass spectrometry to identify the suite of proteins that control metabolic activities in microbial communities. *Anal. Chem.* 85: 4203–4214.

Hogeweg, P. (2011). The roots of bioinformatics in theoretical biology. *PLoS Comput. Biol.* 7 (3): e1002021.

Horton, M.W., Bodenhausen, N., and Beilsmith, K. (2014). Genome-wide association study of *Arabidopsis thaliana* leaf microbial community. *Nat. Commun.* 5: 5320.

Hug, L.A., Baker, B.J., Anantharaman, K. et al. (2016). A new view of the tree of life. *Nat. Microbiol.* 1: 16048.

Huson, D.H., Auch, A.F., Qi, J., and Schuster, S.C. (2007). MEGAN analysis of metagenomic data. *Genome Res.* 17: 377–386.

Ideker, T., Galitski, T., and Hood, L. (2001). A new approach to decoding life: systems biology. *Annu. Rev. Genomics Hum. Genet.* 2 (1): 343–372.

Innerebner, G., Knief, C., and Vorholt, J.A. (2011). Protection of *Arabidopsis thaliana* against leaf pathogenic *Pseudomonas syringae* by *Sphingomonas* strains in a controlled model system. *Appl. Environ. Microbiol.* 77: 3202–3210.

Jones, D.T. (2000). Protein structure prediction in the postgenomic era. *Curr. Opin. Struct. Biol.* 10 (3): 371–379.

Juretic, D., Lucic, B., and Trinajstic, N. (2005). Why focusing on bioinformatics? *Period. Biol.* 107 (4): 379.

Kierul, K., Voigt, B., Albrecht, D. et al. (2015). Influence of root exudates on the extracellular proteome of the plant growth-promoting bacterium *Bacillus amyloliquefaciens* FZB42. *Microbiology* 161: 131–147.

Koberl, M., Schmidt, R., Ramadan, E.M. et al. (2013). The microbiome of medicinal plants: diversity and importance for plant growth, quality, and health. *Front. Microbiol.* 4: 400.

Korte, A., Vilhj-Aalmsson, B.J., Segura, V. et al. (2012). A mixed-model approach for genome-wide association studies of correlated traits in structured populations. *Nat. Genet.* 44: 1066–1071.

Lam, H., Deutsch, E.W., Eddes, J.S. et al. (2007). Development and validation of a spectral library searching method for peptide identification from MS/MS. *Proteomics* 7: 655–667.

Lambais, M.R., Crowley, D.E., Cury, J.C. et al. (2006). Bacterial diversity in tree canopies of the Atlantic forest. *Science* 312: 1917.

Lambais, M.R., Barrera, S.E., Santos, E.C. et al. (2017). Phyllosphere metaproteomes of trees from the Brazilian Atlantic forest show high levels of functional redundancy. *Microb. Ecol.* 73: 123–134.

Levy, A., Conway, J.M., Dangl, J.L., and Woyke, T. (2018). Elucidating bacterial gene functions in the plant microbiome. *Cell Host Microbe* 24: 475–485.

Lin, W., Wu, L., Lin, S. et al. (2013). Metaproteomic analysis of ratoon sugarcane rhizospheric soil. *BMC Microbiol.* 13: 135.

Lindow, S.E. and Brandl, M.T. (2003). Microbiology of the phyllosphere. *Appl. Environ. Microbiol.* 69: 1875–1883.

Liu, J., Dehbi, M., Moeck, G. et al. (2004). Antimicrobial drug discovery through bacteriophage genomics. *Nat. Biotechnol.* 22 (2): 185.

Lu, K., Xiao, Z., and Jian, H. (2016). A combination of genome-wide association and transcriptome analysis reveals candidate genes controlling harvest index-related traits in *Brassica napus*. *Sci. Rep.* 6 (36): 452.

Ma, B., Zhang, K., Hendrie, C. et al. (2003). PEAKS: powerful software for peptide de novo sequencing by tandem mass spectrometry. *Rapid Commun. Mass Spectrom.* 17: 2337–2342.

Mansfield, J., Genin, S., and Magori, S. (2012). Top 10 plant pathogenic bacteria in molecular plant pathology. *Mol. Plant Pathol.* 13: 614–629.

Marasco, R., Rolli, E., Ettoumi, B. et al. (2012). A drought resistance-promoting microbiome is selected by root system under desert farming. *PLoS One* 7: e48479.

Markley, J.L., Bruschweiler, R., Edison, A.S. et al. (2017). The future of NMR-based metabolomics. *Curr. Opin. Biotechnol.* 43: 34–40.

McDaniel, A.M., Benson, P.L., Roesener, G.H., and Martindale, J. (2005). An integrated computer based system to support nicotine dependence treatment in primary care. *Nicotine Tob. Res.* 7: S57–S66.

Mesuere, B., Van der Jeugt, F., Willems, T. et al. (2018). High-throughput metaproteomics data analysis with Unipept: a tutorial. *J. Proteomics* 171: 11–22.

Morella, N.M., Gomez, A.L., Wang, G. et al. (2018). The impact of bacteriophages on phyllosphere bacterial abundance and composition. *Mol. Ecol.* 27: 2025–2038.

Nakashima, N., Mitani, Y., and Tamura, T. (2005). Actinomycetes as host cells for production of recombinant proteins. *Microb. Cell Fact.* 4 (1): 7.

O'Leary, N.A., Wright, M.W., Brister, J.R. et al. (2016). Reference sequence (RefSeq) database at NCBI: current status, taxonomic expansion, and functional annotation. *Nucleic Acids Res.* 44: D733–D745.

Ouzounis, C. (2002). Bioinformatics and the theoretical foundations of molecular biology.

Pearson, W.R. and Lipman, D.J. (1988). Improved tools for biological sequence comparison. *Proc. Natl. Acad. Sci. U. S. A.* 85 (8): 2444–2448.

Peiffer, J.A., Spor, A., Koren, O. et al. (2013). Diversity and heritability of the maize rhizosphere microbiome under field conditions. *Proc. Natl. Acad. Sci. U. S. A.* 110: 6548–6553.

Petrosino, J.F., Highlander, S., Luna, R.A. et al. (2009). Metagenomicpyro sequencing and microbial identification. *Clin. Chem.* 55 (5): 856–866.

Pollard, M.O., Gurdasani, D., Mentzer, A.J. et al. (2018). Long reads: their purpose and place. *Hum. Mol. Genet.* 27: R234–R241.

Prabha, R., Singh, D.P., Gupta, S.K., and Rai, A. (2014). Whole-genome phylogeny of *Prochlorococcus marinus* group of cyanobacteria: genome alignment and overlapping gene approach. *Interdiscipl. Sci. Comput. Life Sci.* 6 (2): 149–157.

Pundir, S., Martin, M.J., and O'Donovan, C. (2017). UniProt protein knowledgebase. *Methods Mol. Biol.* 1558: 41–55.

Rai, A., Saito, K., and Yamazaki, M. (2017). Integrated omics analysis of specialized metabolism inmedicinal plants. *Plant J.* 90: 764–787.

Rastogi, G., Coaker, G.L., and Leveau, J.H.J. (2013). New insights into the structure and function of phyllosphere microbiota through high-throughput molecular approaches. *FEMS Microbiol. Lett.* 348: 1–10.

Riesenfeld, C.S., Schloss, P.D., and Handelsman, J. (2004). Metagenomics: genomic analysis of microbial communities. *Annu. Rev. Genet.* 38: 525–552.

Robinson, W.H., Fontoura, P., Lee, B.J. et al. (2003). Protein microarrays guide tolerizing DNA vaccine treatment of autoimmune encephalomyelitis. *Nat. Biotechnol.* 21 (9): 1033.

Robinson, M.D., McCarthy, D.J., and Smyth, G.K. (2010). edgeR: a bioconductor package for differential expression analysis of digital gene expression data. *Bioinformatics* 26: 139–140.

Rogozin, I.B., Makarova, K.S., Natale, D.A. et al. (2002). Congruent evolution of different classes of non-coding DNA in prokaryoticgenomes. *Nucleic Acids Res.* 30 (19): 4264–4271.

Ruiz-Perez, C.A., Restrepo, S., and Zambrano, M.M. (2016). Microbial and functional diversity within the phyllosphere of *Espeletia* species in an Andean high mountain ecosystem. *Appl. Environ. Microbiol.* 82: 1807–1817.

Sapp, M., Ploch, S., Fiore-Donno, A.M. et al. (2018). Protists are an integral part of the *Arabidopsis thaliana* microbiome. *Environ. Microbiol.* 20: 30–43.

Sasaki, E., Zhang, P., Atwell, S. et al. (2015). "Missing" G x E variation controls flowering time in *Arabidopsis thaliana*. *PLoS Genet.* 11: e1005597.

Scheuring, I. and Yu, D.W. (2012). How to assemble a beneficial microbiome in three easy steps. *Ecol. Lett.* 15: 1300–1307.

Schlapfer, P., Zhang, P., Wang, C. et al. (2017). Genome-wide prediction of metabolic enzymes, pathways, and gene clusters in plants. *Plant Physiol.* 173: 2041–2059.

Searls, D.B. (2000). Bioinformatics tools for whole genomes. *Annu. Rev. Genomics Hum. Genet.* 1 (1): 251–279.

Serin, E.A., Nijveen, H., Hilhorst, H.W., and Ligterink, W. (2016). Learning from co-expression networks: possibilities and challenges. *Front. Plant Sci.* 7: 444.

Shi, Y., Tyson, G.W., Eppley, J.M., and DeLong, E.F. (2011). Integrated metatranscriptomic and metagenomic analyses of stratified microbial assemblages in the open ocean. *ISME J.* 5 (6): 999.

Simon, C. and Daniel, R. (2011). Metagenomic analyses: past and future trends. *Appl. Environ. Microbiol.* 77 (4): 1153–1161.

Smith, K.P. and Goodman, R.M. (1999). Host variation for interactions with beneficial plant-associated microbes. *Annu. Rev. Phytopathol.* 37: 473–491.

Spence, C., Alff, E., Johnson, C. et al. (2014). Natural ricerhizospheric microbes suppress rice blast infections. *BMC Plant Biol.* 14: 130.

Spicer, R., Salek, R.M., Moreno, P. et al. (2017). Navigating freely-available software tools for metabolomics analysis. *Metabolomics* 13: 106.

Teeling, H., Fuchs, B.M., Becher, D. et al. (2012). Substrate-controlled succession of marine bacterioplankton populations induced by aphytoplankton bloom. *Science* 336 (6081): 608–611.

Turnbaugh, P.J. and Gordon, J.I. (2008). An invitation to the marriage of metagenomics and metabolomics. *Cell* 134 (5): 708–713.

Turner, T.R., Ramakrishnan, K., Walshaw, J. et al. (2013). Comparative metatranscriptomics reveals kingdom level changes in the rhizosphere microbiome of plants. *ISME J.* 7: 2248–2258.

Uroz, S., Buée, M., Deveau, A. et al. (2016). Ecology of the forest microbiome: highlights of temperate and boreal ecosystems. *Soil Biol. Biochem.* 103: 471–488.

Varshney, R.K., Graner, A., and Sorrells, M.E. (2005). Genomics-assisted breeding for crop improvement. *Trends Plant Sci.* 10 (12): 621–630.

Vassilev, D., Leunissen, J., Atanassov, A. et al. (2005). Application of bioinformatics in plant breeding. *Biotechnol. Biotechnol. Equip.* 19 (Suppl. 3): 139–152.

Venter, J.C., Adams, M.D., Myers, E.W. et al. (2001). The sequence of the human genome. *Science* 291 (5507): 1304–1351.

Vorholt, J.A., Vogel, C., Carlstrom, C.I., and Muller, D.B. (2017). Establishing causality: opportunities of synthetic communities for plant microbiome research. *Cell Host Microbe* 22: 142–155.

Waterman, M.S. (2018). *Introduction to Computational Biology: Maps, Sequences, and Genomes.* Chapman and Hall/CRC.

Wink, M. (2003). Evolution of secondary metabolites from an ecological and molecular phylogenetic perspective. *Phytochemistry* 64: 3–19.

Zagursky, R.J., Olmsted, S.B., Russell, D.P., and Wooters, J.L. (2003). Bioinformatics: how it is being used to identify bacterial vaccine candidates. *Expert Rev. Vaccines* 2 (3): 417–436.

Zerbino, D.R., Achuthan, P., Akanni, W. et al. (2018). Ensembl 2018. *Nucleic Acids Res.* 46: D754–D761.

14

Microbial Mediated Zinc Solubilization in Legumes for Sustainable Agriculture

Pawan Saini[1,2], *Sharon Nagpal*[3], *Pooja Saini*[4], *Arun Kumar*[5], *and Mudasir Gani*[6]

[1] *Department of Plant Breeding and Genetics, Punjab Agricultural University (PAU), Ludhiana, Punjab, India*
[2] *CSB-Central Sericultural Research & Training Institute (CSR&TI), Pampore, Jammu-Kashmir (UT), India*
[3] *Department of Microbiology, Punjab Agricultural University (PAU), Ludhiana, Punjab, India*
[4] *Department of Plant Breeding, Genetics and Biotechnology, Dr. Khem Singh Gill College of Agriculture, Eternal University (EU), Baru Sahib, Himachal Pradesh, India*
[5] *Department of Agronomy, Punjab Agricultural University (PAU), Ludhiana, Punjab, India*
[6] *Faculty of Agriculture, Sher-e-Kashmir University of Agricultural Sciences and Technology Kashmir (SKUAST-K) Wadura, Sopore, Kashmir, Jammu-Kashmir (UT), India*

14.1 Introduction

During the period of green revolution major attention was given to increased crop productivity, and protein malnutrition shifted the interest from conventional micronutrient rich crops to more productive and profitable starchy crops resulting in a bumper enhancement in food grain production at a global level to fight against hidden hunger (Pingali 2012). But simultaneously it resulted in the depletion of the soil micronutrient pool. Similar to cereals, pulses also have profound cultivation potential and are the main constituents of a human diet (Fuller et al. 2001; Caracuta et al. 2015). Besides cereals, pulses are an important group of food crops belonging to the family Fabaceae. Chickpea (*Cicer arietinum*), mung bean (*Vigna radiata*), pigeonpea (*Cajanus cajan*), cowpea (*Vigna unguiculata*), urdbean (*Vigna mungo*), llentil (*Lens esculnaris*), french bean (*Phaseolus vulgaris*), lathyrus (*Lathyrus sativus*), field pea (*Pisum sativum*), horse gram (*Macrotyloma uniflorum*) and moth bean (*Vigna aconitifolium*) are the chief pulses produced in world. The highest pulse production takes place in Asia where India sustains first position by contributing 20% to the worldwide production of pulses (Tan et al. 2017). Maximum production in India among pulses is attributed to chickpea (48%) followed by pigeonpea (15%), mung bean (7%), urdbean (7%), lentil (5%) and field pea (5%). Two reasons for cultivation of pulse crops at the global level are their agronomic characteristics and nutritional value (Tan et al. 2017). Pulses are the major source of dietary protein and known as "poor man's meat" in developing countries. Pulses have an excellent profile of protein ranging from 18 to 28% (Singh et al. 2015). They are also enriched with micronutrients such as riboflavin, niacin, thiamine, folate, iron, zinc,

Phytomicrobiome Interactions and Sustainable Agriculture, First Edition. Edited by Amit Verma, Jitendra Kumar Saini, Abd El-Latif Hesham, and Harikesh Bahadur Singh.

calcium, and magnesium (Phillips 1993; Iqbal et al. 2006; Jukanti et al. 2012). For a long time, the awareness of increasing deficiency of vitamins, primarily vitamin A in spite of intake of enough calories, led to an uplift in demand and supply of food rich in micronutrients such as Fe and Zn.

To improve the quality and quantity of plant products, the plants should be given adequate nutrients. Nutrient essentiality criteria suggest that there are 17 essential nutrient elements. Of which, nine are macronutrients (C, H, O, N, P, K, S, Ca, and Mg) and eight micronutrients (Fe, Mn, B, Zn, Co, Cu, Cl, and Ni) (Arnon and Stout 1939). Micronutrients are crucial for the adequate growth and enhanced productivity of crops. They are usually required in trace amounts. Of the micronutrients, zinc (Zn) is an important nutrient required in 5–100 $mg\,kg^{-1}$ in plant tissues (5–100 $mg\,kg^{-1}$) (Iqbal et al. 2010; Goteti et al. 2013). It is a bluish-white metallic element and present in earth's crust in an 0.008% concentration (Sharma et al. 2014). It has three basic functions – catalytic, structural, and regulatory (Cousins 2006). Zinc is an essential element for crop production. Zinc is associated with >50 different metallo-enzymes of distinct functions (Cousins 1996). Zinc also plays a pivotal role in cellular differentiation/proliferation, signal transduction, transcription, and metabolism (Singh et al. 2011). It is required in various key plant physiological pathways, enzyme reactions, oxidation–reduction reactions and metabolic processes (Alloway 2002; Mousavi 2011; Yosefi et al. 2011; Hafeez et al. 2013). Various physiological phenomenons, such as metabolism of carbohydrates, proteins, auxin, photosynthesis, production of starch from sugars, chlorophyll and pollen formation, biological membrane integrity and resistance to infectious pathogens, are all regulated by the zinc ion (Alloway 2008).

Zinc is an immobile element, thus it cannot move to the plant parts that are lacking (Mousavi et al. 2013). Deficiency of zinc worldwide resulted in a cessation of all the physiological functions of crop plants ultimately inhibiting growth and development. Zinc is essential for normal metabolism in plants, humans, and even microorganisms (King 2011; Hafeez et al. 2013; Sharma et al. 2013). It occurs as both available and unavailable zinc compounds in soil. It becomes limiting due to neutral/alkaline and calcareous soil (Sharma et al. 2014). In addition to this, its usability by plants is decreased by various physical factors (temperature, light) in the soil. Zinc deficiency in crop plants becomes a major problem attributed toward low availability of zinc in soil rather than low zinc content. Thus, there is a need to find the solutions to make unavailable forms of Zn as available forms for plant growth and development. In the nature, few Zn solubilizing microorganisms, viz. bacteria and fungi, are known to solubilize the complex forms of Zn to simple forms for crop plants.

14.2 Chronological Events of Zinc Biology

A student of renowned microbiologist Louis Pasteur, Jules Raulin in 1869, reported the utility of zinc for *Aspergillus niger* causal agent of black mold grapes, onions, and peanuts. These findings were later confirmed by Bertrand and Javillier (1911). Mazé (1914) documented the role of zinc in maize in hydroponic. During 1920, researchers hypothesized and investigated the presence of zinc in human tissues. Later, studies at Harvard School of Public Health, Boston, working on zinc concentration in different tissues in humans, revealed that zinc presents endogenously and serves many functions of biological importance (Lutz 1926; Drinker and Collier 1926). Studies in sunflower and barley showed that zinc was required for the growth and development of the plant, documenting its essentiality in productivity (Sommer and Lipman 1926). Mendel and Bradley (1905) reported that zinc might be required by animals. Birckner (1919) first suggested that zinc may have some role to play in higher animals. After vigorous testing trials starting from Raulin's original observation, zinc was observed to be essential for rats in 1934 at the University of Wisconsin (Todd et al. 1934) and for humans in 1961 (Prasad et al. 1961).

14.3 Role of Zinc in Living System

Zinc constitutes a crucial part as a micronutrient for both eukaryotes and prokaryotes. It is part of major organs, tissues, and fluids of body. There exists approximately 1.5–2.0 g or about 0.003% zinc of total body weight of an adult. Recommended daily intake of zinc was estimated to be 3–16 mg day^{-1}. It depends on gender, age, diet type and many other factors. The estimated physiological requirement of zinc is 1.4 mg day^{-1} and 1.0 mg day^{-1} for male and female respectively (WHO 1996). The optimum dietary consumption for human adults is 15 mg day^{-1}. According to the reports, the population suffering from dietary zinc deficiency is more than 30% (Hotz and Brown 2004).

Zinc being a type II nutrient element is required for multiple general metabolic functions like nitrogen, essential amino acids, magnesium, and potassium metabolisms (Golden 1989). The essential trace nutrient element zinc is appraised to be crucial for cell division as well as DNA and protein synthesis. It is considered important for the metabolic activity of more than 300 metalloenzymes (Vallee 1991). Zn is required by many principal proteins of body. The involvement of eukaryotic zinc proteins in gene expression is 36% and in eukaryotic cell zinc proteins make up approximately 10% of the proteome (Jagana et al. 2019). Zinc is found to be present in all enzyme classes, viz., *oxido-reductases, transferases, hydrolases, lyases, isomerases, and ligases* (Webb 1992; Barak and Helmke 1993), as designated by the International Union of Biochemistry and Molecular Biology (IUBMB). As evident from the chronological events of the zinc biochemistry, the three phases of discovery can be perceived as zinc recognition, catalytic and structural roles aiding zinc detection and vitality of zinc in regulation of cell. The Copenhagen Consensus determined the deficiency of both zinc and vitamin A together as the most crucial universal problem and suggested that eradicating zinc deficiency will show direct high effects and returns for mankind around the globe (Jagana et al. 2019).

14.3.1 Essentiality of Zinc in Humans

Until the 1960s, no cases in humans regarding zinc deficiency were detected (Brown et al. 1993; Welch 1993). In 1961, the essentiality of zinc for humans was recognized (Prasad et al. 1961). According to the nutritionist, zinc deficiency affects a large number of women and children across the world and is a risk factor for immunity and is cause of major infection in elder peoples. Billions of lives have been affected by malnutrition of micronutrients, mostly iron deficiency (5 billion) and zinc starvation (2.7 billion) across the globe (Anonymous 2004). In India, 27% of population is dealing with Zn deficiency ailments like poor immune system and physical growth, diarrhea and mental retardation (World Health organization 2004). In a study conducted in Haryana, India it was found that nearly 65% of the pregnant women were deficient in zinc, mainly due to low intake of dietary Zn (less than 15 mg Zn per day) (Singh et al. 2011). In elderly people, inadequate zinc storage is a risk factor leading to pneumonia (Meydani et al. 2007). A low level of zinc in the body causes zinc deficiency. Patients are diagnosed with several diseases, such as diabetes mellitus, burns, hemolytic anemia, liver cirrhosis, malabsorption or intestinal fistula. The human body shows problems such as hair and memory loss, skin problems and weakness in the body muscles. Further, zinc deficit during pregnancy causes abnormalities in fetus like retarded brain development and even infertility in the male child. Zn deficiency can be cause of various congenital diseases like *Acrodermatitis enteropathica* (Lukaski 2004; Morlet 2004; Zimmermann 2001) and reduces the serum testosterone level which mainly affects T-helper cells and also cause hyperammonemia, neurosensory disorders and decrease in lean body mass (Prasad 2008). Zn plays a crucial role as immunity modulator through its accessibility, which is strongly delimited by some transporters and regulators. Zn deficiency affects monocytes, polymorphonuclear-, natural killer-, T-, and B-cells involved in both innate and adaptive immunity at the survival, proliferation and maturation levels (Bonaventura et al. 2015).

14.3.2 Essentiality of Zinc in Plants

Role of zinc for the growth and development of crop plants was established by Mazé (1914) by using hydroponics method of raising maize plants. Crops usually require zinc concentration of 0.05 $kg\,ha^{-1}\,ton^{-1}$ of grain yield, but the fixation of zinc in soil leads to a higher application of more zinc fertilizers (Mohammad et al. 2018). It plays crucial role in plant metabolism as a structural constituent or regulatory co-factor in various enzymes by influencing various enzymatic activities, stabilizing ribosomal fraction, synthesizing cytochrome, metabolism of carbohydrate, lipid, and nucleic acids (Tisdale et al. 1984; Cakmak 2000; Palmer and Guerinot 2009). Zinc is involved in the activation of various enzymes namely, RNA polymerase, alcohol dehydrogenase, carbonic anhydrase, etc. The transition metal zinc has unpaired electrons that encourage it to participate in oxidation–reduction reactions. Similarly, zinc is actively involved in the metabolism of carbohydrate, synthesis of protein, cellular membranes maintenance, auxin synthesis, pollen formation, early development of fruits and provides resistance to infection by certain pathogens (Figure 14.1) (Marschner 1995; Alloway 2008). Impaired physiological functions, leading to severe reduction in growth, yields and poor quality of crop are the result of zinc limitation (Brown et al. 1993). Zn^{2+} ions are constituents of many transcription factors which regulates cell proliferation and differentiation such as those belonging to the zinc finger family (Vallee and Falchuk 1993; Lin et al. 2005; Palmer and Guerinot 2009). It plays a role in the structural stability and functioning of genetic material and is essential for curtaining chromatin structure, metabolism of DNA and RNA and regulation of genes (Ho 2004). Zinc deficiency results in a decrease in protein synthesis process due to the degradation of RNA (Ciais et al. 2004) and a lower activity of RNA polymerase (Marschner 1995). Transcription factors of zinc finger family aid in development and functioning of flower

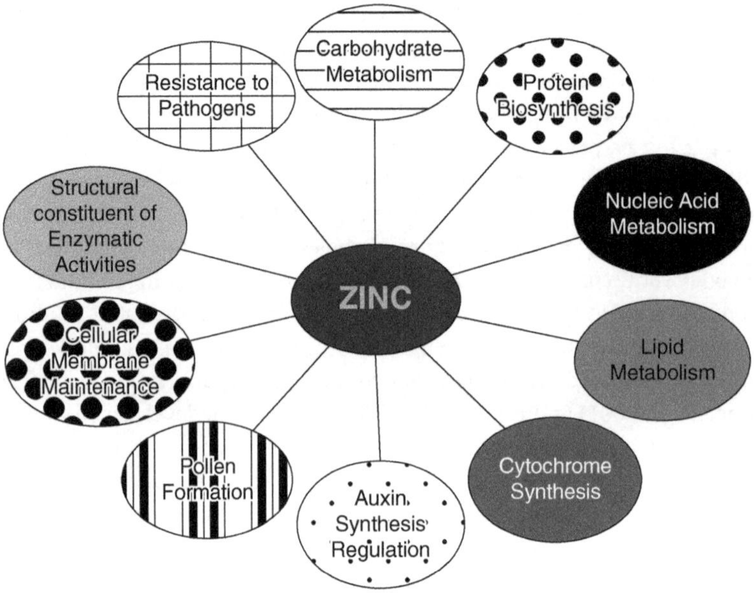

Figure 14.1 Role of zinc in plants. *Sources:* Marschner (1995); Alloway (2008).

tissues such as anther, tapetum, pollen, and pistil secretary tissues (Kobayashi et al. 1998). It has role in chloroplast formation and is involved in activity of SPP peptidase and the damage repair of photosystem-II mediated by turning over photo-damaged D1 protein (Hansch and Mendel 2009). Zn ion is the component of oxidative stress enzymes like superoxide dismutase and catalase which protect plants. It is also involved in the generation and detoxification of reactive oxygen species (ROS) (Cakmak and Marschner 1988b; Cakmak 2000), maintenance of integrity and working of cell membranes (Pinton et al. 1994). The regulatory gene expression at the time of environment stresses is mostly Zn dependent in plants (Cakmak 2000). Synthesis of amino acid tryptophan a substrate for indole acetic acid (IAA) (growth hormone) requires Zn ions (Alloway 2004; Brennan 2005). Zinc limitation accumulates inactive RNases and starch in plants indicating its role degradation of RNA (Zargar et al. 2015).

Zinc deficiency cause abnormalities in plants as deficiency symptoms, viz., diminished growth, yellowing of the leaves between veins, bronzing of chlorotic leaves, small and abnormally shaped leaves, stunted and rosetted leaves where leaves form a whorl or shortened stems and spikelet sterility. Zinc deficiency symptoms mainly occur in the new/terminal parts, as zinc is not a mobile element in plants. Because of its low mobility in plants, a constant supply of zinc is essential for optimum growth. Zn deficiency also resulted in the deterioration of the quality of harvested products, plant susceptibility to abiotic (temperature, light) and biotic factors (diseases) (Marschner 1995; Cakmak 2000). It also affects water absorption and movement in plants (Tavallali et al. 2010). Zinc is closely related to N_2 metabolism pathway in plants which interferes with protein synthesis (Fageria 2004). Flowering and fruiting processes also get affected under severe Zn deficiency condition in plants (Epstein and Bloom 2005).

14.4 Zinc Deficiency vs. Zinc Toxicity in Crop Plants

Zinc deficiency is a serious micronutrient problem occurring in plants and animals including humans (Saravanan et al. 2011; Welch 2002). But to date no deficiency symptoms were observed in microorganisms. When there is an inadequate supply of zinc from soil solution to plants, it results in debilitated physiological functions of zinc and adversely affects the plant growth. When the zinc concentration in soil solution goes down from 0.5 mg kg^{-1} dry soil, the limiting symptoms appear. The zinc concentration in the soil solution is usually unsubstantial (0.002–0.196 mg l^{-1}) in spite of 50–100 mg l^{-1} available zinc (Srivastava and Gupta 1996). Estimation of leaf zinc content is an alternate way of evaluating the zinc deficiency. The decrease in leaf zinc content from 20 mg kg^{-1} in plant leads to deficiency symptoms (Alloway 2008). Zinc deficiency is a threat to food production worldwide (Alloway 2001; Welch and Graham 2004) as it is widespread among different crops including legume crops (Naik and Das 2008).

According to Singh et al. (2005) "one of the widest ranging abiotic stresses in world agriculture arises from low zinc availability in calcareous soils, particularly in cereals." Zinc deficiency occurs in a wide range of soil types and different climatic zones, i.e. semi-arid areas with calcareous soils, tropical areas with highly weathered soils and sandy textured soils (Alloway 2008; Akay 2011). Sandy soils with high calcium carbonate is prone to zinc and iron limitation. In eroded soils, organic matter deficiency causes zinc deficiency (Mousavi 2011).

Metals are needed in traces by the biota and their higher amount results in toxicity problems and affects both macro- as well as microflora (Rout and Das 2003). The general trend of higher effect of metal toxicity is more prominent in plants, followed by soil fauna and microorganisms (Alloway 2008). Both plants and microbial populations exhibit zinc toxicity symptoms which are required for growth and development. But when zinc concentration exceeds more than its normal level, it results in growth inhibition. Zinc toxicity symptoms such as shoot stunting, deformation of leaves and ultimately death and yellowing of leaves are clearly visible in affected plants. Many researchers have reported the cases of zinc toxicity (Beckett and Davis 1977, 1978; Davis and Parker 1993; Taylor et al. 1991; Webber 1981; Wheeler and Power 1995; White et al. 1974). In zinc-sensitive plants the threshold total zinc varies from 150 to 200 mg kg^{-1} soil and it is 100–500 mg kg^{-1} soil at which zinc toxicity occurs and 25% yield penalty may be observed (Kabata-Pendias 2001). There are various reasons behind the Zn pollution in soils, such as industrial pollution, wastes from Zn mines, polluted flood waters and immoderate use of Zn-rich sewage sludge (Alloway 2008). Among the physical factors, the most important is pH of the soil because with an increase in the pH the basicity of zinc also increases which directly affects the growth. Zinc toxicity affects the physiological processes like photosynthesis, cell organelles, cellular membranes, chromatin material and nucleoli (Sresty and Madhava Rao 1999). Plasma membrane is the most important cell compartment as it permits the metal ions to get entry into a cell that may be affected due to zinc toxicity. Increase in Zn levels in soil solution results in the decrease in shoot-to-root ratio as well as translocation of other essential elements like Fe, Mg, K, P, and Ca (Godbold and Huttermann 1985). Zinc toxicity mainly affects the root physiology by inhibiting cell elongation and its division (Barcelo and Poschenrieder 1990). Zinc toxicity levels in diverse range of crops have been reported (Chardonnes et al. 1999; Staker and Cummings 1941).

Generally, soil microorganisms are more susceptible to higher zinc concentration. Among them, the nitrifying bacteria are most sensitive to soil contamination with excessive zinc (Baath 1989). Perhaps *Rhizobium leguminosarum,* the nitrogen-fixing bacterium, which is most sensitive to high Zn content performs nodulation in the roots of legume crops (Alloway 2008). Nitrification is the process of conversion of ammonia into nitrite and nitrate and is a most crucial phase in nitrogen cycle. In the presence of excessive Zn in the soil, nitrification inhibition takes place resulting in hampered nitrogen metabolism of plants as well as nitrifying bacterium. Increased Zn content in soil may modify the composition of bacteria. Soil microorganisms are the source of soil enzymes (Kucharksi 1997). Moffett et al. (2003) reported that zinc toxicity causes reduction in the microorganism pool. Excess levels of heavy metals affect soil enzymes directly or indirectly. In small amounts, heavy metals stimulate enzymatic activity while larger amount inhibits the activity of microbes and extracellular enzymes (Frankenberger et al. 1983). In a study conducted by Kucharski (1992) a higher dose of zinc (400 mg kg^{-1}) reduced the activity of dehydrogenase and acid phosphatase. According to Januszek (1999), zinc is more toxic to soil microbes and enzymes than lead (Pb).

14.5 Availability of Zinc in Soil Environment

Soil serves as growth medium for different communities of living organisms, a natural body indulged in a vigorous interaction with the atmosphere influencing climate and hydro-cycle of the planet. It is the outer layer of the earth's surface weathered and fragmented from the terrestrial surface with a wide range of attributes. Zn is an essential element for the normal growth and is required in trace amounts. Zinc content is present in various fixed forms such as hopeite ($Zn_3[PO_4]_2{\cdot}4\ H_2O$), zincite (ZnO), wellemite (Zn_2SiO_4), franklinite ($ZnFe_2O_4$) and smithsonite ($ZnCO_3$). These fixed forms of Zn are sparingly soluble (Dhaked et al. 2018). Indian soils are deficient in available zinc content. With the increasing concentration of carbonates ($pH > 7$), fixation of zinc increases in soil, thereby being unavailable which becomes available with help of Zn solubilizing microorganisms. Zinc concentration in soil is categorized into five proportions, i.e. water-soluble fraction, soil particles bound form, organic acid bound form via adsorption, chelation or complexation, and nonexchangeable forms such as clay minerals and metallic oxides (Alloway 2008). The total zinc content of soil depends upon the parent rock, weathering (type and intensity), organic matter, texture, pH and climate (Saeed and Fox 1977; Chesworth 1991; Sharma et al. 2013). According to World Food Organization (FAO) near about 30% world cultivable soils are zinc-deficient (Sillanpaa 1990) and many Indian soils exhibit the zinc deficiency due to a low level of available zinc in the soil (Tiwari and Dwivedi 1994). In most agricultural soils, the total zinc content ranges from 10 to 300 mg kg^{-1} (Barber 1995). In a study conducted by Katyal and Sharma (1991) it was found that the total zinc content in Indian soils varies among the eight soil orders, i.e. entisols (47 mg kg^{-1}), inceptisols (60 mg kg^{-1}), aridisols (61 mg kg^{-1}), vertisols (63 mg kg^{-1}), alfisols (44 mg kg^{-1}), altisols (43 mg kg^{-1}), mollisols (30 mg kg^{-1}) and oxisols (72 mg kg^{-1}). Spodosols and luvisols were known to have the least zinc concentration corresponding to 28 mg kg^{-1} and 35 mg kg^{-1}, respectively, over those of fluvisols and histosols with 60 mg kg^{-1} and 58 mg kg^{-1} zinc concentration, respectively (Kiekens 1995). Alkaline rocks (basalt) produced soils contain more amount of zinc as compared to soils formed from acidic rocks (granite and gneisses) (Vinogradov 1959). Lighter and heavier soils have low and high zinc, respectively (Frank et al. 1976). Availability of soil zinc to plants depends on many soil factors such as zinc content in soil, clay content, $CaCO_3$ content, redox reactions, soil pH, rhizospheric microbes, moisture content of soil, other micro-element concentration and macronutrient concentration. Interaction of zinc with other nutrient elements also affects the Zn availability.

14.6 Factors Affecting Zinc Availability to Plants

The term availability describes the potential of plants to absorb nutrients from the surrounding soil (Tiller 1983). Several researchers have enlisted the factors affecting solubility of Zn in soils and ultimately its availability (Marschner 1993; Catlett et al. 2002; Zaborowska et al. 2006; Alloway 2008, 2009; Sadeghzadeh and Rengel 2011; Hafeez et al. 2013; Sharma et al. 2013). The soil factors which affect the zinc availability to plants are those which thereafter alter the amount from/into the soil solution. The main soil factors affecting the availability of zinc to plants include parent material of soils and total Zn, pH, organic matter, texture, flooding, temperature, microbial inactivated zinc, plant species and genotypes, phosphate fertilizers and effect of nitrogen (Figure 14.2) (Lindsay 1972; Takkar and Randhawa 1978; Pendias and Pendias 1992; Alloway 2008).

Low soil zinc content that leads to deficiencies occurring in plants is known as primary zinc deficiency, on the other hand, due to higher zinc content, the effects arising are known

Figure 14.2 Causes of zinc deficiency in crops. *Sources:* Lindsay (1972); Takkar and Randhawa (1978); Pendias, A.K. and Pendias, H. (1992); Alloway (2008).

as secondary zinc deficiencies (zinc toxicity) (Alloway 2009). Acidic soils derived from gneiss and granite and sandy soils generally have low zinc content which makes zinc unavailable to plants. The total zinc content varies in soils from 10 to 300 μg and average soil zinc content is found to be 50 μg (Lindsay 1972). Soil pH is an important factor as zinc solubility and availability is dependent on pH. With the increase in the soil pH the adsorptive capacity of soil increases which decreases the zinc availability. Simultaneously increasing the pH of soil decreases the zinc solubility. Alkaline soils with high carbonate content absorb zinc and make it unavailable (Udo et al. 1970). Zinc deficiency is common in alkaline, calcareous and heavily limed soils than in neutral or acidic soils. Addition of organic material like farmyard manure or compost increases the available zinc. Humic substances such as humic and fulvic acid showed a strong affinity for binding with Zn^{2+} ion to form chelate compounds which increases zinc solubility as well as mobility over a range of pH (Kiekens 1995). Heavy textured soils like clay can retain more quantities of zinc over light textured soils due to their high cation exchange capacity (CEC) and reactive sites (Shukla and Mittal 1979). Consequently, zinc deficiency is more likely to occur in sandy soils than clay soils. Adsorption of zinc is controlled by pH and CEC in clay soils (Ellis and Knezek 1972). Zinc availability is associated with phosphorous content in soil because heavy application of phosphorous fertilizers causes reduction in zinc uptake by plants (Olsen 1972). An increase in carbonate content and pH results in zinc deficiency due to formation of zinc sulphide (ZnS) under flooding conditions (Mikkelsen and Shiou 1977). Zinc fertilizers applied as zinc sulfate at pH 7.7 and 9.1 may form unavailable compounds viz., $Zn(OH)_2$ and $ZnCO_3$ hence, displaying Zn deficiency symptoms in crops at higher pH (Sunithakumari et al. 2016). Maintaining adequately high amounts of accessible soil Zn can occur by soil fertilization application in soil and/or sustaining sufficient Zn fraction in the plant vegetative tissue throughout the proliferative growth phases through foliar sprays and biofortification of plants with Zn (Cakmak et al. 2017). Plant scientists are framing diverse policies to cater to the Zn deficits by fertilizers or by adapting breeding approaches to increase the uptake and/or biological availability of zinc to crop plants.

14.7 Response of Legume Crops to Zinc

Pulse crops belong to seed legumes under the family Fabaceae. They occupy a unique position in Indian agriculture acting as an important source of dietary vegetable proteins in daily life for poor people. India produces various types of pulses crops, among them lentil, chickpea, cowpea, blackgram, greengram, fieldpea, and grasspea are important ones. Pulses are known to maintain and restore by processes such as biological nitrogen fixation (BNF) as well as maintaining and improving soils physical parameters through the deep root system and leaf fall, thus playing important part in sustainable agriculture. India being the largest producer and consumer of pulses in the world accounts for 34.40% of the world area and 24.30% of the world production (Joshi and Rao 2017). The area of pulse crops is persistently increasing with decreased productivity. During the last two decades pulse crop production has been stagnated between 11 and 14 million tons. This is primarily due to a decreased infertility of soil especially macro- and micronutrient content, disproportionate use of fertilizers, physiological disorders during crop growth causing loss in productivity.

To meet the present requirement and to fulfill the projected demands of 32 million tons of people by 2030, an annual growth rate of 4.2 is recommended (Singh et al. 2011). To reach this annual growth rate, micronutrient including Zn is required to negate hidden hunger. Zinc plays a dual role in increasing the yield as well as crop quality. Among the various methods of zinc application, soil zinc application is most effective for correcting the zinc deficiency as well as provides more beneficial results. Application of zinc showed a wide range of differences among the different pulse crops according to their sensitivity.

Zinc deficiency problem is universal and it can be corrected through agronomy and genetic improvement. Several researchers reported a response of zinc fertilization in pulse crops on growth and yield parameters (Togay et al. 2004; Poshtmasari et al. 2008; Shanti et al. 2008; Valenciano et al. 2010; Akay 2011; Patel et al. 2011; Singh et al. 2011; Valenciano et al. 2011; Salehin and Rahman 2012; Quddus et al. 2014; Rahman 2015; Ayalew et al. 2016; Hossain et al. 2016; Ahmad and Tahir 2017; Balai et al. 2017; Hidoto et al. 2017; Maheshwari and Karthik 2017; Begum et al. 2018; Moshiul Islam et al. 2018). Fertilization is used to provide macronutrients as well as micronutrients like zinc to promote root and shoot development and uptake of nutrients. Application of zinc fertilizers can possibly counteract the zinc deficiency and ensure optimum growth of plants as well as enhance the yield owing to the increased zinc content in grain. Zinc sulphate ($ZnSO_4{\cdot}7H_2O$) is the most common as well as cheapest source of zinc for correcting zinc deficiency. In addition to this, other zinc-containing fertilizers are also very efficient for combating zinc deficiency through soil or foliar application (Singh et al. 2011). Pulse crops (chickpea, mung bean, urdbean, lentil, pigeonpea, rajmash, and pea) respond well to zinc application and it is found that the balanced zinc application enhances the nodulation, nitrogen fixation, whole zinc content in grain, harvest index, seedling vigor index and grain yield. Foliar application of zinc is more effective in food crops biofortification as in field-grown wheat and rice with varied soil and climatic conditions (Cakmak et al. 2017). Thus, balanced zinc fertilization in pulses is found to be very effective for enhancing the yield as well as quality which ultimately provide benefits in terms of monetary return.

14.8 Microbial Mediated Zinc Solubilization in Legume Crops

In soil, zinc exists in the form of zinc sulfide (ZnS), smithsonite ($ZnCO_3$), sphalerite (ZnFe), zincite (ZnO), willemite ($ZnSiO_4$) and zinc silicates ($ZnSiO_3$). Soil microbes (bacteria and fungi) play a crucial role in the availability of soil zinc by a solubilization process to crop plants. The soil rhizosphere is the ideal habitat for the growth of microbes as well as plants. Rhizospheric microflora plays an important role in converting unavailable metal forms into available ones (Lovely 1991). Soil rhizosphere microflora plays a vital part in increasing the growth of plants through various mechanisms, i.e. organic acid and plant growth hormones production, nutrient movement, atmospheric nitrogen fixation, changes in plant growth hormones level and abiotic stress tolerance. In addition to this, soil microbes can be used as biocontrol agents against the pathogens (Khalid et al. 2009; Verma et al. 2015). High pH, $CaCO_3$, organic matter, PO_4, and Cu can fix zinc in the soil. Acidification and low pH plays an important role in zinc solubilization process

(Cunningham and Kuiack 1992; Alexander 1997). Plants take up zinc as divalent ion Zn^{2+}. Production of organic acids (acetic, citric, lactic, propionic, gluconic acid), decrease pH which consequently increases the zinc solubility (Baruah and Barthakur 1999). Other than acidification and low pH, Zn content in soil, CEC, organic matter content and soil mineralogy are associated with zinc solubility (Martinez and Motto 2000). Microorganisms can solubilize the complexes by protonation, chelation and redox reactions in the exterior or interior of their cell. For the adsorption of zinc by clays and organic matter there exist two different mechanisms. The first works under acidic conditions and is comparable to cation exchange system. On the other hand, the second mechanism under alkaline condition uses chemisorption and organic acid complexation (Kiekens 1980). Microbial organic acids increase availability of zinc in soil by exudation with protons and as counter ions and consequently reducing pH of soil and complexing with anions increasing Zn solubility and making it available to plants (Kumar 2018). Both bacteria and fungi among the microbial diversity play an important role in making available Zn by plant–microbes interaction in the soil rhizosphere. The role of both zinc-solubilizing bacteria (ZnSB) and zinc-solubilizing fungi (ZnSF) in improving the zinc solubilization and making Zn available to legume crops is comprehensively discussed below.

14.8.1 Zinc-Solubilizing Bacteria (ZnSB)

The rhizosphere-associated bacteria has the potential of transforming metals like zinc into available form through solubilization (Cunningham and Kuiack 1992; Iram et al. 2016; Sunithakumari et al. 2016). There are varied mechanisms of solubilization or mobilization of Zn which includes secretion of organic acids, chelation, root architecture changes, exopolysaccharide production and lowering down of pH of rhizosphere ZnSB. Soil microorganisms make use of single or many varied mechanisms to enhance zinc availability in soil aiding in the zinc acquisition/uptake to plants (Baruah 2018; Hussain et al. 2018). Zinc solubilizers act as enhancers making zinc available to crops reducing the malnourishment of the micronutrients to human. Nutrient solubilization, siderophores, and exopolysaccharides production are the main growth-promoting traits of these microorganisms and through these traits they have potential to enhance the quality of crops (Ahmad et al. 2011). Zinc-solubilizing potential of few bacterial genera has been studied in cereals and other crops (Hafeez et al. 2002; Tariq et al. 2007; Biari et al. 2008; Rana et al. 2012; Goteti et al. 2013; Fageria et al. 2014: Ramesh et al. 2014; Vaid et al. 2014; Wang et al. 2014; Hussain et al. 2015; Prasad et al. 2016; Kumar et al. 2016; Dhaked et al. 2017; Jaivel et al. 2017; Jerlin et al. 2017) but not many studies deal with zinc-solubilizing bacteria (ZnSB) in legume crops globally (Iqbal et al. 2010; Sharma et al. 2012, 2014; Gopalkrishanan et al. 2016; Navsare et al. 2018).

The bacteria widely used for solubilizing zinc are *Bacillus* and *Pseudomonas* sp. The identified few bacteria which can solubilize Zn viz., *Pseudomonas fluorescens* (Di Simine et al. 1999); *Microbacterium saperdae, Psuedomonas monteilli* and *Entrobacter cancerogens* (Whiting et al. 2001); *Pseudomonas aeruginosa* (Fasim et al. 2002); *Thiobacillus thioxidans, Thiobacillus ferroxidans* (Saravanan et al. 2003); *Gluconacetobacter diazotrophicus* (Saravanan et al. 2007); *Stenotrophomonas maltpohilia, Mycobacterium brisbanense, Entrobacter aerogenes, P. aeruginosa, and Xanthomonas retroflexus* (Sunithakumari et al. 2016). Nitrogen fixer (*Rhizobium, Azospirillum*, and *Azotobacter*) do not solubilize zinc but few endophytes including *Actinobacter* are good zinc solubilizers (Saravanan et al. 2007). On evaluating five bacterial isolates (U, 8M, 36, 102, and 111) in *Vigna radiata* found maximum plant growth parameters at seedling stage with zinc phosphate ($Zn_3[PO_4]_2{\cdot}4H_2O$) (Iqbal et al. 2010). A total of 134 putative *Bacillus* isolates from the soybean field examined for zinc solubilization on Tris-minimal medium amended with 0.1% of different zinc sources in vitro, KHBD-6 and KHBAR-1 were excellent zinc solubilizers aiding zinc availability in soybean (Sharma et al. 2012). *Bacillus cereus* and *Pseudomonas* sp. have excellent ability for solubilization of zinc minerals (Muhammad Shakeel et al. 2015). Among the endophytic bacteria *Klebsiella, Enterococcus, Bacillus, Pseudomonas*, and *Paenibacillus* sp., isolated from soybean and mungbean, *Klebsiella* and *Pseudomonas* sp., have been found to be promising zinc solubilizers on Tris-minimal medium supplemented with zinc oxide (ZnO) and zinc phosphate ($Zn_3[PO_4]_2{\cdot}4H_2O$) (Sharma et al. 2014). Potassium- and zinc-solubilizing microorganisms were tested in five treatments in mung bean and out of five treatments, treatment5 (T5) {RDF + *Rhizobium* + PSB + KSB + ZnSB} resulted in an increase in various growth parameters improving soil fertility in mung bean (Navsare et al. 2018). Various crops have been advocated by extrinsic application of zinc substrates equivalent to fertilizer application, leading to fixation of ~96 to 99% of applied zinc. This problem can be solved using bacterial isolates able to solubilize insoluble zinc forms (Saravanan et al. 2003). Zinc-solubilizing bacteria have played crucial role in enhancing productivity in *Zea mays, Glycine max* and *Triticum aestivum* (Khande et al. 2017). Since, zinc being a limiting factor in plant growth necessitates zinc solubilization by ZnSB.

14.8.2 Zinc-Solubilizing Fungi (ZnSF)

It becomes evident from Jules Raulin (1869) work that zinc is indispensable for the growth of *A. niger* fungus. Other fungi viz., *Penicillium simplicissimum* (Franz et al. 1993); *Abisidia cylindrospora, Arctostaphylos glauca* and *Absidia spinosa* (Coles et al. 2001); *Glomus macrocarpus* (Swaminathan and Verma 1979); *Glomus intraradices* (Subramanian et al. 2009); *Phomopsis, Aspergillus* sp. (Sutjaritvorakul et al. 2013) showed their sensitivity toward zinc and are able to solubilize zinc to make it available to crop plants for their growth and development. Arbuscular mycorrhizal fungi (AMF) is an important class of fungi which provides essential nutrients to plants, associating with the majority of terrestrial plants. They are effective making the immobile nutrients available to plants (Liu et al. 2000). Among the AMF, ericoid mycorrhizae (*Paxillus involutus, Hymenoscyphus ericae, Oidiodendron maius, Suillus luteus*, and *Beauveria caledonica* (Fomina et al. 2005; Schöll et al. 2006; Gadd 2007) have strong potential for Zn solubilization. According to Subramanian et al. (2009) AMF improves bioavailability of Zn to crop plants and has been proved to be more efficient acid-tolerant than bacteria and better phosphate mobilizers. Several fungi have been described to produce siderophores, plant growth hormones such as indole-3-acetic acid (IAA) and gibberellins (Kumar et al. 2018). There are some good reports available on fungi being used as bio-inoculants for Zn solubilization in cereals and other food crops, but so far, no report is available for legumes crops. But there are many opportunities to utilize these fungi in legume crops which act as zinc solubilizers in cereals and other crops. Kumar et al. (2018) documented an entomopathogenic fungus with direct and indirect plant growth–promoting traits. The plant growth promoters (PGP) traits of the fungus resemble that of PGPR and independent of the insect species.

14.9 Conclusion and Future Prospects

Zinc and legume crops both itself act as double-edged sword in all the living systems including plants, animals, humans, and microorganisms. Zinc is a vital element which is required in trace amounts and result in both deficiency as well as toxicity when its concentration deviates from the optimum required concentration. Legume crops are rich source of protein which is an important dietary component necessary for everyone. Zinc is involved in the biosynthesis of protein, carbohydrate metabolism, nucleic acid metabolism and other metabolic pathways essential for growth and development. Soil is a natural reservoir of all the essential minerals both macro- as well as micronutrient element. There are several factors which are limiting the bioavailability of zinc which affects the nutrient use and uptake efficiency of crop plants from soil to crop plants. Mobilization of nutrients is the most critical phenomenon related to bioavailability of zinc to plants. Microbial-mediated solubilization of zinc and other macro- and micro-elements is an important process which helps in the bioavailability of zinc to crop plants for their overall growth and development. It is one of the methods of fortifying the crop plants through the application of zinc-solubilizing bacteria (ZnSB) and fungi (ZnSF) as biofertilizers to enhance the grain zinc content. Among microbes, ZnSB and ZnSF are reported by various researchers in cereals and other crops. But with respect to legume crops only few zinc-solubilizing bacteria (ZnSB) are documented in the literature. In this area, considerable research focus is required for the identification of ZnSB and ZnSF in legume crops for the enhancement of grain zinc content to fight against hidden hunger.

References

Ahmad, A. and Tahir, M. (2017). Role of Zinc Sulphate for Maize (Zea mays L.) and Mungbean (*Vigna radiata* L. *Wilczek*) yield improvement: a review. *Bull. Biol. Allied Sci. Res.* 2: 1.

Ahmad, A.M.A., Ahmad, G., Magda, M.H., and Tawfik, M.M. (2011). Integrated effect of organic and biofertilizers on wheat productivity in new reclaimed soils. *Res. J. Agric. Biol. Sci.* 7: 105–114.

Akay, A. (2011). Effect of zinc fertilizer applications on yield and element contents of some registered chickpeas varieties. *Afr. J. Biotechnol.* 10 (60): 12892–12896.

Alexander, M. (1997). *Introduction to Soil Microbiology*, 33–399. New York: Wiley.

Alloway, B.J. (2001). *Zinc – The Vital Micronutrient for Healthy, High-Value Crops*. Brussels: International Zinc Association.

Alloway, B.J. (2002). *Zinc – The Vital Micronutrient for Healthy, High Value Crops*. International Zinc Association (IZA).

Alloway, B.J. (2004). *Zinc in Soils and Crop Nutrition*. Brussels, Belgium: International Zinc Association Communications. IZA Publications.

Alloway, B.J. (2008). *Zinc in Soils and Crop Nutrition*, 2e. Brussels, Belgium and Paris, France: IZA and IFA.

Alloway, B.J. (2009). Soil factors associated with zinc deficiency in crops and humans. *Environ. Geochem. Health* 31: 537–548.

Anonymous (2004). *Fifth Report on the World Nutrition Situation*. Geneva: United Nations, Administrative Committee on Coordination/Sub-Committee on Nutrition.

Arnon, D.I. and Stout, P.R. (1939). The essentiality of certain elements in minute quantity for plants with special reference to copper. *Plant Physiol.* 14: 371–375.

Ayalew, A., Beyene, S., and Walley, F. (2016). Evaluation of zinc and iron availabilities in soils of selected areas of Ethiopia for haricot bean production under greenhouse condition. *J. Resour. Develop. Manage.* 16: 46–55.

Baath, E. (1989). Effects of heavy metals in soil on microbial processes and populations (a review). *Water Air Soil Pollut.* 47: 335–379.

Balai, K., Jajoria, M., Verma, R. et al. (2017). Nutrient content, uptake, quality of chickpea and fertility status of soil as influenced by fertilization of Phosphorus and Zinc. *J. Pharmacogn. Phytochem.* 6 (1): 392–298.

Barak, P. and Helmke, P.A. (1993). The chemistry of zinc. In: *Zinc in Soils and Plants* (ed. A.D. Robson), 90–106. Dordrecht: Kluwer Academic Publishers.

Barber, S.A. (1995). *Soil Nutrient Bioavailability*. New York: Wiley.

Barcelo, J. and Poschenrieder, C. (1990). Plant water relations as affected by heavy metal stress: a review. *J. Plant Nutr.* 13: 1–37.

Baruah, R. (2018). Towards the bioavailability of zinc in agricultural soils. In: *Role of Rhizospheric Microbes in Soil* (ed. V.S. Meena), 99–136. Springer Nature Singapore Pvt. Ltd.

Baruah, T.C. and Barthakur, H.P. (1999). *A Text Book of Soil Analysis*. New Delhi: Vikas Publishing House Pvt. Ltd.

Beckett, P.H.T. and Davis, R.D. (1977). Upper critical levels of toxic elements in plants. *New Phytol.* 79: 95–106.

Beckett, P.H.T. and Davis, R.D. (1978). The additivity of the toxic effects of Cu, Ni and Zn in young barley. *New Phytol.* 81: 155–173.

Begum, R., Swain, S.K., and Mohanty, S.K. (2018). Effect of micronutrients on plant growth and yield in green gram. *Int. J. Chem. Studies* 6 (2): 1671–1673.

Bertrand, G. and Javillier, M. (1911). The combined influence of zinc and manganese on the development of *Aspergillus niger*. *C. R. Acad. Sci.* 152: 900–902.

Biari, A., Gholami, A., and Rahmani, H.A. (2008). Growth promotion and enhanced nutrient uptake of maize (*Zea mays* L.) by application of plant growth promoting rhizobacteria in arid region of Iran. *J. Biol. Sci.* 8: 1015–1020.

Birckner, V. (1919). The zinc content of some food products. *J. Biol. Chem.* 38: 191–203.

Bonaventura, P., Benedetti, G., Albarède, F., and Miossec, P. (2015). Zinc and its role in immunity and inflammation. *Autoimmun. Rev.* 14 (4): 277–285.

Brennan, R.F. (2005). Zinc Application and Its Availability to Plants. Ph. D. dissertation, School of Environmental Science, Division of Science and Engineering, Murdoch University.

Brown, P.H., Cakmak, I., and Zhang, Q. (1993). Form and function of zinc in plants. In: *Zinc in Soils and Plants* (ed. A.D. Robson), 93–106. Dordrecht: Kluwer Academic Publishers.

Cakmak, I. (2000). Possible roles of zinc in protecting plant cells from damage by reactive oxygen species. *New Phytol.* 146: 185–205.

Cakmak, I. and Marschner, H. (1988b). Increase in membrane permeability and exudation in roots of zinc deficient plants. *J. Plant Physiol.* 132: 356–361.

Cakmak, I., McLaughlin, M.J., and White, P. (2017). Zinc for better crop production and human health. *Plant and Soil* 411: 1–4.

Caracuta, V., Barzilai, O., Khalaily, H. et al. (2015). The onset of faba bean farming in the Southern Levant. *Sci. Rep.* 5: 14370.

Catlett, K.M., Heil, D.M., Lindsay, W.L., and Ebinger, M.H. (2002). Soil chemical properties controlling Zinc2+ activity in 18 Colorado soils. *Soil Sci. Soc. Am. J.* 66: 1182–1189.

Chardonnes, A.N., Koevoets, P.L.M., van Zanten, A. et al. (1999). Properties of enhanced tonoplast zinc transport in naturally selected zinc-tolerant silene vulgaris. *Plant Physiol.* 120: 779–785.

Chesworth, W. (1991). Geochemistry of micronutrients. In: *Micronutrients in Agriculture*, 2e (eds. J.J. Mortvedt, F.R. Cox, L.M. Shuman and R.M. Welch), 1–30. Madison, WI: Soil Science Society of America Inc.

Ciais, D., Cherradi, N., Bailly, S. et al. (2004). Destabilization of vascular endothelial growth factor mRNA by the zinc – finger protein TIS11b. *Oncogene* 23: 8673–8680.

Coles, K.E., David, J.C., Fisher, P.J. et al. (2001). Solubilization of zinc compounds by fungi associated with the hyper accumulator *Thlaspi caerulescens*. *Bot. J. Scotl.* 51: 237–247.

Cousins, R.J. (1996). Zinc. In: *Present Knowledge in Nutrition*, 7e (eds. E.E. Ziegler and L.J. Filer Jr.), 293–306. Washington, DC: International Life Sciences Institute (ILSI) Press.

Cousins, R.J. (2006). Zinc. In: *Present Knowledge in Nutrition* (eds. B.A. Bowman and R.M. Russell), 4454–4457. Washington, DC: International Life Sciences Institute (ILSI) Press.

Cunningham, J.E. and Kuiack, C. (1992). Production of citric and oxalic acid and solubilization of calcium phosphate by *Penicillium billai*. *Appl. Environ. Microbiol.* 58: 1451–1458.

Davis, J.G. and Parker, M.B. (1993). Zinc toxicity symptom development and portioning of biomass and zinc in peanut plants. *J. Plant Nutr.* 16: 2352–2369.

Dhaked, B.S., Trivani, S., Reddy, R.S., and Padmaja, G. (2017). Isolation and screening of potassium and zinc solubilizing bacteria from different rhizosphere soil. *Int. J. Curr. Microbiol. App. Sci.* 65 (8): 1271–1281.

Dhaked, B.S., Trivani, S., Reddy, R.S. et al. (2018). Potassium and zinc solubilizing efficiency assessment of microorganisms from different rhizosphere soils. *Int. J. Agric. Sci.* 10 (6): 5437–5442.

Di Simine, C.D., Sayer, J.A., and Gadd, G.M. (1999). Solubilization of zinc phosphate by a strain of Pseudomonas fluorescens isolated from a forest soil. *Biol. Fertil. Soils* 28: 87–94.

Drinker, K.A. and Collier, E.S. (1926). The significance of zinc in the living organism. *J. Ind. Hyg.* 8: 257.

Ellis, B.G. and Knezek, B.D. (1972). Adsorption reactions of micronutrients in soils. In: *Micronutrients in Agriculture* (eds. J. Mortvedt, J. Giordano and W.L. Lindsay), 59–78. Madison, Wisconsin: Soil Science Society of America.

Epstein, E. and Bloom, A.J. (2005). *Mineral Nutrition of Plants: Principles and Perspectives*, 2e, 1–380. Oxford University Press.

Fageria, N.K. (2004). Dry matter yield and nutrient uptake by lowland rice at different growth stages. *J. Plant Nutr.* 27 (6): 947–958.

Fageria, N.K., Baligar, V.C., and Elson, M.K. (2014). Zinc requirements of tropical legume cover crops. *Am. J. Plant Sci.* 5: 1721–1732.

Fasim, F., Ahmed, N., Parsons, R., and Gadd, G.M. (2002). Solubilization of zinc salts by a bacterium isolated from the air environment of a tannery. *FEMS Microbiol. Lett.* 213: 1–6.

Fomina, M., Alexander, I.J., Colpaert, J.V., and Gadd, G.M. (2005). Solubilization of toxic metal minerals and metal tolerance of mycorrhizal fungi. *Soil Biol. Biochem.* 37: 851–866.

Frank, R., Ishida, K., and Suda, P. (1976). Metals in agricultural soils of Ontario. *Can. J. Soil Sci.* 56: 191–196.

Frankenberger, W.T., Johanson, J.B., and Nelson, C.O. (1983). Urease activity in sewage sludge amended soils. *Soil Biol. Biochem.* 15 (5): 543–549.

Franz, A., Burgstaller, W., Muller, B., and Schinner, F. (1993). Influence of medium components and metabolic inhibitors on citric acid production by *Penicillium simplicissimum*. *J. Gen. Microbiol.* 139: 2101–2107.

Fuller, D.Q., Korisetter, R., and Venkatasubbaiah, P.C. (2001). Southern Neolithic cultivation systems: a reconstruction based on archaebotanical evidence. *South Asian Stud.* 17: 171–187. https://doi.org/10.1080/02666030.2001.9628599.

Gadd, G.M. (2007). Geomycology: biogeochemical transformations of rocks, minerals, metals and radionuclides by fungi, bioweathering and bioremediation. *Mycol. Res.* 111: 3–49.

Godbold, D.L. and Huttermann, A. (1985). Effect of zinc, cadmium and mercury on root elongation of *Piceaabies* (Karst.) seedlings and significance of these metals to forest die-back. *Environ. Pollut. Ser. A* 38: 375–381.

Golden, M.H.N. (1989). The diagnosis of zinc deficiency. In: *Zinc in Human Biology* (ed. C.F. Mills), 173–181. London, UK: Springer-Verlag.

Gopalkrishanan, S., Vadlamudi, S., Samineni, S., and Sameer Kumar, C.V. (2016). Plant growth-promotion and biofortification of chickpea and pigeonpea through inoculation of biocontrol potential bacteria, isolated from organic soils. *SpringerPlus* 5 (1): 1882.

Goteti, P.K., Emmanuel, L.D.A., Desai, S., and Shaik, M.H.A. (2013). Prospective zinc solubilising bacteria for enhanced nutrient uptake and growth promotion in maize (*Zea mays* L.). *Int. J. Microbiol.* 2013: 1–7.

Hafeez, F.Y., Hameed, S., Zaidi, A.H., and Malik, K.A. (2002). Biofertilizer for sustainable agriculture. In: *Techniques for sustainable agriculture* (eds. F. Azam, M.M. Iqbal, C. Inayatullah and K.A. Malik), 67–73. Faisalabad: ISBN/NIAB.

Hafeez, B., Khanif, Y.M., and Saleem, M. (2013). Role of zinc in plant nutrition – a review. *Am. J. Exp. Agric.* 3 (2): 374–391.

Hansch, R. and Mendel, R.R. (2009). Physiological functions of mineral micronutrients (Cu, Zn, Mn, Fe, Ni, Mo, B, Cl). *Curr. Opin. Plant Biol.* 12: 259–266.

Hidoto, L., Worku, W., Hussein, M., and Taran, B. (2017). Effects of zinc application strategy on zinc contenet and productivity of chickpea grown under zinc deficient soils. *J. Soil Sci. Plant Nutr.* 17 (1): 112–126.

Ho, E. (2004). Zinc deficiency, DNA damage and cancer risk. *J. Nutr. Biochem.* 15: 572–578.

Hossain, M.B., Hasan, M.M., Sultana, R., and Abdul Bari, A.K.M. (2016). Growth and yield response of chickpea to different levels of boron and zinc. *Fundam. Appl. Agric.* 1 (2): 82–86.

Hotz, C. and Brown, K.H. (2004). Assessment of the risk of zinc deficiency in populations and options for its control. *Food Nutr. Bull.* 25: S91–S204.

Hussain, A., Arshad, M., Zahir, Z.A., and Asghar, M. (2015). Prospects of zinc solubilizing bacteria for enhancing growth of maize. *Pak. J. Agric. Sci.* 52: 915–922.

Hussain, A., Zahir, Z.A., Asghar, H.N. et al. (2018). Zinc solubilizing bacteria for zinc biofortification in cereals: a step towards sustainable nutritional security. In: *Role of Rhizospheric Microbes in Soil* (ed. V.S. Meena), 203–227. Springer Nature Singapore Pvt. Ltd.

Iqbal, A., Khalil, I.A., Ateeq, N., and Sayyar Khan, M. (2006). Nutritional quality of important food legumes. *Food Chem.* 97: 331–335.

Iqbal, U., Jamil, N., Ali, I., and Hasnain, S. (2010). Effect of zinc-phosphate-solubilizing bacterial isolates on growth of *Vigna radiata*. *Ann. Microbiol.* 60: 243–248.

Iram, N., Ahmad, H., Khokkar, S.N. et al. (2016). Impact of zinc solubilizing bacteria on zinc contents of wheat. *Am. Eurasian J. Agric. Environ. Sci.* 16 (3): 449–454.

Jagana, C.S., Baba, Z.A., Krishnanand, S.I. et al. (2019). Isolation and characterization of zinc solubilizing bacteria from Kashmir Himalayas, India. *Int. J. Curr. Microbiol. App. Sci.* 8 (6): 1248–1258.

Jaivel, N., Sivakumar, U., and Marimuthu, P. (2017). Characterization of zinc solubilization and organic acid detection in *Pseudomonas* sp. RZ1 from rice phyllosphere. *Int. J. Chem. Studies* 5 (6): 272–277.

Januszek, K. (1999). Enzymatic activity of some forest soils in southern Poland in the light of field investigations. *Zsez. Nauk. AR Krakow, Rozpr.* 250: 1–130.

Jerlin, B., Sharmila, S., Kathiresan, K., and Kayalvizhi, K. (2017). Zinc solubilizing bacteria from rhizospheric soil of mangroves. *Int. J. Microbiol. Biotechnol.* 2 (3): 148–155.

Joshi, P.K. and Rao, P.R. (2017). Global pulses scenario: status and outlook. *Ann. N. Y. Acad. Sci.* 1392: 6–17.

Jukanti, A.K., Gaur, P.M., Gowda, C.L.L., and Chibber, R.N. (2012). Nutritional quality and health benefits of chickpea (*Cicer arietinum* L.): a review. *Braz. J. Nutr.* 108: S11–S26.

Kabata-Pendias, A. (2001). *Trace Elements in Soils and Plants*, 3e, 1–413. Boca Raton, FL: CRC Press.

Katyal, J.C. and Sharma, B.D. (1991). DTPA extractable and total Zn, Cu, Mn and Fe in Indian soils. *Geoderma* 49: 165–179.

Khalid, A., Arshad, M., Shaharoona, B., and Mahmood, T. (2009). Plant growth promoting rhizobacteria and sustainable agriculture. In: *Microbial Strategies for Crop Improvement* (eds. M.S. Khan, A. Zaidi and J. Musarrat), 133–160. Berlin/Heidelberg: Springer.

Khande, R., Sushil, K.S., Ramesh, A., and Mahaveer, P.S. (2017). Zinc solubilizing *Bacillus*strains that modulate growth, yield and zinc biofortification of soybean and wheat. *Rhizosphere* 4: 126–138.

Kiekens, L. (1980). Absorptieverschijnselen van zwaremetalen in gronden. Doctoral Thesis, University of Gent.

Kiekens, L. (1995). Zinc in heavy metals. In: *Soils* (ed. B.J. Alloway). London: Blackie Academic and Professional.

King, J.C. (2011). Zinc: an essential but elusive nutrient. *Am. J. Clin. Nutr.* 94 (Suppl): 679S–684S.

Kobayashi, A., Sakamoto, A., Kubo, K. et al. (1998). Seven zinc-finger transcription factors are expressed sequentially during the development of anthers in petunia. *Plant J.* 13: 571–576.

Kucharksi, J. (1997). Relationships between soil enzymatic activity and fertility. In: *Microorganisms in the Natural Environment: Occurrence, Activity and Significance* (eds. W. Baranasz and A.R. Krakow), 327–342.

Kucharski, J. (1992). The effect of zinc on yellow lupine development and on activity of soil microorganisms. *Pol. J. Soil Sci.* 25 (2): 177–183.

Kumar, V.V. (2018). Biofertylizers and biopesticides in sustainable agriculture. In: *Role of Rhizospheric Microbes in Soil* (ed. V.S. Meena), 377–398. Springer Nature Singapore Pvt. Ltd.

Kumar, A.S., Meenakumari, K.S., and Anith, K.N. (2016). Screening of zinc solubilization potential of soil bacteria from Zn deficient soils of Kerala. *J. Trop. Agric.* 54 (2): 194–200.

Kumar, C.M.S., Jacob, T.K., Devasahayam, S. et al. (2018). Multifarious plant growth promotion by an entomopathogenic fungus *Lecanicillium psalliotae*. *Microbiol. Res.* 207: 150–163.

Lin, C.W., Chang, H.B., and Huang, H.J. (2005). Zinc induces mitogen-activated protein kinase activation mediated by reactive oxygen species in rice roots. *Plant Physiol. Biochem.* 43: 963–968.

Lindsay, W.L. (1972). Zinc in soil and plant nutrition. *Adv. Agron.* 24: 147–188.

Liu, A., Hamel, C., Hamilton, R.I. et al. (2000). Acquisition of Cu, Zn, Mn and Fe by mycorrhizal maize (*Zea mays* L.) grown in soil at different P and micronutrient levels. *Mycorrhiza* 9: 331–336.

Lovely, D.R. (1991). Dissimilarity Fe(III) and Mn(IV) reduction. *Microbial. Rev.* 55: 259–287.

Lukaski, H.C. (2004). Vitamin and mineral status: effects on physical performance. *Nutrition* 20: 632–644.

Lutz, R.E. (1926). The normal occurrence of zinc in biologic material. *J. Ind. Hyg.* 8: 127.

Maheshwari, U. and Karthik, A. (2017). Effect of foliar nutrition on growth, yield attributes and seed yield of pulse crops. *Adv. Crop Sci. Technol.* 5 (3): 278. https://doi.org/10.4172/2329-8863.1000278.

Marschner, H. (1993). Zinc uptake from soils. In: *Zinc in Soils and Plants* (ed. A.D. Robson), 59–77. Dordrecht, The Netherlands: Kluwer Academic Publishers.

Marschner, H. (1995). *Mineral Nutrition of Higher Plants*, 2e. London: Academic Press.

Martinez, C.E. and Motto, H.L. (2000). Solubility of lead, zinc and copper added to mineral soils. *Environ. Pollut.* 107: 153–158.

Mazé, P. (1914). Influences respective des elements de la solution minérale sur le development du mais. *Ann Inst Pasteur (Paris)* 28: 21–68.

Mendel, L.B. and Bradley, H.C. (1905). Experimental studies on the physiology of the molluscs—second paper. *American Journal of Physiology-Legacy Content* 14 (4): 313–327.

Meydani, S.N., Barnett, J.B., Dallal, G.E. et al. (2007). Serum zinc and pneumonia in nursing home elderly. *Am. J. Clin. Nutr.* 86 (4): 1167–1173.

Mikkelsen, D.S. and Shiou, K. (1977). Zinc fertilization and behaviour in flooded soils. In: *Approaches to Crop Improvement* (eds. A.R. Yeo and T.J. Flowers), 175–200). Spec. Publ. No. 5. Farnham Royal: Comm. Agric. Bur.

Moffett, B.F., Nicholson, F.A., Uwakwe, N.C. et al. (2003). Zinc contamination decreases the bacterial diversity of agricultural soil. *FEMS Microbiol. Ecol.* 43: 13–19.

Mohammad, Y.K., Ricciuti, P., Allegretta, I. et al. (2018). Solubilization of insoluble zinc compounds by zinc solubilizing bacteria (ZSB) and optimization of their growth conditions. *Environ. Sci. Pollut. Res.* 25 (26): 25862–25868.

Morlet, J.E. (2004). The top 10 hot topics in ahing. *J. Gerontol.* 59: 24–33.

Moshiul Islam, M., Razul Karim, M., Oliver, M.M.H. et al. (2018). Impacts of trace element addition on lentil (*Lens culinaris* L.) agronomy. *Agronomy* 8: 100.

Mousavi, S.R. (2011). Zinc in crop production and interaction with phosphorous. *Aus. J. Basic Appl. Sci.* 5: 1503–1509.

Mousavi, S.R., Galavi, M., and Rezaei, M. (2013). Zinc (Zn) importance for crop production – a review. *Int. J. Agron. Plant Prod.* 4 (1): 64–68.

Naik, S.K. and Das, D.K. (2008). Relative performance of chelated zinc and zinc sulphate for lowland rice (Oryza sativa L.). *Nutr. Cycl. Agroecosyst.* 81: 219–227.

Navsare, R.I., Mane, S.S., and Supekar, S.J. (2018). Effect of potassium and zinc solubilizing microorganism on growth, yield and quality of mungbean. *Int. J. Chem. Stud.* 6 (1): 1996–2000.

Olsen, S.R. (1972). Micronutrient interactions. In: *Micronutrients in Agriculture* (eds. J. Mortvedt, J. Giordano and W.L. Lindsay), 243–264. Madison, Wisconsin: Soil Science Society of America.

Palmer, C.M. and Guerinot, M.L. (2009). Facing the challenges of Cu, Fe and Zn homeostasis in plants. *Nat. Chem. Biol.* 5: 333–340.

Patel, M.M., Patel, I.C., Patel, R.I., and Acharya, S. (2011). Effect of zinc and iron on yield and yield attributes of rainfed cowpea (*Vigna unguiculata* L. *Walp*). *Ann. Arid Zone* 50 (1): 17–19.

Pendias, A.K. and Pendias, H. (1992). *Trace Elements in Soil and Plants*, 2e. Boca Raton, FL: CRC Press.

Phillips, R.D. (1993). Starchy legumes in human nutrition, health and culture. *Plant Foods Hum. Nutr.* 44: 195–211.

Pingali, P.L. (2012). Green revolution: impacts, limits and the path ahead. *Proc. Natl. Acad. Sci. U. S. A* 109: 12302–12308.

Pinton, R., Cakmak, I., and Marschner, H. (1994). Zinc deficiency enhanced NAD(P) H – dependent superoxide radical production in plasma membrane vesicles isolated from roots of bean plants. *J. Exp. Bot.* 45: 45–50.

Poshtmasari, H.K., Bahmanyar, M.A., Pirdashti, H., and Ahmadi Shad, M.A. (2008). Effect of Zn rates and application forms on protein and some micronutrients accumulation in common bean (Phaseolus vulgaris L.). *Pak. J. Biol. Sci.* 11 (7): 1042–1046.

Prasad, A.S. (2008). Zinc in human health: effect of zinc on immune cells. *Mol. Med.* 14: 353–357.

Prasad, A.S., Halsted, J.A., and Nadimi, M. (1961). Syndrome of iron deficiency anemia, hepatosplenomegaly, hypogonadism, dwarfism and geophagia. *Am. J. Med.* 31: 532–546.

Prasad, S., Syed, I., and Anuradha, P. (2016). Effect of zinc solubilizing microorganisms in enhancing enzyme activity and nutrient availability in groundnut grown on vertisol. *Int. J. Agric. Sci.* 8 (49): 2099–2102.

Quddus, M.A., Naser, H.M., Hossain, M.A., and Hossain, M.A. (2014). Effect of zinc and boron on yield and yield contributing characters of lentil in low Ganges river flood plain soil at Madaripur, Bangladesh. *Bangladesh J. Agric. Res.* 39 (4): 591–603.

Rahman, Md. A. (2015). Effect of zinc and boron on the growth and yield of mungbean. M.Sc. Thesis. Sher-E-Bangla Agricultural University, Dhaka. pp. 1–70.

Ramesh, A., Sharma, S.K., Sharma, M.P. et al. (2014). Inoculation of zinc solubilizing *Bacillus aryabhattai* strains for improved growth, mobilization and biofortification of zinc in soybean and wheat cultivated in Vertisols of Central India. *Appl. Soil Ecol.* 73: 87–96.

Rana, A., Joshi, M., Prasanna, R. et al. (2012). Biofortification of wheat through inoculation of plant growth promoting rhizobacteria and cyanobacteria. *Eur. J. Soil Biol.* 50: 118–126.

Raulin, J. (1869). Études cliniques sur la vegetation (Chemical studies on vegetation). *Ann. Sci. Nat. Bot. Biol. Veg. Ser.* 11: 93–99.

Rout, G.R. and Das, P. (2003). Effect of metal toxicity on plant growth and metabolism: I. Zinc. *Agronomie* 23 (1): 3–11.

Sadeghzadeh, B. and Rengel, Z. (2011). Zinc in soils and crop nutrition. In: *The Molecular and Physiological Basis of Nutrient Use Efficiency in Crops* (ed. M.J. Hawkesford), 335–375. New York: Wiley.

Saeed, M. and Fox, R.L. (1977). Relation between suspension pH and Zn solubility in acid and calcareous soils. *Soil Sci.* 124: 199–204.

Salehin, F. and Rahman, S. (2012). Effects of zinc and nitrogen fertilizer and their application method on yield and yield components of *Phaseolus vulgaris* L. *Agric. Sci.* 3 (1): 9–13.

Saravanan, V.S., Subramonianm, S.R., and Anthony Raj, S. (2003). Assessing in vitro solubilization potential of different zinc solubilizing bacterial (ZSB) isolates. *Braz. J. Microbiol.* 34: 121–125.

Saravanan, V.S., Madhaiyan, M., and Thangaraju, M. (2007). Solubilization of zinc compounds by the diazotrophic, plant growth promoting bacterium *Gluconacetobacter diazotrophicus. Chemosphere* 66: 1794–1798.

Saravanan, V.S., Kumar, M.R., and Sa, T.M. (2011). Microbial zinc solubilization and their role in plants. In: *Bacteria in Agrobiology: Plant Nutrient Management* (ed. D.K. Maheshwari), 47–63. Berlin Heidelberg: Springer-Verlag.

Schöll, L.V., Smits, M.M., and Hoffland, E. (2006). Ectomycorrhizal weathering of the soil minerals muscovite and hornblende. *New Phytol.* 171: 805–814.

Shakeel, M., Rais, A., Hassan, M.N., and Hafeez, F.Y. (2015). Root associated Bacillus sp. improves growth, yield and zinc translocation for basmati rice (Oryza sativa) varieties. *Frontiers in Microbiology* 6: 1286.

Shanti, M., Peda Babu, B., Rajendra Prasad, B., and Minhas, P.S. (2008). Effect of zinc on blackgram in rice- blackgram cropping system of coastal saline soils. *Legume Res.* 31 (2): 79–86.

Sharma, S.K., Sharma, M.P., Ramesh, A., and Joshi, O.P. (2012). Characterization of zinc-solubilizing Bacillus isolates and their potential to influence zinc assimilation in soybean seeds. *J. Microbiol. Biotechnol.* 22 (3): 352–359.

Sharma, A., Patni, B., Shankhdhar, D., and Shankhdhar, S.C. (2013). Zinc – An indispensable micronutrient. *Physiol. Mol. Biol. Plants* 19 (1): 11–20.

Sharma, P., Kumawat, K.C., Kaur, S., and Kaur, N. (2014). Assessment of zinc solubilization by endophytic bacteria in legume rhizosphere. *Indian J. Appl. Res.* 4 (6): 439–441.

Shukla, U.C. and Mittal, S.B. (1979). Characterization of zinc application in some soils of *India. J. Soil Sci. Soc. Am.* 43: 905–908.

Sillanpaa, M. (1990). Micronutrient assessment at the country level: an international study. FAO Soils Bulletin No. 48, Rome, Food and Agriculture Organization, United Nations.

Singh, B., Natesan, S.K.A., Singh, B.K., and Usha, K. (2005). Improving zinc efficiency of cereals under zinc deficiency. *Curr. Sci.* 88: 36–44.

Singh, K.K., Praharaj, C.S., Choudhary, A.K. et al. (2011). Zinc response in pulses. *Indian J. Fert.* 7 (10): 118–126.

Singh, U., Praharaj, C.S., Singh, S.S., Bohra, A., and Shivay, Y.S. (2015). Biofortification of pulses: strategies and challenges. In: 2nd International Conference on Bio-Resource and stress management, Hyderabad, India, January 07–10, 2015, Hyderabad. pp. 50–55.

Sommer, A.L. and Lipman, C.B. (1926). Evidence on the indispensable nature of zinc and boron for higher green plants. *Plant Physiol.* 1: 231–249.

Sresty, T.V.S. and Madhava Rao, K.V. (1999). Ultrastructural alterations in response to zinc and nickel stress in the root cells of pigeon-pea. *Environ. Exp. Bot.* 41: 3–13.

Srivastava, P.C. and Gupta, U.C. (1996). *Trace Elements in Crop Production*, 356. New Delhi: Oxford & IBH Publishers.

Staker, E.V. and Cummings, R.W. (1941). The influence of zinc on the productivity of certain New York peat soils. *Soil Sci. Soc. Am. Proc.* 6: 2017–2214.

Subramanian, K.S., Tenshia, V., Jayalakshmi, K., and Ramachandran, V. (2009). Role of arbuscular mycorrhizal fungus (*Glomus intraradices*) – (fungus aided) in zinc nutrition of maize. *Agric. Biotechnol. Sustain. Dev.* 1: 29–38.

Sunithakumari, K., Padma Devi, S.N., and Vasandha, S. (2016). Zinc solubilizing bacterial isolates from agricultural fields of Coimbatore, Tamil Nadu, India. *Curr. Sci.* 110 (2): 196–205.

Sutjaritvorakul, T., Gadd, G.M., Suntornvongsagul, K. et al. (2013). Solubilization and transformation of insoluble zinc compounds by Fungi isolated from a zinc mine. *Environ. Asia* 6 (2): 42–46.

Swaminathan, K. and Verma, B.C. (1979). Response of three crop species to vesicular arbuscular mycorrhizal infection on zinc deficient Indian soils. *New Phytol.* 114: 1–38.

Takkar, P.N. and Randhawa, N.S. (1978). Micronutrients in Indian agriculture. *Fert. News* 23: 3–26.

Tan, G.Z.H., Das Bhowmink, S.S., Hoang, T.M.L. et al. (2017). Finger on the pulse: pumping iron into chickpea. *Front. Plant Sci.* 8: 1755.

Tariq, M., Hameed, S., Malik, K.A., and Hafeez, F.Y. (2007). Plant root associated bacteria for zinc mobilization in rice. *Pak. J. Bot.* 39: 245–253.

Tavallali, V., Rahemi, M., Eshghi, S. et al. (2010). Zinc alleviates salt stress and increases antioxidant enzyme activity in the leaves of pistachio (*Pistacia vera* L. 'Badami') seedlings. *Turk. J. Agric. For.* 34 (4): 349–359.

Taylor, G.J., Stadt, K.J., and Dale, M.R.T. (1991). Modeling the phytotoxicity of aluminium, cadmium, copper, manganese, nickel and zinc using the Weibull frequency distribution. *Can. J. Bot.* 69: 267–359.

Tiller, K.G. (1983). Micronutrients. In: *Soils: An Australian Viewpoint* (eds. J. Lenaghan and G. Katsntoni), 365–387. Melbourne and London: Division of Soils, CSIRO and Academic Press.

Tisdale, S.L., Nelson, W.L., and Beaten, J.D. (1984). Zinc. In: *Soil Fertility and Fertilizers*, 4e, 382–391. New York: Macmillan Publishing Company.

Tiwari, K.N. and Dwivedi, B.S. (1994). Available zinc status of soils and delineation of the areas of zinc deficiency in Uttar Pradesh. *Fert. News* 39 (3): 31–39.

Todd, W.R., Elvehjem, C.A., and Hart, E.B. (1934). Zinc in the nutrition of the rat. *Am. J. Physiol.* 107: 146–156.

Togay, N., Ciftci, V., and Togay, Y. (2004). The effects of zinc fertilization on yield and yield components of dry bean (*Phaseolus vulgaris* L.). *Asian J. Plant Sci.* 3 (6): 701–704.

Udo, E.J., Bhon, L.H., and Tukker, T.C. (1970). Zinc adsorption by calcareous. *J. Soil Sci. Soc. Am. Proc.* 34: 405–407.

Vaid, S.K., Kumar, B., Sharma, A. et al. (2014). Effect of zinc solubilizing bacteria on growth promotion and zinc nutrition of rice. *J. Soil Sci. Plant Nutr.* 14: 889–910.

Valenciano, J.B., Boto, J.R., and Marcelo, V. (2010). Response of chickpea (*Cicer arietinum* L.) yield to zinc, boron and molybdenum application under pot conditions. *Spanish J. Agric. Res.* 8 (3): 797–807.

Valenciano, J.B., Boto, J.R., and Marcelo, V. (2011). Chickpea (*Cicer arietinum* L.) response to zinc, boron and molybdenum application under field conditions. *New Zealand J. Crop Hort. Sci.* 39 (4): 217–229.

Vallee, B.L. (1991). Introduction to metallothionein. *Methods Enzymol.* 205: 3–7.

Vallee, B.L. and Falchuk, K. (1993). The biochemical basis of zinc physiology. *Physiol. Rev.* 73: 79–118.

Verma, J.P., Jaiswa, D.K., Meena, V.S., and Meena, R.S. (2015). Current need of organic farming for enhancing sustainable agriculture. *J. Clean. Prod.* 102: 545–547.

Vinogradov, A.P. (1959). *The Geochemistry of Rare and Dispersed Chemical Elements in Soils.* New York: Consultants Bureau.

Wang, Y., Yang, X., Zhang, X. et al. (2014). Improved plant growth and Zn accumulation in grains of rice (*Oryza sativa* L.) by inoculation of endophytic microbes isolated from a Zn hyperaccumulator, *Sedum alfredii* H. *J. Agric. Food Chem.* 62: 1783–1791.

Webb, E.C. (1992). Enzyme nomenclature 1992: recommendations of the Nomenclature Committee of the International Union of Biochemistry and Molecular Biology on the nomenclature and classification of enzymes. San Diego: Published for the International Union of Biochemistry and Molecular Biology by Academic Press. ISBN 0-12-227164-5.

Webber, J. (1981). Trace elements in Agriculture. In: *Effect of Heavy Metal Pollution on Plants*, vol. 2 (ed. N.W. Lepp), 159–184. London: Applied Science Publishers.

Welch, R.M. (1993). Zinc concentrations and forms in plants for humans and animals. In: *Zinc in Soils and Plants* (ed. A.D. Robson), 183–196). Dordrecht: Kluwer Academic Publishers.

Welch, R.M. (2002). The impact of mineral nutrients in food crops on global human health. *Plant and Soil* 247: 83–90.

Welch, R.M. and Graham, R.D. (2004). Breeding for micronutrients in staple food crops from a human nutrition perspective. *J. Exp. Bot.* 55: 353–364.

Wheeler, D.M. and Power, I.L. (1995). Comparison of plant uptake and plant toxicity of various ions in wheat. *Plant Soil.*: 167–173.

White, M.C., Chaney, R.L., and Decker, A.M. (1974). Differential varietal tolerance in soyabean to toxic levels of zinc in sassafras sandy loam. *Agron. Abstr.*: 144–145.

Whiting, S.N., de Souza, M.P., and Terry, N. (2001). Rhizosphere bacteria mobilize Zn for hyperaccumulation by *Thalaspi caerulescens. Environ. Sci. Technol.* 35: 3144–3150.

WHO, UNICEF (2004). *WHO–UNICEF Joint Statement on the Clinical Management of Acute Diarrhea.* Geneva: World Health Assembly.

World Health Organization (WHO) (1996). Zinc. In: *Trace Elements in Human Nutrition and Health*, 72–105. Geneva: World Health Organization.

Yosefi, K., Galavi, M., Ramrodi, M., and Mousavi, S.R. (2011). Effect of bio-phosphate and chemical phosphorous fertilizer accompanied with micronutrients foliar application on growth, yield and yield components of maize (Single Cross 704). *Aust. J. Crop Sci.* 5 (2): 175–180.

Zaborowska, M., Wyszkowska, J., and Kucharski, J. (2006). Microbiological activity of zinc-contaminated soils. *J. Elementol.* 11 (4): 543–557.

Zargar, S.M., Mahajan, R., Farhat, S. et al. (2015). Understanding the role of iron and zinc in animals and crop plants from genomics perspective. *Curr. Trends Biotechnol. Pharm.* 9 (2): 181–196.

Zimmermann, M. (2001). *Micronutrients in Health and Diseases*. Stuttgart: Georg Thieme Verlag.

15

Composition and Interconnections in Phyllomicrobiome

Meghmala Waghmode[1], *Aparna Gunjal*[2], *Neha Patil*[1], *and Sonali Shinde*[3]

[1] *Department of Microbiology, PDEA's Annasaheb Magar Mahavidyalaya, Pune, Maharashtra, India*
[2] *Department of Microbiology, Dr. D.Y. Patil, Arts Commerce and Science College, Pimpri, Pune, Maharashtra, India*
[3] *Annasaheb Kulkarni, Department of Biodiversity, MES Abasaheb Garware College, Pune, Maharashtra, India*

15.1 Introduction

The co-evolution of the microorganisms and their plant hosts has helped to understand many interactions between microorganisms and plants. This has helped to improve the environmental stability. One such interface above-ground surface or aerial habitat in plants affected by microbes is termed as the phyllosphere and its microbial inhabitants as epiphytes and endophytes (Lindow and Brandl 2003).

The members of *Proteobacteria*, *Firmicutes*, and *Actinobacteria* are majorly present in the plants phyllosphere region. Reports are available on the fungal community found in the phyllosphere of oak tress (*Quercus macrocapra*). The microbial community in the phyllosphere of oak tress include *Devriesia*, *Mycosphaerella*, *Ramularia*, *Stenella*, *Diosszegia*, *Paraphaeosphaeria*, *Phaeosphaeria*, *Sphaceloma*, *Davidiella*, *Microsphaeropsis*, etc. (Jumpponen and Jones 2009).

Plant connects two enormously different universes, above- and below-ground and serves as a channel to flow the energy and nutrient. Phyllosphere microbiota forms the micro-ecosystem where chemical communication and compounds get exchanged between microbial communities that include bacteria, viruses, archaea, and eukaryotic taxa (Christian et al. 2015; Penuelas and Terradas 2014). The phyllosphere microbes colonization arises from propagules dispersed from the air, rain herbivores, and, leaf litter (Beattie and Lindow 1999; Clay et al. 2005). Plant surfaces harbor huge number of beneficial microbes, whereas others act as pathogens, symbionts, and saprobes for host plants (Busby et al. 2016; Mejia et al. 2014). Most studies revolve around the pathogenic microbes, because of severe economic crop loss (Dik et al. 1991). A stronger pathogenic effect denotes negative feedback, indicating lack of co-adaptation, coexistence and, reduced fitness (Mordecai 2011; Rua et al. 2016). Instead, host species which supports defensive response under high parasite pressure leads to good response (Whitaker

Phytomicrobiome Interactions and Sustainable Agriculture, First Edition. Edited by Amit Verma, Jitendra Kumar Saini, Abd El-Latif Hesham, and Harikesh Bahadur Singh.

et al. 2017; Clay et al. 2005). Thus, it would be an exciting fact to study the nonpathogenic microbes which are the inhabitants of the phyllosphere. These microbes can be used as biocontrol agents against plant diseases. Use of *Baculovirus* as biocontrol agent is well known, similarly, bacterial agents to control bean rust is also well known (Yuen et al. 2001). Utilizing biocontrol agents have helped in sustainable agriculture potentially reducing the use of chemical fertilizers (Adesemoye and Kloepper 2009; Pati 1992). With growing recognition of food pathogen and chemical fertilizers on the fresh food product, different sanitation schemes are proposed that reduce the risk of exposure of pathogens. One such sanitation scheme is the use of antagonistic organism that decreases the effect of the pathogens (Gardener and Fravel 2002).

Plant microbiome serves as the boundary between humans and other environmental microbiomes. They also connect a vital pathway for human exposure to environmental antimicrobial resistance (Chen et al. 2019). Organisms can enter the food chain via contaminated crops and also poses a possible threat due to the transfer of antibiotic resistance genes to human pathogens via horizontal gene transfer (HGT) (Chen et al. 2018; Xiang et al. 2019).

Genetic factors of plants are important for the associated microbiota and its significant role in the identification of novel plant factors (Bodenhausen et al. 2014). Understanding what makes the microbes choose their niche completely depends upon what superpower they hae been bestowed that can be either presence of some plasmids or specific genes that can confer resistance to be present on a nutrient-deficient condition or out-compete existing microbes (Remus-Emsermann and Schlechter 2018). Availability of nutrients like iron and fructose and biotic and abiotic factors determine the habitats on the leaf surface that provides the information on the distribution of microbes (Beattie and Lindow 1999).

Plants and microbes communicate via a signaling molecule, stimulating expression of genes. Leguminous plants have established a symbiotic relationship with nitrogen-fixing bacteria, determined by the oligosaccharide signal. This association is known to regulate the degree of nodulation and nitrogen-fixation. The nodulation factor switches on the nodule organogenesis and defense reactions in plants. Nodulation factors could be considered as the first oligosaccharin of bacterial origin as receptors that plants could perceive (Gottfert 1993). Studies by Ikeda et al. (2011) shown in a soybean plant, show remarkable insights of underground events such as nodulation and fertilization which affect the microbial diversity as compared to the microbial diversity which is above-ground. *Methylobacterium* sp. and *Aurantimonas* are the examples of bacteria which responds to nodulation. Regulation of soil nitrogen level in soybean is extensively regulated in leaves than in stems due to the microbial activity (Ikeda et al. 2010). Salicyclic acid is yet another chemical stimulus to understand innate immunity and plant microbiome. Studies on soil-derived microbial communities promote seedling growth. The ratio of Gammaproteobacteria and Bacteriodetes is reproducibly shifted upon manipulation of salicylic acid signaling (Doornbos et al. 2011).

A plant can overcome the stress of nutrients, UV, water, and temperature by producing secondary metabolites like alkaloids, flavonoids, etc. There is another mechanism which increases the plant fitness by harboring beneficial microbes, and a mutation in these plants reduces fitness. Molecular tools are exceptionally useful to determine the composition and interactions of the phyllosphere. Pioneering work by Yang et al. (2001) on culture-independent microbes from the phyllosphere has opened avenues to identify microbes formerly not identified by culture-dependent method, making them to sustain under stressful environmental conditions in plants. However, the communities are highly variable and may undergo substantial compositional changes during the growth phase and season (Ellis et al. 1999). Nutrient resources in a plant are patchy and chemical analysis reveals spatial sequestered of sugars from epiphytes (Mercier and Lindow 2000). Morris et al. (1998) showed depending upon the sugar availability like sucrose, fructose, etc., microbes are either found in aggregates or in solitary. Such forms can be utilized as biofilms for plant disease management. As phyllosphere communities can be ready manipulated, they are a very effective and eco-friendly means of plant protection. Thus, phyllosphere forms an important ecosystem, both an economical and ecological way that hosts a diverse and large microbial community (Cohen and Yamasaki 2003).

In this chapter, we intend to highlight the current advances in understanding the phyllosphere microbiota and composition to increase the yield of the crop. We illustrate the plant–microbe interactions and provide the basis for a comprehensive data of the microbial ecology of the phyllosphere.

15.2 Significance of Phyllospheremicrobiota

The phyllospheremicrobiota is very important. The significance of phyllosphere microbiota in agriculture is represented in Figure 15.1.

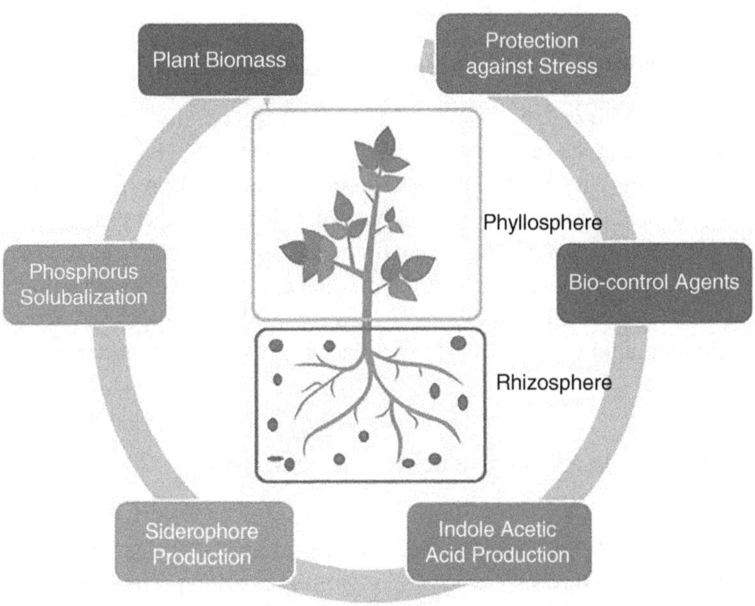

Figure 15.1 The significance of phyllosphere microbiota in agriculture.

15.3 Phyllosphere Microorganisms as Plant Growth Regulator

The phyllosphere is rich in diversity of microorganisms which help in sustainable agriculture. The microorganisms present in the phyllosphere region help in the plant growth by various ways.

15.3.1 Plant Growth Hormones Production by Phyllosphere Microorganisms

Study has been done on *Sphingomonas* sp. from phyllosphere to produce IAA (indole-acetic-acid) which resulted in enhancement of plant growth (Enya et al. 2007; Tsavkelova et al. 2007; Venkatachalam et al. 2016). The phyllosphere microorganisms have improved the plant productivity due to IAA production (Romero et al. 2016). The cytokinins produced by phyllosphere microorganisms can enhance nitrogen transport to above-ground plant tissues (Holland 2011), thus increasing above-ground biomass and crop yield. Report is available on a phyllosphere colonizing *Methylobacterium* for the production of cytokinins and IAA (Madhaiyan et al. 2005).

15.3.2 Phosphorus Solubilization by Phyllosphere Microorganisms

Phosphate-solubilizing microorganisms are also found in the phyllosphere (Batool et al. 2016; Mohammadi 2012). The phosphate solubilizers synthesize organic acids which causes the chelation of cations present in phosphate. The acidic conditions lead to the maximum efficiency of phosphorus incorporation in plants cells and tissues.

15.3.3 Siderophores Production by Phyllosphere Microorganisms

Siderophores are low-molecular-weight compounds which have affinity for iron-based nutrients. Phyllosphere microorganisms also produce siderophores. There is a report where 15 strains were isolated from the phyllosphere and were found to have siderophore-producing potential (Fu et al. 2016). The siderophores produced by the phyllosphere microorganisms also plays a role in the biological control mechanism (Beneduzi et al. 2012). There is a report where pyoverdines were found to inhibit the growth of phytopathogens in maize and peanuts (Pal et al. 2001). Siderophores produced by phyllosphere bacteria *Bacillus subtilis* has the potential to serve as biocontrol agent for *Fusarium* wilt of pepper (Yu et al. 2011).

15.3.4 Phyllosphere Microorganisms as Biocontrol Agents Against the Phytopathogens

The phyllosphere microorganisms also suppress tissue infection caused by plant pathogens (Lindow and Brandl 2003). The phyllosphere microorganisms provide protection against the soil-borne diseases (Mendes et al. 2011). There are reports on phyllosphere bacteria which are biocontrolagents, viz. *Methylobacterium*, *Sphingomonas*, *Pseudomonas*, etc., which inhibit growth of pathogenic microorganisms onto plant surface (Lindow and

Table 15.1 Phyllosphere microorganisms as biocontrol agents.

Plant	Pathogen	Antagonist	Reference
Arabidopsis thaliana (Mustard)	*Pseudomonas syringae* pv. *Syringae* DC 3000	*Sphingomonas*	Innerebner et al. (2011)
Lactuca sativa L. (Lettuce)	*Xanthomonas campestris* pv. *Vitians* (causative agent of leaf spot disease)	*Bacillus*	Rastogi et al. (2013)
Malus domestica (Apple)	*Erwinia amylovora*	*Pantoea agglomerans*	Pusey et al. (2011)

Leveau 2002). The phyllosphere bacteria have antagonistic activity against the pathogens to control diseases in plants. In this, the pathogen and antagonist have overlap of nutritional niche (Ji and Wilson 2002).Antagonistic action relies upon the production of antimicrobials, lytic enzymes, siderophores and induction of systemic resistance (ISR) in host plants (Julakha et al. 2017; Zohara et al. 2016). *Paraburkholderia graminis* PHS1 has been reported to have antifungal activity which is found to suppress disease (Carrion et al. 2018). The phyllosphere microorganisms as biocontrol agents are represented in Table 15.1.

The volatile organic compounds (VOCs) produced by the phyllosphere microorganisms have also shown to have biocontrol properties against the fungal phytopathogens. VOCs produced by endophytic fungi of phyllosphere have shown antifungal activity against *Penicillium expansum*, *Botrytis cinerea* and *Monilinia fructicola* (Mercier and Jimenez 2004). The VOCs produced by the phyllopshere microorganisms are also important indicators of fungal growth in agriculture, monitoring spoilage, chemotaxonomy purposes, for use in biofilters and biodiesel, etc. (Hung et al. 2015; Stinson et al. 2003).

Flowers are more susceptible to pathogen attack due to lack of physical barriers like lignified cell wall, impermeable cuticle and high moisture as well as nutrient content. Floral volatiles are one of the topics of major interests. *Arabidopsis thaliana* flowers reported to have the ability to produce the antimicrobial VOC, sesquiterpene (E)-β-*Caryophyllene* as defense against bacterial pathogen (Huang et al. 2012). Petals of *Saponaria officinalis* (soapweed) and *Lotus corniculatus* are reported to reduce the bacterial diversity by VOCs (Junker et al. 2011b). Leaves harbor bacteria belonging to *Pseudomonadaceae* and *Microbacteriaceae* families, whereas members of *Enterobacteriaceae* family dominate the floral region.

Quorum quenchers (e.g. *Serratia*, *Pseudomonas*, *Bacillus*, *Lysinibacillus*, *Myoides* and *Acinetobacter*) have the potential to produce lactonase enzyme by which they degrade AHL (N-acylhomoserine lactone, component of quorum sensing) and disrupts quorum sensing by which pathogen proliferates. Quorum quenchers can serve as biocontrol agent against tobacco pathogens (Ma et al. 2013).

15.3.5 Phyllosphere Microorganisms to Reduce Biotic and Abiotic Stress

Stress tolerance is recorded in plants harboring *Methylobacterium extorquens* AM1 due to production of proteins (catalase, superoxide dismutase, alkyl hydroperoxidereductase, thioredoxins, organic hydroperoxide resistance protein, heat shock proteins and DNA protection protein) in response to stress (Gourion et al. 2009; Parasuraman et al. 2019).

15.3.6 Synthesis of 1-Aminocyclopropane-1-Carboxylate Deaminase (ACC)

The microorganisms present in the phyllosphere produce ACC deaminase enzyme and provides protection to the plants from excess ethylene levels (Compant et al. 2019). The phyllosphere microorganisms which produces ACC deaminase enzyme reduces ethylene levels by two to fourfold (Grichko and Glick 2001; Mayak et al. 2004; Penrose et al. 2001). ACC deaminase is a multimeric which has molecular mass of 42 kDa. This enzyme cleaves ACC and forms α-ketobutyrate and ammonia (Glick 2007). There is a report on the study of production of ACC deaminase enzyme by *Methylobacteria* isolated from the phyllosphere of rice and role of ACC deaminase in ethylene regulation (Balachandar et al. 2009).

15.3.7 Phyllosphere Microorganisms in Nitrogen-Fixation

The phyllosphere microorganisms also help in the plant growth by fixing nitrogen (Lambais et al. 2017).There is a report where phyllosphere nitrogen-fixing microorganisms were found to enhance the dry weight, nitrogen content, grain weight and yield of rice in comparison to application of urea pathogens (Lindow and Brandl 2003). Phyllosphere microorganisms help in the oxidization of ammonia to nitrate by the process of nitrification (Guerrieri et al. 2015; Krupa 2003). Chemoautotrophic nitrifiers like archaea have been reported in the phyllosphere of many plants (Watanabe et al. 2016). The importance of phyllosphere microbiota is shown in Figure 15.2.

15.3.8 Frost Injury and Frost Control by Altering the Phyllosphere Microbiota

Population of bacteria residing on the outside surface of the plant with the ice nucleation activity (i.e. ice+bacteria) is important in the control of damage caused by frost. It is

Figure 15.2 Importance of phyllosphere microbiota.

Figure 15.3 Frost injury on leaf surface due to ice nucleation activity.

important to lower the number of these bacteria on the surface of plants growing under cold temperature to reduce the damage due to ice + bacteria. Replacement of ice + bacteria with ice-bacteria is the strategy by which we can reduce the damage of the plants by frost. There is a report on spray application of lyophilized *Pseudomonas fluorescens* strain A506 on plants to reduce the damage incurred by frost and bacteria with ice nucleation activity (Lindow and Brandl 2003). The frost injury on leaf surface due to ice nucleation activity is represented in Figure 15.3.

15.3.9 Remediation of Toxic Pollutants

Plants growing under oil-rich environmental conditions include the microorganisms utilizing oil in which *Microbacterium* sp., *Rhodococcus* sp., *Citrobacter freundii* are predominant genera (Ali et al. 2012). Indoor air pollutant, toluene was found to reduce due to the application of bacteria of phyllosphere (De Kempeneer et al. 2004). Phyllosphere of legume plant (peas and beans) are reported to have a greater number of oil-utilizing bacteria as compared with those of non-legume plant (tomato and sunflower), which proved the application of legume plants to remediate volatile hydrocarbon compounds (Ali et al. 2012). The phyllosphere region is composed of many distinct genera with bacteria of genus *Pseudomonas*, *Burkholderia gladioli*, etc. (Benizri et al. 2005). The genus *Pseudomonas* and *Burkholderia gladioli*, etc., present in the phyllosphere can degrade the organic pollutant like phenol. The *Arthrobacter* genus present in the phyllosphere has been reported to degrade phenol, chlorophenol, BTEX, phenanthrene), *s*-triazines, and pesticides (Scheublin and Leveau 2013).

15.3.10 Plant Probiotics

The plant probiotics are the bacteria beneficial to the plants (Berlec 2012). *Stenotrophomonas* sp. and *Ochrobacterum* sp. formulated with chitosan (250–500 ppm) showed a higher yield of strawberry (Muktaa et al. 2017). The phyllosphere microorganisms *Stenotrophomonas* sp. and *Ochrobacterum* sp. along with chitosan can be used as natural bioregulators for strawberry fruits (Muktaa et al. 2017).

15.3.11 Role of Phyllosphere Microorganisms in Climate Change

Plants emit VOCs which get intercepted, biocaptured or metabolized by the microorganisms associated with phyllosphere (Bringel and Couée 2015). In this, methylotrophic microorganisms play very important role by utilizing methanol, formaldehyde and chloromethane. Greenhouse gas, methane, is emitted from intact plant as well as detached leaves (Keppler et al. 2006). Phyllosphere microorganisms have the potential to convert plant VOCs into other useful volatile compounds.

15.3.12 Phyllosphere Microorganisms in Nutrient Yield and Increase of Plant Growth

The microorganisms in the phyllosphere significantly influence the plant growth and productivity. Microorganisms in the phyllosphere increase the plant growth by production of plant growth hormones, viz. IAA and cytokinins, siderophores production, by phosphorus solubilization, etc. There is a report on phyllosphere bacteria for an increase in the yield of potato (*Solanum tuberosum* L.) (Kumar et al. 2018). There is a report on improved growth in *Helianthus annuus* L. due to phyllospheric bacteria (Iqbal et al. 2018). Also, there is a report on improvement in plant biomass, seed germination and vigor index of tomato due to phyllosphere *Methylobacterium* sp. isolated from tomato leaves (Senthilkumar and Krishnamoorthy 2017).

15.3.13 Plant Hormones as Colonization Mediators of the Plant Leaves

The exogenous wax layer of aerial plant surface (cuticle) has long-chain fatty acids, pentacyclic triterpenoids and sterols. The cuticle represents 15% of leaf dry weight. The endophytes present in the phyllosphere region which are leaf epiphytic cross the cuticle and other superficial tissue layer (Hartley et al. 2015). The microbial colonization is reported in the epidermal cell wall junctions, stomata and trichomes. At 10^7 cells/g plant material, bacteria are abundant colonizers of the phyllosphere. The population of microorganisms for colonization is more on the upper surface of the leaves. The microorganisms may be present in biofilm-forms or aggregates on the leaf surface. The interaction of the phyllosphere microorganisms and leaf surfaces; and environmental conditions is important for colonization of the interior of plant leaves (Whipps et al. 2008). The phyllosphere microorganisms in plant leaves can be limited by the availability of nutrients. There is a report on the proteomic profiling of *Methylobacterium extorquens* which colonizes the leaves (Gourion et al. 2006). There is a study on colonization of leaves of plants soybean, clover, and *Arabidopsis* where the majority of *Sphingomonas* and *Methylobacterium* are (Rastogi et al. 2013). The bacterial-induced modification of the leaf surface, i.e. the nutrient leakage due to production of IAA enhances the colonization of the leaves by the phyllosphere microorganisms. Bacterial colonization is influenced by different plant species.

The secondary metabolites and glucosinolates improve colonization by beneficial bacteria in the phyllosphere (Schreiner et al. 2009).

15.4 Plant–Pathogen Interactions Mediated by Phyllosphere Microbiome

15.4.1 Interaction Dependent on the Ionome

Ionome is the composition of nutrient, mineral and trace elements found in the organisms. Plants have the potential to absorb the metals. Strategy of high metal uptake has been used by the plants as a self-defense mechanism (Poschenrieder et al. 2006). High metal concentration restricts the pathogen colonization. Increased concentration of calcium, magnesium, manganese, and copper were observed in the resistant variety of pomegranate (*Punica granatum* L.) as compared to the sensitive variety (Maity et al. 2016). The plant–pathogen interaction by phyllosphere microorganisms is represented in Table 15.2.

15.4.2 Role of Secretory Systems and Secretory Products

Pathogens produce exopolysaccharide which protect the bacteria from plant defense system and *Erwinia amylovora* produces mylovoran and levanexopolysaccharide for biofilm formation and as a contributing factor, respectively. Virulence factors reported in *E. amylovora* are siderophore desferrioxamine (Smits and Duffy 2011), metalloproteases,

Table 15.2 Plant–pathogen interaction by phyllosphere microorganisms.

Pathogen	Virulence factor	Mechanism	Reference
Pseudomonas sp.	Biosurfactant	Increases the wettability of leaf surface by which substrates will be solubilized	Bunster et al. (1989)
Pseudomonas syringae pv. *Syringae*	Toxin Syringomycin	Cell necrosis and disease by affecting ion flux	Hutchison et al. (1995)
Xanthomonas campestris pv. *Vitians* (causative agent of leaf spot disease)	—	Affect the nutrient uptake by changing the lettuce ionome	Nicolas et al. (2019)
Erwinia amylovora (Fire blight)	Exopolysaccharides (amylovoran, levan) and the use of the type III secretion system (T3SS), biofilm formation and motility	Penetration of the host	Pique et al. (2015)

plasmids (Llop et al. 2011), histone-like proteins (Hildebrand et al. 2006) and two-component signal transduction systems (Wang et al. 2009; Zhao et al. 2009).

15.4.3 Quorum Sensing

Quorum sensing has been employed by plant pathogens like N-acylhomoserine lactone (AHL)-based quorum sensing (QS) system found in gram-negative bacteria. Tobacco crop is reported to have quorum quenching mechanism due to phyllosphere bacteria (Ma et al. 2013).There are exchange signals which occur between the phyllosphere microorganisms. These signals create immune responses against pathogens or allow entry of useful microorganisms (Hartmann et al. 2014). Microorganisms present in the phyllosphere communicate with each other by quorum sensing.

15.5 Conclusion and Future Prospects

Commercial formulations of microbes can be employed as an eco-friendly method for the management of plant from diseases and frost injury. There is a need to study the structure and regulation of microbial communities. Owing to the changes in environmental parameters like humidity, moisture, temperature, soil type, and plant developmental stages, the microbial inhabitants can be unpredictable with time. There are spare reports on such study and can be an area that needs to be worked upon. To understand such unpredictability, the use of biosensors can be monitored to provide extensive data on these microbes and their complex associations with habitat. Designing biosensors like that of a whole bacterial biosensor, iron biosensor, fructose biosensor, etc., to understand microbial ecology around plants leads to effective means of plant protection. Such strains will be useful in biotechnological applications. These are the hotspots areas like understanding the cell-to-cell signaling via small molecule produced by microbes regulates the plant fitness bringing in different avenues for managing the microbial colonization and plant health. Grain quality and yield can be manipulated by spraying the beneficial microbes. Soil microbe and plant interactions are much studied, however, aerial plant part and microbe interactions need to be studied, with the growing technological advances to measure microbial activity, and identity. A synthetic microbial community can be built if one understands the microbiome fingerprint associated with the plant. There is a severe gap in our understanding of compositional variation concerning the spatio-temporal scale. With technological advancement such as several "omics" may help to gain new insights into the structural and functional challenges in phyllosphere microorganisms.

Thus, the use of phyllosphere microbiome for the plant growth and nutrient cycling will be eco-friendly and economical. The plant–microbe interactions in the phyllosphere will help to understand the microbial ecology. The phyllosphere microorganisms will have biotechnological applications and also in environmental clean-up.

References

Adesemoye, A.O. and Kloepper, J.W. (2009). Plant-microbes interactions in enhanced fertilizer-use efficiency. *Appl. Microbiol. Biotechnol.* 85: 1–12.

Ali, N., Sorkhoh, N., Salamah, S. et al. (2012). The potential of epiphytic hydrocarbon-utilizing bacteria on legume leaves for attenuation of atmospheric hydrocarbon pollutants. *J. Environ. Manage.* 93: 113–120.

Balachandar, D., Chinnadurai, C., Binnie, J., and McManus, M. (2009). Characterization of 1-aminocyclopropane-1-carboxylatedeaminase producing methylobacteria from phyllosphere of rice and their role in ethylene regulation. *World J. Microbiol. Biotechnol.* 25: 1403–1411.

Batool, F., Rehman, Y., and Hasnain, S. (2016). Phylloplane associated plant bacteria of commercially superior wheat varieties exhibit superior plant growth promoting abilities. *Front. Life Sci.* 9: 313–322.

Beattie, G.A. and Lindow, S.E. (1999). Bacterial colonization of leaves: a spectrum of strategies. *Phytopathology* 89: 353–359.

Beneduzi, A., Ambrosini, A., and Passaglia, L.M. (2012). Plant growth-promoting rhizobacteria (PGPR): their potential as antagonists and biocontrol agents. *Genet. Mol. Biol.* 35: 1044–1051.

Benizri, E., Piutti, S., Verger, S. et al. (2005). Replant diseases: bacterial community structure and diversity in peach rhizosphere as determined by metabolic and genetic fingerprinting. *Soil Biol. Biochem.* 37: 1738–1746.

Berlec, A. (2012). Novel techniques and findings in the study of plant microbiota: search for plant probiotics. *Plant Sci.* 193–194: 96–102.

Bodenhausen, N., Bortfeld-Miller, M. et al. (2014). A synthetic community approach reveals plant genotypes affecting the phyllosphere microbiota. *PLoS Genet.* 10: 1–12.

Bringel, F. and Couée, I. (2015). Pivotal roles of phyllosphere microorganisms at the interface between plant functioning and atmospheric trace gas dynamics. *Front. Microbiol.* 6: 1–14.

Bunster, L., Fokkema, N.J., and Schippers, B. (1989). Effect of surface-active *Pseudomonas* spp. on leaf wettability. *Appl. Environ. Microbiol.* 55: 1340–1345.

Busby, P.E., Peay, K.G., and Newcombe, G. (2016). Common foliar fungi of *Populus trichocarpa* modify *Melampsora* rust disease severity. *New Phytol.* 209: 1681–1692.

Carrion, V.J., Cordovez, V., Tyc, O. et al. (2018). Involvement of *Burkholderiaceae* and sulfurous volatiles in disease suppressive soils. *ISME J.* 12: 2307–2321.

Chen, Q.L., An, X.L., Zheng, B.X. et al. (2018). Long-term organic fertilization increased antibiotic resistome in phyllosphere of maize. *Sci. Total Environ.* 645: 1230–1237.

Chen, Q.L., Cui, H.L., Su, J.Q. et al. (2019). Antibiotic resistomes in plant microbiomes. *Trends Plant Sci.* 24: 530–541.

Christian, N., Whitaker, B.K., and Clay, K. (2015). Microbiomes: unifying animal and plant system through the lens of community ecology theory. *Front. Microbiol.* 6: 1–15.

Clay, K., Holah, J., and Rudgers, J.A. (2005). Herbivores cause a rapid increase in hereditary symbiosis and alter plant community composition. *Proc. Natl. Acad. Sci. U. S. A.* 102: 12465–12470.

Cohen, M.F. and Yamasaki, H. (2003). Involvement of nitric oxide synthase in sucrose-enhanced hydrogen peroxide tolerance of *Rhodococcus* sp. strain APG1, a plant-colonizing bacterium. *Nitric Oxide* 9: 1–9.

Compant, S., Samad, A., Faist, H. et al. (2019). A review on the plant microbiome: ecology, functions, and emerging trends in microbial application. *J. Adv. Res.* 19: 29–37.

De Kempeneer, L., Sercu, B., Vanbrabant, W. et al. (2004). Bioaugmentation of the phyllosphere for the removal of toluene from indoor air. *Appl. Microbiol. Biotechnol.* 64: 284–288.

Dik, A.J., Fokkema, N.J., and Van Pelt, J.A. (1991). Consumption of aphid honeydew, a wheat yield reduction factor, by phyllosphere yeasts under field conditions. *Netherlands. J. Plant Pathol.* 97: 209–232.

Doornbos, R.F., Geraats, B.J., Kuramae, E.E. et al. (2011). Effects of jasmonic acid, ethylene, and salicylic acid signaling on the rhizosphere bacterial community of *Arabidopsis thaliana*. *Mol. Plant Microbe Interact.* 24: 395–407.

Ellis, R.J., Thompson, I.P., and Bailey, M.J. (1999). Temporal fluctuations in the Pseudomonad population associated with sugar beet leaves. *FEMS Microbiol. Ecol.* 28: 345–356.

Enya, J., Shinohara, H., Yoshida, S. et al. (2007). Culturable leaf-associated bacteria on tomato plants and their potential as biocontrol agents. *Microb. Ecol.* 53: 524–536.

Fu, S.F., Sun, P.F., Lu, H.Y. et al. (2016). Plant growth-promoting traits of yeasts isolated from the phyllosphere and rhizosphere of *Drosera spatulata* L. *Fungal Biol.* 120: 433–448.

Gardener, B.M. and Fravel, D.R. (2002). Biological control of plant pathogens: research, commercialization, and application in the USA. *Plant Health Prog.* 3: 1–8.

Glick, B.R. (2007). Modulation of plant ethylene levels by the bacterial enzyme ACC deaminase. *FEMS Microbiol. Lett.* 251: 1–7.

Gottfert, M. (1993). Regulation and function of *Rhizobial* nodulation genes. *FEMS Microbiol. Lett.* 104: 39–63.

Gourion, B., Rossignol, M., and Vorhholt, J.A. (2006). A proteomic study of *Methylobacterium extorquens* reveals a response regulator essential for epiphytic growth. *Proc. Natl. Acad. Sci. U. S. A.* 103: 13186–13191.

Gourion, B., Sulser, S., Frunzke, J. et al. (2009). The PhyR-sigma (EcfG) signaling cascade is involved in stress response and symbiotic efficiency in *Bradyrhizobium japonicum*. *Mol. Microbiol.* 73: 291–305.

Grichko, V.P. and Glick, B.R. (2001). Amelioration of flooding stress by ACC deaminase-containing plant growth-promoting bacteria. *Plant Physiol. Biochem.* 39: 11–17.

Guerrieri, R., Vanguelova, E., Michalski, G. et al. (2015). Isotopic evidence for the occurrence of biological nitrification and nitrogen deposition processing in forest canopies. *Glob. Chang. Biol.* 21: 4613–4626.

Hartley, S.E., Eschen, R., Horwood, J.M. et al. (2015). Infection by a foliar endophyte elicits Novel *Arabidopside*-based plant defense reactions in its host, *Cirsium arvense*. *New Phytol.* 205: 816–827.

Hartmann, A., Rothballer, M., Hense, B.A. et al. (2014). Bacterial quorum sensing compounds are important modulators of microbe-plant interactions. *Front. Plant Sci.* 5: 1–4.

Hildebrand, M., Aldridge, P., and Geider, K. (2006). Characterization of *hns* genes from *Erwinia amylovora*. *Mol. Genet. Genomics* 275: 310–319.

Holland, M. (2011). Nitrogen: give and take from phyllosphere microbes. In: *Ecological Aspects of Nitrogen Metabolism in Plants*, 1e (eds. J.C. Ploacco and C.D. Todd), 217–230. New York: Wiley.

Huang, M., Sanchez-Moreiras, A.M., Abel, C. et al. (2012). The major volatile organic compound emitted from *Arabidopsis thaliana* flowers, the sesquiterpene (E)-beta-caryophyllene, is a defense against a bacterial pathogen. *New Phytol.* 193: 997–1008.

Hung, R., Lee, S., and Bennett, J.W. (2015). Fungal volatile organic compounds and their role in ecosystems. *Appl. Biotechnol. Microbiol.* 8: 3395–3405.

Hutchison, M.L., Tester, M.A., and Gross, D.C. (1995). Role of biosurfactant and ion channel-forming activities of syringomycin in trans-membrane ion flux: a model for the mechanism of action in the plant-pathogen interaction. *Mol. Plant Pathog. Interact.* 8: 610–620.

Ikeda, S., Okubo, T., Kaneko, T. et al. (2010). Community shifts of soybean stem-associated bacteria responding to different nodulation phenotypes and N levels. *ISME J.* 4: 315–326.

Ikeda, S., Anda, M., Inaba, S. et al. (2011). Autoregulation of nodulation interferes with impacts of nitrogen fertilization levels on the leaf-associated bacterial community in soybeans. *Appl. Environ. Microbiol.* 77: 1973–1980.

Innerebner, G., Knief, C., and Vorholt, J.A. (2011). Protection of *Arabidopsis thaliana* against leaf-pathogenic *Pseudomonas syringae* by *Sphingomonas* strains in a controlled model system. *Appl. Environ. Microbiol.* 77: 3202–3210.

Iqbal, M., Wagi, S., and Ahmed, A. (2018). Phyllospheric bacterial treatments improve growth in *Helianthus annuus* L. *Rads. J. Biol. Res. Appl. Sci.* 9: 35–40.

Ji, P. and Wilson, M. (2002). Assessment of the importance of similarity in carbon source utilization profiles between the biological control agent and the pathogen in biological control of bacterial speck of tomato. *Appl. Environ. Microbiol.* 68: 4383–4389.

Julakha, A.M., Rahmana, M., Sabira, A. et al. (2017). Chitosan and plant probiotics application enhance growth and yield of strawberry. *Biocatal. Agric. Biotechnol.* 11: 9–18.

Jumpponen, A. and Jones, K.L. (2009). Massively parallel 454 sequencing indicates hyperdiverse fungal communities in temperate *Quercus macrocarpa* phyllosphere. *New Phytol.* 184: 438–448.

Junker, R.R., Loewel, C., Gross, R. et al. (2011b). Composition of epiphytic bacterial communities differs on petals and leaves. *Plant Biol.* 13: 918–924.

Keppler, F., Hamilton, J.T., Brass, M. et al. (2006). Methane emissions from terrestrial plants under aerobic conditions. *Nature* 439: 187–191.

Krupa, S.V. (2003). Effects of atmospheric ammonia (NH_3) on terrestrial vegetation: a review. *Environ. Pollut.* 124: 179–221.

Kumar, S., Chaudhary, D., Rashmi, R. et al. (2018). Exploring phyllosphere bacteria for growth promotion and yield of potato (*Solanum tuberosum* L.). *Int. J. Curr. Microbiol. Appl. Sci.* 7: 1065–1071.

Lambais, M.R., Barrera, S.E., Santos, E.C. et al. (2017). Phyllosphere metaproteomes of trees from the Brazilian Atlantic forest show high levels of functional redundancy. *Microb. Ecol.* 73: 123–134.

Lindow, S.E. and Brandl, M.T. (2003). Microbiology of phyllosphere. *Appl. Environ. Biol.* 69: 1875–1883.

Lindow, S.E. and Leveau, J.J. (2002). Phyllosphere microbiology. *Curr. Opin. Biotechnol.* 13: 238–243.

Llop, P., Cabrefiga, J., Smits, T.M. et al. (2011). *Erwinia amylovora* novel plasmid pEI70: complete sequence, biogeography, and role in aggressiveness in the fire blight phytopathogen. *PLoS One* 6: e28651.

Ma, A., Lv, D., Zhuang, X., and Zhuang, G. (2013). Quorum quenching in culturable phyllosphere bacteria from tobacco. *Int. J. Mol. Sci.* 14: 14607–14619.

Madhaiyan, M., Poonguzhali, S., Lee, H.S. et al. (2005). Pink-pigmented facultative methylotrophic bacteria accelerate germination, growth and yield of sugarcane. *Biol. Fertil. Soils* 41: 350–358.

Maity, A., Sharma, J., Sarkar, A. et al. (2016). Nutrient imbalance indices are closely related with susceptibility of pomegranate to bacterial blight disease. *Sci. Hortic.* 211: 79–86.

Mayak, S., Tirosh, T., and Glick, B.R. (2004). Plant growth-promoting bacteria that confer resistance to water stress in tomato and pepper. *Plant Sci.* 166: 525–530.

Mejia, L.C., Herre, E.A., Sparks, J.P. et al. (2014). Pervasive effects of a dominant foliar endophytic fungus on host genetic and phenotypic expression in a tropical tree. *Front. Microbiol.* 5: 1–16.

Mendes, R., Kruijt, M., de Bruijn, I. et al. (2011). Deciphering the rhizosphere microbiome for disease-suppressive bacteria. *Sciences* 332: 1097–1100.

Mercier, J. and Jimenez, J.I. (2004). Control of fungal decay of apples and peaches by the biofumigant fungus *Muscodor albus*. *Post-harvest Biol. Technol.* 31: 1–8.

Mercier, J. and Lindow, S.E. (2000). Role of leaf surface sugars in colonization of plants by bacterial epiphytes. *Appl. Environ. Microbiol.* 66: 369–374.

Mohammadi, K. (2012). Phosphorus solubilizing bacteria: occurrence, mechanisms and their role in crop production. *Resour. Environ.* 2: 80–85.

Mordecai, E.A. (2011). Pathogen impacts on plant communities: unifying theory, concepts, and empirical work. *Ecol. Monogr.* 81: 429–441.

Morris, C.E., Monier, J.M., and Jacques, M.A. (1998). A technique to quantify the population size and composition of the biofilm component in communities of bacteria in the phyllosphere. *Appl. Environ. Microbiol.* 64: 4789–4795.

Muktaa, J.A., Rahmana, M., Sabira, A.A. et al. (2017). Chitosan and plant probiotics application enhance growth and yield of strawberry. *Biocatal. Agric. Biotechnol.* 11: 9–18.

Nicolas, O., Charles, M.T., Jenni, S. et al. (2019). The Ionomics of lettuce infected by *Xanthomonas campestris* pv. Vitians. *Front. Plant Sci.* 10: 1–10.

Pal, K.K., Tilak, K.V., Saxena, A.K. et al. (2001). Suppression of maize root diseases caused by *Macrophomina phaseolina, Fusarium moniliforme* and *Fusarium graminearum* by plant growth promoting rhizobacteria. *Microbiol. Res.* 156: 209–223.

Parasuraman, P., Pattnaik, S., and Busi, S. (2019). Phyllosphere microbiome: functional importance in sustainable agriculture. In: *New and Future Developments in Microbial Biotechnology and Bioengineering* (eds. J.J. Singh and D.P. Singh), 135–148.

Pati, B.R. (1992). Effect of spraying nitrogen fixing phyllospheric bacterial isolates on rice plants. *Zentralbl. Mikrobiol.* 147: 441–446.

Penrose, D.M., Moffatt, B.A., and Glick, B.R. (2001). Determination of 1-aminocyclopropane-1-carboxylic acid (ACC) to assess the effects of ACC deaminase-containing bacteria on roots of canola seedlings. *Can. J. Microbiol.* 47: 77–80.

Penuelas, J. and Terradas, J. (2014). The foliar microbiome. *Trends Plant Sci.* 19: 278–280.

Pique, N., Minana-Galbis, D., Merino, S. et al. (2015). Virulence factors of *Erwinia amylovora*: a review. *Int. J. Mol. Sci.* 16: 12836–12854.

Poschenrieder, C., Tolra, R., and Barcelo, J. (2006). Can metals defend plants against biotic stress? *Trends Plant Sci.* 11: 288–295.

Pusey, P.L., Stockwell, V.O., Reardon, C.L. et al. (2011). Antibiosis activity of *Pantoea agglomerans* biocontrol strain E325 against *Erwinia amylovora* on apple flower stigmas. *Phytopathology* 101: 1234–1241.

Rastogi, G. and Sani, R.K. (2011). Molecular techniques to assess microbial community structure, function, and dynamics in the environment. In: *Microbes and Microbial Technology* (eds. I. Ahmad, F. Ahmad and J. Pichtel), 29–57. New York: Springer.

Rastogi, G., Coaker, G.L., and Leveau, J.J. (2013). New insights into structure and function of phyllosphere microbiota through high-throughput molecular approaches. *FEMS Microbiol. Lett.* 348: 1–10.

Remus-Emsermann, M.P. and Schlechter, R.O. (2018). Phyllosphere microbiology: at the interface between microbial individuals and the plant host. *New Phytol.* 218: 1327–1333.

Romero, F.M., Marina, M., and Pieckenstain, F.L. (2016). Novel components of leaf bacterial communities of field-grown tomato plants and their potential for plant growth promotion and biocontrol of tomato diseases. *Res. Microbiol.* 167: 222–233.

Rua, M.A., Antoninka, A., Antunes, P.M. et al. (2016). Home-field advantage? Evidence of local adaptation among plants, soil, and arbuscular mycorrhizal fungi through meta-analysis. *BMC Evol. Biol.* 16: 1–15.

Scheublin, T.R. and Leveau, J.H. (2013). Isolation of *Arthrobacter* species from the phyllosphere and demonstration of their epiphytic fitness. *Microbiol. Open* 2: 205–213.

Schreiner, M., Krumbein, A., and Ruppel, S. (2009). Interaction between plant and bacteria: Glucosinolate and phyllospheric colonization of cruciferous vegetables by *Enterobacter radicinicitans* DSM 16656. *J. Mol. Microbiol. Biotechnol.* 17: 124–125.

Senthilkumar, M. and Krishnamoorthy, R. (2017). Isolation and characterization of tomato leaf phyllosphere *Methylobacterium* and their effect on plant growth. *Int. J. Curr. Microbiol. App. Sci.* 6: 2121–2136.

Smits, T.M. and Duffy, B. (2011). Genomics of iron acquisition in the plant pathogen *Erwinia amylovora*: insights in the biosynthetic pathway of the siderophore desferrioxamine E. *Arch. Microbiol.* 193: 693–699.

Stinson, A.M., Zidack, N.K., Strobel, G.A. et al. (2003). Effect of myco-fumigation with *Muscodor albus* and *Muscodor roseus* on seedling 28 diseases of sugar beet and *Verticillium* wilt of eggplant. *Plant dis.* 87: 1349–1354.

Tsavkelova, E.A., Cherdyntseva, T.A., Botina, S.G. et al. (2007). Bacteria associated with orchid roots and microbial production of auxin. *Microbial Res.* 162: 69–76.

Venkatachalam, S., Ranjan, K., Prasanna, R. et al. (2016). Diversity and functional traits of culturable microbiome members, including cyanobacteria in the rice phyllosphere. *Plant Biol.* 18: 627–637.

Wang, D.P., Korban, S.S., and Zhao, Y.F. (2009). The Rcsphosphorelay system is essential for pathogenicity in *Erwinia amylovora*. *Mol. Plant Pathol.* 10: 277–290.

Watanabe, K., Kohzu, A., Suda, W. et al. (2016). Microbial nitrification in throughfall of a Japanese cedar associated with archaea from the tree canopy. *Springerplus* 5: 1–15.

Whipps, J.M., Hand, P., Pink, D. et al. (2008). Phyllosphere microbiology with special reference to diversity and plant genotype. *J. Appl. Microbiol.* 105: 1744–1755.

Whitaker, B.K., Bauer, J.T., Bever, J.D. et al. (2017). Negative plant-phyllosphere feedbacks in native *Asteraceae* hosts – a novel extension of the plant-soil feedback framework. *Ecol. Lett.* 20: 1064–1073.

Xiang, Q., Zhu, D., Giles, M. et al. (2019). Agricultural activities affect the pattern of the resistome within the phyllosphere microbiome in peri-urban environments. *J. Hazard. Mater.* 382: 121068.

Yang, C.H., Crowley, D.E., Borneman, J. et al. (2001). Microbial phyllosphere populations are more complex than previously realized. *Proc. Natl. Acad. Sci. U. S. A.* 98: 3889–3894.

Yu, X., Ai, C., Xin, L. et al. (2011). The siderophore-producing bacterium, *Bacillus subtilis* CAS15, has a biocontrol effect on *Fusarium* wilt and promotes the growth of pepper. *Eur. J. Soil Biol.* 47: 138–145.

Yuen, G.Y., Steadman, J.R., Lindgren, D.T. et al. (2001). Bean rust biological control using bacterial agents. *Crop Prot.* 20: 395–402.

Zhao, Y.F., Wang, D.P., Nakka, S. et al. (2009). Systems level analysis of two-component signal transduction systems in *Erwinia amylovora*: role in virulence, regulation of amylovoran biosynthesis and swarming motility. *BMC Genomics.* 10: 1–16.

Zohara, F., Akanda, M.M., Paul, N.C. et al. (2016). Inhibitory effects of *Pseudomonas* spp. on plant pathogen *Phytophthora capsica in-vitro* and *in planta*. *Biocatal. Agric. Biotechnol.* 5: 69–77.

Index

Phytomicrobiome Interactions and Sustainable Agriculture, First Edition. Edited by Amit Verma, Jitendra Kumar Saini, Abd El-Latif Hesham, and Harikesh Bahadur Singh.

q

r

S